OPTIMIZATION PRINCIPLES

BOOKS IN THE IEEE PRESS SERIES ON POWER ENGINEERING

Analysis of Faulted Power Systems
P. M. Anderson
1995 Hardcover 536pp 0-7803-1145-0

Subsynchronous Resonance in Power Systems
P. M. Anderson, B. L. Agrawal, J. E. Van Ness
1990 Softcover 282pp 0-7803-5350-1

Power System Protection
P. M. Anderson
1999 Hardcover 1344pp 0-7803-3427-2

Power and Communication Cables: Theory and Applications
Edited by R. Bartnikas and K. D. Srivastava
2000 Hardcover 896pp 0-7803-1196-5

Understanding Power Quality Problems: Voltage Sags and Interruptions
Math H. J. Bollen
2000 Hardcover 576pp 0-7803-4713-7

Electric Power Applications of Fuzzy Systems
Edited by M. E. El-Hawary
1998 Hardcover 384pp 0-7803-1197-3

Inspection of Large Synchronous Machines: Checklists, Failure Identification, and Troubleshooting
Kerszenbaum
1996 Hardcover 200pp 0-7803-1148-5

Power System Stability, Volumes I, II, and III
An IEEE Press Classic Reissue Set
E. W. Kimbark
1995 Hardcover 1008pp 0-7803-1135-3

Analysis of Electric Machinery and Drive Systems, Second Edition
Paul C. Krause, Oleg Wasynczuk, and Scott D. Sudhoff
2002 Hardcover 624pp 0-471-14326-X

Power System Control and Stability, Second Edition
P. M. Anderson, A. A. Fouad
2002 Hardcover 672pp 0-471-23862-7

OPTIMIZATION PRINCIPLES

Practical Applications to the Operation and Markets of the Electric Power Industry

NARAYAN S. RAU

IEEE PRESS SERIES ON POWER ENGINEERING

IEEE Press Power Engineering Series
Mohamed E. El-Hawary, *Series Editor*

IEEE

IEEE PRESS

WILEY-INTERSCIENCE

A JOHN WILEY & SONS, INC., PUBLICATION

Published simultaneously in Canada.

Library of Congress Cataloging-in-Publication Data:

Rau, Narayan S.
Optimization principles : practical applications to the operation and markets of the electric power industry / Narayan S. Rau.
p. cm.
Includes bibliographical references and index.
ISBN 0-471-45130-4 (cloth)
1. Electric utilities—Management—Mathematical models. 2. Mathematical optimization. 3. Electronic spreadsheets. I. Title.

HD9685.A2R354 2003
333.793′2′015193—dc21

2003049732

10 9 8 7 6 5 4 3 2 1

I dedicate this book to my deceased parents, who
gave up a lot in their lives to make me who I am.

Na chora haryam, na cha raja haryam;
na brathru bhojyam, na cha bharakari;
vyam krite' vardhate' e'va nityam
vidya dhanam sarva dhana pradhanam.

Cannot thieves steal, cannot even king usurp;
Cannot siblings covet, not even heavy to carry;
As is spent, keeps on increasing continually,
wealth of **vidya** (knowledge) is the supreme of all wealth.

CONTENTS

PREFACE

My career in the electric power industry has spanned four decades during which I have seen several changes, none as drastic as the deregulation of the industry towards the end of the last century. The merits or otherwise of deregulation—if competition in the electricity industry is possible to the extent that it can bring benefits to consumers—is worthy of debate. But my concern is about the acquisition of new skills and proficiency by the professionals in the electric power industry to meet new challenges.

Optimization techniques have been around for quite some time, not to mention that the mathematical underpinnings precede such techniques by a century or more. However, the power industry, except in the field of manufacturing apparatus, has been rather conservative and phlegmatic in embracing these techniques for analyses. An example is that of optimal power flow (OPF) solution, which has been around for at least a quarter of a century. Nevertheless, practicing professionals, by and large, viewed it as the pursuit and intellectual curiosity of some academicians and researchers. Such techniques, with some rare exceptions, had no day-to-day use for decision-making in the power industry.

Suddenly, unaware as the industry was, deregulation changed it all. The market rules and economic principles to price commodities at marginal cost made it necessary to use more and more sophisticated algorithms containing optimization techniques, either for the dispatch of the power system itself or for the settlement of markets. OPF is one such technique essential for the computation of locational marginal prices (LMP) of energy in a network.

Against this backdrop, as the need for sophisticated methods and algorithms were and are being felt, software vendors are rising to the occasion. However, the corresponding universal appreciation of such sophistications appears to be absent among the practicing professionals of the industry.

Given the refractory attitude of the past and that changes are fast, the practicing professional has had no time to catch up. It is not uncommon in meetings to hear senior staff speak of LMP and other matters related to even elementary optimization techniques making their lack of awareness transparent to those who know the field well.

In my experience of teaching university students as an adjunct professor, I have observed that the engineering curriculum in most schools does not address linear programming techniques in the undergraduate level, and on occasion in postgraduate levels as well. However, this is a fundamental requirement in most undergraduate curriculums for Economics majors!

Because of these observations, I thought that a book on optimization techniques that addressed practical applications more than the theory was needed. Additionally, such a book should enable the senior managers to study at their own leisure and privacy using the PC as an aid. Such an education might even be enjoyable and might enthrall the reader to appreciate the beauty of optimization techniques and its application not only to engineering problems, but to social sciences as well.

An in-house course on optimization, which I was privileged to teach during 2001, was organized at ISO New England Inc. This book evolved from the contents of that course. During that course, I noted that the electricity industry is recruiting many nonengineering professionals of late, particularly in departments dealing with markets and trade. During discussions in such an interdisciplinary group, I found that when attention turned to reactive power, there was considerable discomfort among professionals without engineering background. Therefore, although it appears inconsistent to include a discussion on basic principles of electricity, I thought that its inclusion would benefit nonengineers who may want to use optimization techniques to analyze markets and trade. Appendix A.1 discusses such basics culminating in the explanation of active and reactive power concepts. It neither is a rigorous treatment of the subject, nor is it directed to engineers.

The intent of this book is to help practicing engineers as well as the senior staff in the electric power industry to become aware of, if not dexterous in, optimization techniques. Of course, a student who works out all the exercises and understands the underlying programs of spreadsheets will move closer to dexterity.

As the power industry has changed, so has the availability of personal computers and algorithms. In my earlier days of profession, engineers spent several days developing algorithms to solve a problem. Today, algorithms of all types and from multiple sources are available at very reasonable prices, making it unattractive for any organization to develop its own algorithms. For example, some mathematical and statistical packages are available from NAG, IMSL, GAMS, and several other sources based on the fruits of research of the world's best engineers and mathematicians. Presumably, there are several other sources, all of which one cannot know in one's own lifetime. For example, Wall Street financial professionals routinely use linear program software packages that solve for thousands of choice variables with several thousands of constraints. On top of that, on the ubiquitous PC, one has Excel, MATLAB, and other software, which have optimization packages that are excellent for small to even fairly large problems.

Wiley's ftp site located at ftp://ftp.wiley.com/public/sci_tech_med/electric_power/ contains the spreadsheets and solutions to the problems explained in the text. The spreadsheets are programmed to solve the problems contained therein. Therefore, it is necessary for the reader to have access to a personal computer with Microsoft Excel software. The advanced reader may prefer to use MATLAB instead of the Excel Solver to confirm the solution to examples in the book.

This book serves well as a text for postgraduate or senior undergraduate students in engineering for a one-semester course. A rigorous curriculum can extend this to a two-semester course by making exercise problems compulsory.

The part of thanking people, in most prefaces, is left until the end, as it is in this case. I must say that I would not have undertaken the travails of writing this book in the first place had it not been for the insistence and encouragement of Mr. John Lally and Dr. Alex Bykhovsky, both of whom participated in the course I gave at ISO New England. Consequently, I consider it my obligation to thank them first. Of course, this book would not have been possible without the understanding and patience of my wife, Kusum, a companion and friend for 40 years.

NARAYAN S. RAU

Easthampton, Massachusetts
May 2003

CHAPTER 1

INTRODUCTION

The mathematics related to optimization is not new. Even its application to solve practical problems has already spanned well over half a century. During the World War II, the use of Operation Research techniques to optimize the use and movement of men and material was not uncommon. The application to the power industry is also over 40 years old. With the advent of digital computers, system dispatch to minimize cost came into vogue. Electrical engineering science also borrowed mathematical models that were popular in management sciences. Such models are now used in planning system expansion, system operation, and ratemaking.

The emphasis of this book is on the electrical industry. Despite this, it is not at all surprising that the mathematical formulation of the problems for solution in this industry bears a remarkable resemblance to those in other industries. Consequently, notwithstanding the emphasis on the electric power industry, we digress occasionally into problems in other fields. Such an excursion not only illustrates the common themes in the problems of industries, but also will demonstrate the beauty of this branch of applied mathematics.

The electric power industry has undergone colossal changes in the last decade. It appears that such changes in the future are also inevitable. Because of such changes, the incorporation of optimization into decision-making has also become inevitable. By and large, the engineering curriculum has not permitted teaching undergraduate students optimization theory. It is not uncommon to see graduates not having been exposed even to linear optimization such as linear programs. In the evolving deregulated electric power industry, models used for system dispatch, auctions of rights and hedging instruments, and models for settlement of markets, all

Optimization Principles, by Narayan S. Rau
ISBN 0-471-45130-4

use optimization of one type or another. In particular, Optimal Power Flow (OPF), which has been around for some 30 plus years, has come into prominence for system dispatch, security-constrained unit commitment, declaration of locational marginal prices in a transmission network, and a myriad of other tasks. The primary intent of this book is to familiarize those in the electric power industry with the principles and the development of such models. The emphasis is more toward the practical applications of optimization principles. However, since a mere practical use of an algorithm without a theoretical understanding would not develop an engineer in his or her profession, some theoretical background is also included. However, a student might prefer to glean only the practical applications by studying the solved problems of chapters.

The primary intent of this book is to provide a practical insight into optimization technique via demonstration programs in Microsoft Excel. The book is organized essentially into three parts and appendices. The first part addresses mathematical preliminaries and shows examples of some mathematical techniques via solved problems. The second part deals with linear optimization techniques. This part is in consists of chapters, Chapter 3 addresses theoretical material followed by solved problems in Chapter 4 demonstrating practical applications. The third part addresses some mathematical theory of nonlinear optimization in Chapter 5, unconstrained nonlinear optimization in Chapter 6, and constrained optimization in Chapter 7. Solved problems of practical interest are in Chapter 8. The first appendix deals with basic principles of electricity directed toward those who are not engineers. Additional appendices deal with background material addressing network theory and other matters related to optimization.

This book has a secondary purpose as well. Because of the changes in the industry, there are an increasing number of professionals with some or no engineering background at all. Such professionals usually are engaged in markets, trading, and the settlement activities of this evolving deregulated industry. There is a need for them to understand the inner workings of the mathematical models that they use in their day-to-day tasks. Such an understanding will make their tasks more interesting, and, possibly, more pleasant. To such readers, Appendix A.2 provides some basic principles of electricity, particularly as applied to networks. Clearly, in one book of this nature a reader cannot be made into an electrical engineer. The reason for outlining the basic principles is to impart to such readers an appreciation of network equations, in particular the concept of reactive power. Clearly, the outline in Appendix A.2 is too simplistic and redundant for engineers who are familiar with network equations.

1.1 DEREGULATED ELECTRICITY MARKETS – TERMINOLOGY AND ACRONYMS

There are several acronyms and terminologies associated with the deregulated electricity industry. It is impossible to list them all. However, the following is a brief discussion of a few of them as they pertain to the subject matter of the remaining chapters of this book.

In the past, an electric utility had the responsibility to install generation and transmission to serve load. In this structure, called the vertically integrated structure, utilities were permitted by their respective regulatory agencies to set rates that recovered prudent costs plus some return on investment. The cost of energy, by and large, was based on the average cost of providing it.

In the deregulated scenario, the entities that generate energy, build and own transmission, and serve load may all be different. The entity generating energy is called the Generator (note the capitalization), and the entity serving load is called the Load Serving Entity (LSE) or Load. Similarly, Transmission providers have different labels. Depending on the organizational structure and tariff, some associated names are: Independent Transmission Company (ITC), Independent Transmission Provider (ITP), Transmission Company (TRANSCO), Grid Company (GRIDCO), Regional Transmission Organization (RTO), and Transmission Provider (TP). The Independent System Operator (ISO) operates the generation and transmission network, and does not earn profits.

There are different rules and the structure of markets at different locations. Consequently, it is not possible to discuss the nuances and rules of different markets. However, the essential feature of all markets is that the Generators submit offer bids (called offers) to supply energy (prices and quantities) and also submit other commodities such as reserve capacity and regulating capacity. The System Operator dispatches generation to minimize cost, making sure that the reliability of the system is not jeopardized. Some markets accept demand side bids in which loads or LSE submit price quantity bids for energy withdrawals from the grid. The System Operator reconciles offers and bids, which essentially resembles an auction. Examples in this book indicate this process.

In 1979, Bohn and co-workers (Caramanis et al., 1982) introduced the concept of Spot Prices. This is now called Locational Marginal Price (LMP) or Location-Based Marginal Price (LBMP). In this method, the cost of supplying the next unit of energy is calculated at each major node (bus bar) of the electrical network. This, therefore, represents the marginal cost of supplying the energy at those locations. The basis of computing LMP is by using a procedure called Optimal Power Flow (OPF). The OPF program optimizes the system dispatch to minimize cost.

In markets that adopt LMP pricing, withdrawal of energy from the network is charged at the marginal rate at that node (LMP times consumption), and injection of energy by Generators is paid at this rate at the node of injection.

Examples in subsequent chapters outline the OPF procedure and show how to compute LMP from the results of OPF.

1.2 STUDY PLAN

There are other books that discuss the application of optimization techniques to power systems (Mamoh, 2000; Song, 2002). The emphasis of this book is on using Excel spreadsheets to educate the reader about practical aspects of optimization. Of course, some theoretical background material on optimization cannot be avoided. The expectation is that by setting up Excel spreadsheets, the reader becomes somewhat of a programmer gaining an understanding of the theoretical basis of algorithms. Readers who are well-versed in the theoretical underpinnings of algorithms and want to pursue solution techniques further may consult Mamoh (2000), which uses MATLAB to obtain solutions.

It is obvious that the reader is expected to have access to a personal computer and be familiar with the use of Microsoft Excel. The examples of this book, in the interest of clear exposition, are of limited size, generally restricted to less than five or six design variables. That is not to say that larger problems cannot be solved by using such spreadsheets, limited only by the capability of Excel program (help menu of Excel outlines that there cannot be more than a certain number of variables and other such limits on its use). The question then is—How can one write larger spreadsheets? The answer to this is that as the reader gains confidence in setting-up small spreadsheets, their extension to larger problems becomes readily obvious. Writing larger programs is a process of increasing the number of choice variables and the number of constraints—but the fundamental requirement is that one should be able to conceive and formulate optimization problems.

For the reader familiar with the mathematics of matrices and wishing to learn practical applications of linear optimization first, it is suggested that he or she proceed directly to Chapters 4 and 8. In a similar vein, readers familiar with the mathematics of nonlinear optimization may skip Chapter 5 and proceed directly to Chapters 6, 7, and 8 to study the applications of nonlinear optimization. Of course, those who want a refresher of matrix operations and solution of linear inequalities will do well to read Chapters 2 and 3. The philosophy of linear optimization, in particular the simplex method, is the topic of discussion in Chapter 3. Chapters 6 and 7 describe the mathematics behind the algorithms for unconstrained and constrained nonlinear optimization methods.

Appendix A.1 is directed to readers without training in electrical engineering. Consequently, electrical engineers can well avoid this rudimentary section of the book. Appendix B is devoted to the development of ac network flow equations. The equations derived there are used in the chapters of this book to solve optimal load flow problems. Appendix C develops the mathematics behind least-square error techniques leading to state estimation.

While it is expedient to gain a practical understanding of optimization techniques, it is rewarding to understand some theoretical underpinnings to demystify the process of obtaining a solution. Hence, for those that take the practical route first, it is suggested that during the course of time, they revisit the chapters that explain the theory behind the methods of optimization. Understandably, in view of many excellent books on the mathematical theory of these methods, this book's intent is to give a quick sketch of popular procedures rather than that of mathematical rigor. Of course, the punctilious reader will do well to read all chapters in their serial order, as well as the reference materials listed in the bibliography.

1.3 ORGANIZATION AND CONVENTIONS

The marginal notes in the text indicate the appropriate Excel file to be opened by the reader that corresponds to the text. These files can be downloaded from Wiley's ftp site at ftp://ftp.wiley.com/public/sci_tech_med/electric_power/. It is expected of the student to study the spreadsheet carefully, checking the formulas associated with the cells. After some exercises of that nature, it may become either faster or unnecessary for the student to examine details of formulas associated with the cells. Additionally, it is clear that spreadsheets can be set in different fashions depending on the preference of the user. In the spreadsheets associated with this book, no attempt has been made to set up an efficient procedure because their intent is only to illustrate the solution procedure. Consequently, a reader wishing to extend the solution to a larger system is cautioned against a mere reproduction and expansion of the spreadsheets. Although such a procedure may be acceptable in some cases, it is better to examine an efficient way of setting up spreadsheets for larger problems.

The conventions used in this book are as follows. Lowercase letters signify scalars. For example, x_1 and x_2 are scalars. However, lowercase bold letters represent vectors. For example, **y** is a vector of dimension 3 (a three-vector) whose elements are y_1, y_2, y_3. The elements of a vector are normally written vertically as

$$\mathbf{y} = \begin{pmatrix} y_1 \\ y_2 \\ y_3 \end{pmatrix},$$

or as $\mathbf{y} = [y_1, y_2, y_3]^T$.

Uppercase bold letters represent matrices. For example, [**A**] or (**A**) represents a matrix. Sometimes, when there is no confusion with network parameters (see below), a matrix is simply represented by uppercase bold letters as **A**. If **A** is a 4×3 matrix, **A** can be written as $[\mathbf{A}] = [\mathbf{a}_1, \mathbf{a}_2, \mathbf{a}_3]$, where $\mathbf{a}_1$, $\mathbf{a}_2$, and $\mathbf{a}_3$ are "four-vectors." A four-vector implies that the vector contains four scalar components. Each vector, for example $\mathbf{a}_1$, can be expressed in terms of its scalar components as $\mathbf{a}_1 = [a_{11}, a_{21}, a_{31}, a_{41}]^T$. The "row-column" notation is used for the subscripts of matrix elements. Thus, a_{13} implies the element in the first row and the third column of matrix **A**.

The exceptions to this convention are symbols related to network equations. The general convention used by power engineers is that voltage, current, power, and reactive power are expressed in uppercase letters. Thus V, I, S, P, and Q represent the voltage, current, complex power, real, and reactive powers at a single node while bold uppercase letters **V**, **I**, **S**, **P**, and **Q** represent a vectors of the same variables at several nodes. The power angle at a node δ is always represented in lowercase, while $\boldsymbol{\delta}$ represents a vector of angles at several nodes. The admittance of branches is also written in uppercase such as Y_{34} to represent the admittance of branch 3–4. However, an uppercase bold **Y** with associated braces represents the **Y** matrix.

To represent matrices in the network equations, bold letters with the associated braces is used. For example, [**A**] or (**Y**) represents an A matrix or Y matrix. These conventions will be obvious from their context.

PART I

MATHEMATICAL BACKGROUND

CHAPTER 2

FUNDAMENTALS OF MATRIX ALGEBRA

2.1 SCALARS, VECTORS, AND MATRICES

A *scalar* is a single quantity or a measurement. It is expressed in terms of real numbers. For example, the height of a person is 190 cm, the dimensions of a box are 30, 20, 40 indicating its length, width and height.

A *vector* is an ordered collection of scalars. The numbers of scalars and the ordering rules are crucial. The ordered set of scalars describes an ordered set of quantities or attributes. For example, when we say that a box has a height of 30 cm, the real number 30 measures a single aspect of the box. However, if the box has property 1 = length, property 2 = width, property 3 = height, then

$$\begin{bmatrix} 30 \\ 20 \\ 40 \end{bmatrix} \tag{2.1}$$

indicates three attributes of the box as a vector. The vector is normally written in a vertical fashion as in (2.1) and is denoted in this text by lowercase bold letter as

$$\mathbf{v} = [v_1, v_2, \ldots, v_n]^T, \tag{2.2}$$

in which the superscript T implies that the numbers should be transposed to obtain the vector as a column.

Two vectors are equal if their components are equal.

Optimization Principles, by Narayan S. Rau
ISBN 0-471-45130-4

A vector is the generalization of the notion of scalars ordered as a set. The idea of a matrix is to carry this one step further by aggregating two or more such orderings. Therefore, a matrix can be viewed as a multidimensional collection of *scalars*. A matrix is represented by boldfaced capital letters. Any element of a matrix is referenced by the name of the matrix (uppercase or lowercase) with subscripts indicating the row and column position of the element. Thus, X_{ij} or x_{ij} refer to the matrix element in the row i and column j. A matrix **X** with three rows and two columns (3×2) is written as

$$\mathbf{X} = \begin{bmatrix} x_{11} & x_{12} \\ x_{21} & x_{22} \\ x_{31} & x_{32} \end{bmatrix}.$$

If the number of rows and columns in a matrix are equal, the matrix is said to be square.

A column vector is a special case of a matrix with only one column.

Two matrices are equal if there is element-by-element equality for the same ordering. The transpose of a matrix results from interchanging the row and column positions of the elements:

$$(\mathbf{A^T})_{ij} = (\mathbf{A})_{ji}. \tag{2.3}$$

For example,

$$\mathbf{A} = \begin{bmatrix} 1 & 2 & 3 \\ 4 & 5 & 6 \end{bmatrix}, \tag{2.4}$$

and

$$\mathbf{A^T} = \begin{bmatrix} 1 & 4 \\ 2 & 5 \\ 3 & 6 \end{bmatrix}. \tag{2.5}$$

A symmetric matrix **X** is one in which $X_{ij} = X_{ji}$. For example,

$$\mathbf{X} = \begin{bmatrix} 1 & 2 & 3 & 4 \\ 2 & -5 & -2 & 1 \\ 3 & -2 & 7 & 6 \\ 4 & 1 & 6 & -1 \end{bmatrix}$$

is a symmetric matrix. Hence, for a symmetric matrix we have $\mathbf{X} = \mathbf{X}^T$.

By convention, x_{ij} are the diagonal elements, $x_{ij},\ j \geq i$ are superdiagonal elements, and $x_{ij},\ j \leq i$ are subdiagonal elements.

Let **i**, **j**, and **k** be mutually perpendicular vectors of unit length in three coordinate axis of Euclidean space. The coordinates of any vector **v** and **z** can be expressed in terms of the three unit vectors as $\mathbf{v} = \mathbf{i}a_1 + \mathbf{j}a_2 + \mathbf{k}a_3$ and $\mathbf{z} = \mathbf{i}b_1 + \mathbf{j}b_2 + \mathbf{k}b_3$.

Then, the scalar product or the dot product of the two vectors is $\gamma = \mathbf{v}^\mathrm{T}\mathbf{z} = \mathbf{z}^\mathrm{T}\mathbf{v} = a_1b_1 + a_2b_2 + a_3b_3$.

The scalar product of the vector itself—that is, $(\mathbf{x}.\mathbf{x})^{1/2}$—is the Euclidean length or the norm of the vector.

2.2 OPERATIONS ON MATRICES

Multiplication by a scalar is the simplest of operations in which each element of a matrix or vector is multiplied by the scalar. Then, for any scalar α we have

$$\alpha \begin{bmatrix} x_1 \\ x_2 \\ x_3 \end{bmatrix} = \begin{bmatrix} \alpha x_1 \\ \alpha x_2 \\ \alpha x_3 \end{bmatrix}.$$

For example, we have

$$-4 \begin{bmatrix} 1 & -2 \\ -3 & 4 \end{bmatrix} = \begin{bmatrix} -4 & 8 \\ 12 & -16 \end{bmatrix}.$$

The addition of two matrices (vectors) is simply the sums of corresponding elements. Evidently, addition requires that the dimensions of the two matrices should match. For a vector case, we have

$$\begin{bmatrix} x_1 \\ x_2 \\ x_3 \end{bmatrix} + \begin{bmatrix} y_1 \\ y_2 \\ y_3 \end{bmatrix} = \begin{bmatrix} x_1 + y_1 \\ x_2 + y_2 \\ x_3 + y_3 \end{bmatrix}.$$

It is easy to see that matrix addition obeys the laws of *associativity* and *commutativity*; that is, $\mathbf{X} + (\mathbf{Y} + \mathbf{Z}) = (\mathbf{X} + \mathbf{Y}) + \mathbf{Z}$ and $\mathbf{X} + \mathbf{Y} = \mathbf{Y} + \mathbf{X}$. Also, the *scalar product* of vectors obeys the laws of commutativity and distributivity over vector addition since $\mathbf{p}^T\mathbf{y} = \mathbf{y}^T\mathbf{p}$, and $\mathbf{p}^T(\mathbf{y} + \mathbf{z}) = \mathbf{p}^T\mathbf{y} + \mathbf{p}^T\mathbf{z}$.

2.2.1 Product of Matrices

The product $\mathbf{C}$ of matrices $\mathbf{A}$ and $\mathbf{B}$ is obtained as follows. The scalar product of ith row of $\mathbf{A}$ and jth column of $\mathbf{B}$ is the (i, j)th element of $\mathbf{C}$. To obtain the scalar product requires that the number of columns of $\mathbf{A}$ (column dimension of $\mathbf{A}$) must be equal to the number of rows (row dimension) of $\mathbf{B}$. The row and column dimensions of $\mathbf{C}$ correspond to the row dimension of $\mathbf{A}$ and the column dimension of $\mathbf{B}$.

If $\mathbf{x}_i^T$ denotes the ith row of $\mathbf{X}$, and $\mathbf{y}_j$ the jth column of $\mathbf{Y}$, the product of $\mathbf{X}$ and $\mathbf{Y}$ is

$$\mathbf{Z}=\mathbf{XY}=\begin{bmatrix} \mathbf{x}_1^T \\ \mathbf{x}_2^T \\ \vdots \\ \mathbf{x}_n^T \end{bmatrix} \begin{bmatrix} \mathbf{y}_1 & \mathbf{y}_n & \cdots & \mathbf{y}_n \end{bmatrix} = \begin{bmatrix} \mathbf{x}_1^T\mathbf{y}_1 & \mathbf{x}_1^T\mathbf{y}_2 & \cdots & \mathbf{x}_1^T\mathbf{y}_n \\ \mathbf{x}_2^T\mathbf{y}_1 & \mathbf{x}_2^T\mathbf{y}_2 & \cdots & \mathbf{x}_2^T\mathbf{y}_n \\ \mathbf{x}_n^T\mathbf{y}_1 & \mathbf{x}_n^T\mathbf{y}_2 & \cdots & \mathbf{x}_n^T\mathbf{y}_n \end{bmatrix}.$$

For example, if

$$\mathbf{X}=\begin{bmatrix} 2 & 3 \\ 4 & -2 \end{bmatrix}$$

and

$$\mathbf{Y}=\begin{bmatrix} 4 & 3 \\ 1 & 0 \end{bmatrix},$$

then

$$\mathbf{XY}=\begin{bmatrix} 11 & 6 \\ 14 & 12 \end{bmatrix}.$$

It is essential to note that the order of matrices during multiplication is critical because of the clear distinction between the treatment of rows and columns to produce the product. Hence, $\mathbf{XY} \neq \mathbf{YX}$.

In the product $\mathbf{XY}$, the first matrix $\mathbf{X}$ is said to multiply the second matrix $\mathbf{Y}$ or to *premultiply* $\mathbf{Y}$. Similarly, the second matrix $\mathbf{Y}$ is said to *postmultiply* $\mathbf{X}$. Multiplication of matrices satisfies laws of associativity and distributivity over addition. That is, $(\mathbf{XY})\mathbf{Z} = \mathbf{X}(\mathbf{YZ})$ and $\mathbf{X}(\mathbf{Y}+\mathbf{Z}) = \mathbf{XY}+\mathbf{XZ}$.

2.2.2 Special Matrices

The identity matrix $\mathbf{I}$ and the null matrix $\mathbf{N}$ are defined as

$$\mathbf{I}=\begin{bmatrix} 1 & 0 & \cdots & 0 \\ 0 & 1 & 0 & \vdots \\ \vdots & \ddots & \ddots & \vdots \\ 0 & \cdots & \cdots & 1 \end{bmatrix}, \qquad \mathbf{N}=\begin{bmatrix} 0 & 0 & \cdots & 0 \\ 0 & \ddots & \ddots & \vdots \\ \vdots & \vdots & \vdots & \vdots \\ 0 & \cdots & \cdots & 0 \end{bmatrix}.$$

A diagonal matrix denoted by $\mathbf{D}$ has the property that all the off-diagonal elements are zero; for example,

$$\mathbf{D}=\begin{bmatrix} 4 & 0 & 0 \\ 0 & -3 & 0 \\ 0 & 0 & 9 \end{bmatrix}, \tag{2.6}$$

which is written in short notation as $\mathbf{D}=\text{diag}(d_1, d_2, \ldots, d_n)$.

If $a_{ij} = 0$ for $i > j$, a matrix is said to be upper triangular (or right triangular) and all its subdiagonal elements are zero. An example of an upper triangular matrix, represented generally by $\mathbf{R}$, is

$$\mathbf{R} = \begin{bmatrix} 1 & 4 & -6 \\ 0 & -7 & 2 \\ 0 & 0 & 3 \end{bmatrix}.$$

The corresponding definition of lower triangular matrix $\mathbf{L}$ is $L_{ij} = 0$, if $i < j$.

A nonsquare matrix (more columns than rows) that has the property $T_{ij} = 0$, if $i > j$ is said to be a (upper) trapezoidal matrix; for example,

$$\mathbf{T} = \begin{bmatrix} 1 & 2 & 6 & 7 \\ 0 & -1 & -2 & 4 \\ 0 & 0 & 5 & 1 \end{bmatrix}. \tag{2.7}$$

The corresponding property of a nonsquare lower trapezoidal matrix $\mathbf{X}$ is $x_{ij} = 0$, if $i < j$.

2.2.3 Division of Matrices

Unlike numbers, matrices cannot be divided. That is, we cannot write $\mathbf{X}/\mathbf{Y}$.

For two numbers a and b, the quotient a/b, $b \neq 0$ can be alternatively written as ab^{-1} or $b^{-1}a$. For matrices the case is different. Applying the concept of inverse matrices (to be discussed below), one can, in certain cases, obtain $\mathbf{X}^{-1}$ as the inverse of matrix $\mathbf{X}$ to obtain the product $\mathbf{X}^{-1}\mathbf{Y}$ or $\mathbf{Y}^{-1}\mathbf{X}$. However, there is no guarantee that either $\mathbf{X}^{-1}$ or $\mathbf{Y}^{-1}$are defined. Further observe that $\mathbf{X}^{-1}\mathbf{Y} \neq \mathbf{Y}\mathbf{X}^{-1}$. Therefore, the expression $\mathbf{X}/\mathbf{Y}$ cannot be used without ambiguity. Consequently, depending on their existence, the correct representation of operation is $\mathbf{X}^{-1}\mathbf{Y}$ or $\mathbf{Y}^{-1}\mathbf{X}$.

2.2.4 Orthogonality

Any two vectors $\mathbf{x}$ and $\mathbf{y}$ are said to be orthogonal if $\mathbf{x}^T\mathbf{y} = 0$. For example,

$$\mathbf{a} = \begin{bmatrix} 1 \\ -1 \\ 2 \end{bmatrix}$$

and

$$\mathbf{b} = \begin{bmatrix} 1 \\ -1 \\ -1 \end{bmatrix}$$

are *orthogonal* because $\mathbf{a}^T\mathbf{b} = 1 + 1 - 2 = 0$. Any two vectors $\mathbf{a}$ and $\mathbf{b}$ are said to be *orthonormal* if they are orthogonal and have unit Euclidean length, that is, $\mathbf{a} \cdot \mathbf{a}^T = 1$. The above orthogonal vectors **a** and **b** normalized by their Euclidean length are orthonormal. That is, vectors $\mathbf{a}' = (1/\sqrt{6}, -1/\sqrt{6}, 2/\sqrt{6})$ and $\mathbf{b}' = (1/\sqrt{3}, -1\sqrt{3}, -1/\sqrt{3})$ are orthonormal.

A square matrix is said to be an orthogonal matrix if its columns are orthogonal. For example,

$$\mathbf{Q} = \begin{bmatrix} 1 & 1 & 2 \\ -1 & -1 & 2 \\ 2 & -1 & 0 \end{bmatrix}$$

is an orthogonal matrix since its columns are orthogonal vectors.

2.3 LINEAR DEPENDENCE AND INDEPENDENCE

A vector $\mathbf{x}_{r+1}$ is said to be linearly dependent on a set of vectors $\mathbf{x}_1, \mathbf{x}_2, \ldots, \mathbf{x}_r$ if $\mathbf{x}_{r+1}$ can be written as a linear combination of the latter set of vectors. For example, the vector

$$\mathbf{x}_1 = \begin{bmatrix} 15 \\ 30 \\ 2 \end{bmatrix}$$

is a linear combination of

$$\mathbf{x}_2 = \begin{bmatrix} 3 \\ 6 \\ 2 \end{bmatrix}, \qquad \mathbf{x}_3 = \begin{bmatrix} 1 \\ 2 \\ -2 \end{bmatrix},$$

because $\mathbf{x}_1 = 4\mathbf{x}_2 + 3\mathbf{x}_3$. Consequently, $\mathbf{x}_1$ is linearly dependent on the set of vectors $\mathbf{x}_2$ and $\mathbf{x}_3$.

Alternatively, a set of vectors (for example, the set $\mathbf{x}_1, \mathbf{x}_2, \ldots, \mathbf{x}_r$) is said to be linearly dependent if a nontrivial linear combination of the vectors can be found to result in a null vector. In the above, since

$$4\mathbf{x}_2 + 3\mathbf{x}_3 - \mathbf{x}_1 = 0,$$

vectors $\mathbf{x}_1, \mathbf{x}_2, \ldots, \mathbf{x}_3$ are linearly dependent. A corresponding definition addresses the opposite situation of *linear independence* when a vector $\mathbf{x}_{r+1}$ *cannot* be written as a linear combination of a set of vectors $\mathbf{x}_1, \mathbf{x}_2, \ldots, \mathbf{x}_r$.

A set of vectors is said to be linearly independent only if a *trivial* linear combination of the vectors can result in a null vector. For instance, a vector

$$\mathbf{x}_4 = \begin{bmatrix} 1 \\ 2 \\ -3 \end{bmatrix}$$

cannot be written as a linear combination of vectors $\mathbf{x}_1$ and $\mathbf{x}_2$. Therefore, $\mathbf{x}_1$, $\mathbf{x}_2$, and $\mathbf{x}_4$ are linearly independent.

If the columns of a matrix $\mathbf{X}$ are linearly independent, $\mathbf{X}$ is said to have *full column rank*. Similarly, $\mathbf{X}$ is said to have *full row rank* if the rows are linearly independent.

If the columns of $\mathbf{X}$ are independent, the relation $\mathbf{X}\boldsymbol{v} = \mathbf{0}$ indicating a set of linear equations requires that $\boldsymbol{v} = \mathbf{0}$ since, by the above definition of linear independence, a nontrivial linear combination of the columns cannot be zero.

2.4 VECTOR SPACES

Let us represent the n columns of a $(m \times n)$ matrix $\mathbf{X}$ as m-vectors with subscript 1 to n; that is, $\mathbf{x}_1, \mathbf{x}_2, \ldots, \mathbf{x}_n$. Then given a set of n scalars $(\alpha_1, \alpha_2, \cdots, \alpha_n)$ we obtain the linear combination by multiplying jth vector by the jth scalar to get

$$\mathbf{y} = \alpha_1\mathbf{x}_1 + \alpha_2\mathbf{x}_2 + \cdots + \alpha_n\mathbf{x}_n, \tag{2.8}$$

which is called a linear combination of vectors x.

If $\sum_i \alpha_i = 1$, the linear combination is called a *convex combination*. For example, with $\alpha_1 = 0.2$ and $\alpha_2 = 0.8$, we get the convex linear combination of $\mathbf{x}_1 = [1, 2, 3]^T$, and $\mathbf{x}_2 = [3, 4, 5]^T$ as $\mathbf{y} = [0.2, 0.4, 0.6]^T + [2.4, 3.2, 4.0]^T = [2.6, 3.6, 4.6]^T$. The significance of the convex combination is that the vector $\mathbf{y}$ terminates on the line joining $\mathbf{x}_1$ and $\mathbf{x}_2$.

The process of forming a linear combination can be written as a matrix–vector product. For example, for a matrix $\mathbf{X}$ with ith column $\mathbf{x}_i$, (2.8) can be written as

$$\mathbf{y} = [x_1, x_2, \ldots, x_n] \begin{bmatrix} \alpha_1 \\ \alpha_2 \\ \vdots \\ \alpha_n \end{bmatrix}. \tag{2.9}$$

Consequently, a matrix–vector product gives linear combinations of columns of matrix with the components of the vector as the coefficients. The following reinforces this:

$$\mathbf{A}\boldsymbol{v} = \begin{bmatrix} a_{11} & a_{12} & a_{13} \\ a_{21} & a_{22} & a_{23} \\ a_{31} & a_{32} & a_{33} \end{bmatrix} \begin{bmatrix} v_1 \\ v_2 \\ v_3 \end{bmatrix} = \begin{bmatrix} a_{11}v_1 + a_{12}v_2 + a_{13}v_3 \\ a_{21}v_1 + a_{22}v_2 + a_{23}v_3 \\ a_{31}v_1 + a_{32}v_2 + a_{33}v_3 \end{bmatrix}, \tag{2.10}$$

or

$$\mathbf{A}\boldsymbol{v} = \boldsymbol{v}_1 \begin{bmatrix} a_{11} \\ a_{21} \\ a_{31} \end{bmatrix} + \boldsymbol{v}_2 \begin{bmatrix} a_{12} \\ a_{22} \\ a_{32} \end{bmatrix} + \boldsymbol{v}_3 \begin{bmatrix} a_{13} \\ a_{23} \\ a_{33} \end{bmatrix}, \tag{2.11}$$

which indicates a linear combination of the three columns of matrix $\mathbf{A}$.

2.4.1 Discussion of Vector Space

From a collection of all possible vectors of dimension n, consider any vector $\mathbf{x}$. Then $\alpha\mathbf{x}$ also a vector of dimension n as is the vector $\mathbf{z} = \mathbf{x} + \mathbf{y}$, where $\mathbf{y}$ is any another n vector from the collection of vectors. Similarly, a linear combination of the two vectors $\mathbf{a} = \alpha_1\mathbf{x} + \alpha_2\mathbf{y}$ is also of dimension n. Because of these properties, the n-dimensional vector set is said to be *closed* with respect to these operations. Therefore, the set of vectors is said to form an n-dimensional *linear vector space* designated by R^n or E_n.

Various linear combinations of two independent 2-vectors generate the totality of the two-dimensional vector space R^2. Any two linearly independent vectors $\mathbf{u}$ and $\mathbf{v}$ are said to span the 2-space. They are also said to constitute a *basis* for the two space. For example, consider two 2-vectors (1,2) and (2,1). Linear combinations of these vectors can span the entire R^2 space. Alternatively, consider two *unit* vectors [1,0], and [0,1] that are also linearly independent. A linear combination of these spans the 2-space as well. That is why we said earlier that any two independent vectors form *a* basis and not *the* basis.

An analogous extension to the totality of 3-vectors is obvious. For example, the three independent unit vectors

$$\mathbf{e}_1 = \begin{bmatrix} 1 \\ 0 \\ 0 \end{bmatrix}, \qquad \mathbf{e}_2 = \begin{bmatrix} 0 \\ 1 \\ 0 \end{bmatrix}, \qquad \mathbf{e}_3 \begin{bmatrix} 0 \\ 0 \\ 1 \end{bmatrix}$$

that are independent span the 3-space R^3. Consequently, the entire 3-space can be generated from these unit vectors which form a basis for the R^3. For example, the vector $[1, 2, -2]^T$ can be considered as the linear combination $\mathbf{e}_1 + 2\mathbf{e}_2 - 2\mathbf{e}_3$.

The extension of this to n-space is obvious. Linearly independent vectors that form a basis span the n-space called the *Euclidean n-space*. The n vectors are n-element vectors expressed as an ordered *n-tuple* that represents a point in the n-space, or an arrow originating at the origin and extending to the said point.

To explain the concept of the Euclidean space, consider the distance between two vectors $\mathbf{a}$ and $\mathbf{b}$. The distance is a real-valued function $d = d(\mathbf{a}, \mathbf{b})$ with the following properties: (1) When $\mathbf{a}$ and $\mathbf{b}$ coincide, the distance is zero; (2) when $\mathbf{a}$ and $\mathbf{b}$ are distinct, the distance between $\mathbf{a}$ and $\mathbf{b}$ is the same as the distance between $\mathbf{b}$ and $\mathbf{a}$ and are positive real numbers; and (3) the distance between $\mathbf{a}$ and $\mathbf{b}$ is less than the distance from $\mathbf{a}$ to $\mathbf{c}$ (a point distinct from $\mathbf{a}$ or $\mathbf{b}$) plus the distance from $\mathbf{c}$ to $\mathbf{b}$. Mathematically, these can be expressed as

$$d(\mathbf{a}, \mathbf{b}) = 0 \quad (\text{for } \mathbf{a} = \mathbf{b}),$$

$$d(\mathbf{a}, \mathbf{b}) = d(b, a) > 0 \quad (\text{for } \mathbf{a} \neq \mathbf{b}),$$

$$d(\mathbf{a}, \mathbf{b}) \leq d(\mathbf{a}, \mathbf{c}) + d(\mathbf{c}, \mathbf{b}) \quad (\text{for } \mathbf{c} \neq \mathbf{a}, \mathbf{b}).$$

If vectors **x** and **y** are two n-tuple vectors with coordinates $(x_1, x_2, \ldots, x_n)$ and $(y_1, y_2, \ldots, y_n)$, the Euclidean distance between the vectors is given by

$$d(\mathbf{x}, \mathbf{y}) = \sqrt{(x_1 - y_1)^2 + (x_2 - y_2)^2 + \cdots + (x_n - y_n)^2}.$$

This result is the generalization of the *Pythagorean Theorem.* The Euclidean distance is used as a measure of convergence of algorithms and optimization procedures.

2.5 LINEAR TRANSFORMATIONS

We have seen earlier that multiplying a vector on the left by a matrix yields a second vector. Thus the matrix can be considered as a tool to *transform* one vector into another. The matrix is also said to be applied to the first vector to obtain the transformed vector.

The transformation of a vector by a matrix is a linear transformation because of properties already observed. With any two vectors **x** and **y** and a transforming matrix **A**, for any scalars α and β, we have

$$\mathbf{A}(\alpha\mathbf{x} + \beta\mathbf{y}) = \mathbf{A}(\alpha\mathbf{x}) + \mathbf{A}(\beta\mathbf{y}) = \alpha(\mathbf{A}\mathbf{x}) + \beta(\mathbf{A}\mathbf{y}).$$

As a result, we see that the transformation of a linear combination of vectors is the same as the linear combination of transformed vectors.

2.5.1 Properties of Transformations

It is of interest to know whether a nonzero vector exists that can be transformed by a matrix **A** into a null vector. That is, for a nonzero **x**, can the relation

$$\mathbf{A}\mathbf{x} = \mathbf{0} \tag{2.12}$$

be satisfied? It has already been said that this is possible only if the columns of **A** are linearly dependent. Such a matrix with linearly dependent columns is said to be *singular.*

If (2.12) holds only when $\mathbf{x} = \mathbf{0}$, it implies that the columns of **A** are linearly independent, and such a matrix is said to be *nonsingular.* The test for linear independence is that the determinant of the matrix is nonzero. A zero determinant indicates a singular matrix with linearly dependent columns. The highest order of the nonzero determinant of a matrix that can be obtained by eliminating columns (or rows) of the matrix indicates the rank of the matrix. Consequently, if the determinant of a $(n \times n)$ matrix is nonzero, the matrix has full rank equal to n.

2.5.2 Inverse of a Matrix

The process of computing the inverse of a matrix can be seen as the reverse process of a transformation. If some n-vector $\mathbf{x}$ has been transformed to a vector $\mathbf{y}$ by a matrix $\mathbf{A}$, we seek a matrix to undo this transformation. Such a matrix termed $\mathbf{A}^{-1}$, the inverse of $\mathbf{A}$, applied to $\mathbf{y}$ will transform $\mathbf{y}$ back to $\mathbf{x}$. Such a procedure is possible only when $\mathbf{A}$ is nonsingular. Consequently, we have

$$\mathbf{A}^{-1}(\mathbf{A}\mathbf{x}) = \mathbf{x}, \tag{2.13}$$

or

$$(\mathbf{A}^{-1}\mathbf{A} - \mathbf{I})\mathbf{x} = \mathbf{0},$$

thus requiring $\mathbf{A}^{-1}\mathbf{A} = \mathbf{I}$, where $\mathbf{I}$ is the identity matrix. Therefore, if $\mathbf{A}^{-1}$ exists, it is nonsingular and satisfies this relation. It is easy to show that $(\mathbf{AB})^{-1} = \mathbf{B}^{-1}\mathbf{A}^{-1}$ because $(\mathbf{AB})\mathbf{B}^{-1}\mathbf{A}^{-1} = \mathbf{I}$.

Certain matrices have special forms of inverses. The inverse of an upper (lower) triangular matrix is an upper (lower) triangular matrix. If $\mathbf{X}$ is an orthogonal matrix, then $\mathbf{X}^{-1} = \mathbf{X}^T$. The inverse of a sparse matrix (most entries being zero) is not a sparse matrix.

2.6 EIGENVALUES AND EIGENVECTORS

We now want to find a vector $\mathbf{u}$ such that when any square matrix $\mathbf{A}$ is applied to it, the transformed vector is $\lambda\mathbf{u}$, where λ is a scalar, which is nothing but a multiple of the original vector. That is,

$$\mathbf{A}\mathbf{u} = \lambda\mathbf{u}. \tag{2.14}$$

The scalar λ is called the *eigenvalue* of $\mathbf{A}$, and $\mathbf{u}$ is an *eigenvector* corresponding to the eigenvalue λ. As an example, 1 and -4 are the eigenvalues of the matrix

$$\mathbf{A} = \begin{bmatrix} 3 & 1 \\ 2 & 2 \end{bmatrix};$$

when considering the eigenvector $\mathbf{u}_1 = [1, -2]^T$, we get

$$[\mathbf{A}]\begin{bmatrix} 1 \\ -2 \end{bmatrix} = \begin{bmatrix} 1 \\ -2 \end{bmatrix} = 1\begin{bmatrix} 1 \\ -2 \end{bmatrix} = 1\mathbf{u}_1.$$

A similar confirmation can be made for the eigenvalue -4 with an eigenvector $\mathbf{u}_2$ of

$$\begin{bmatrix} 2 \\ 2 \end{bmatrix}$$

corresponding to this eigenvalue. Note that the eigenvectors $\mathbf{u}_1$ and $\mathbf{u}_2$ are defined uniquely in direction only because any nonzero multiple of $\mathbf{u}_1$ and $\mathbf{u}_2$ will also satisfy (2.14). Equation (2.14) can also be written as

$$(\mathbf{A} - \lambda \mathbf{I})\mathbf{u} = \mathbf{0}, \tag{2.15}$$

implying that the matrix $\mathbf{A} - \lambda \mathbf{I}$ is singular if λ is an eigenvalue of $\mathbf{A}$. Matrix $\mathbf{A} - \lambda \mathbf{I}$ is singular if its determinant is zero. The resulting polynomial for the determinant in λ is called *characteristic polynomial.*

If π_A is the product of eigenvalues of a square matrix $\mathbf{A}$, we have the property that the product of eigenvalues of the product of the matrices is equal to the product of the products of eigenvalues of the individual matrices. That is,

$$\pi_{AB} = \pi_A \pi_B.$$

Equation (2.15) can be written as

$$(\mathbf{A} - \lambda \mathbf{I})\mathbf{u} = \mathbf{K}\mathbf{u} = \mathbf{0}, \tag{2.16}$$

which represents a set of homogeneous equation associated with each eigenvalue λ that has to be solved for the elements of the eigenvector $\mathbf{u}$. As discussed earlier, if a nontrivial solution for $\mathbf{u}$ exists, the determinant of matrix $\mathbf{K}$—that is, matrix $(\mathbf{A} - \lambda \mathbf{I})$—must vanish. The determinant of $\mathbf{K}$ is termed the *characteristic equation*. In order that the determinant of $\mathbf{K}$ is zero, it is necessary that the eigenvalues λ_1, λ_2, and so on, are the roots of the characteristic equation. The nontrivial solutions obtained from (2.16) are the eigenvectors associated with the eigenvalues used.

From (2.15), the eigenvalues of a matrix can be computed. The procedure involves the formation of a symmetric matrix with λ subtracted from the diagonal terms as in (2.15). The expression for the determinant gives the characteristic polynomial, a polynomial of degree n where n is the order of the matrix. The roots of the characteristic polynomial are the eigenvalues. An example of computing eigenvalues is shown in Section 2.10.2. As said earlier, the eigenvalues of a symmetric real matrix are real and different.

A corollary is that of computing the eigenvector[1] corresponding to a given eigenvalue. For a given eigenvalue λ, (2.15) can be used to compute the eigenvector $\mathbf{u}$. However, since the matrix on the left-hand side of this equation is singular (determinant is zero), (2.15) will have infinite solutions. The eigenvector is specified in direction only, but it can have any magnitude. However, by specifying the norm of the eigenvector (magnitude), a solution can be obtained. Generally, unit magnitude is specified. An example worked out in Section 2.10.4 illustrates the procedure.

[1]The importance of eigenvectors in optimization problems is that they point toward appropriate search directions for nonlinear optimization problems. This is discussed in Chapter 5.

Next we examine the relation between the powers of matrix to the power of its eigenvalues. An $n \times n$ matrix has n eigenvalues, all not necessarily distinct. From (2.14) we have

$$\lambda_i \mathbf{u}_i = \mathbf{A}\mathbf{u}_i, \qquad i = 1, 2, \ldots, n, \tag{2.17}$$

where $\mathbf{u}_i$ is any eigenvector associated with λ_i.

If (2.17) is premultiplied by $\mathbf{A}$, we obtain

$$\lambda_i \mathbf{A}\mathbf{u}_i = \lambda_i^2 \mathbf{u}_i = \mathbf{A}^2 \mathbf{u}_i. \tag{2.18}$$

It can be seen by repeated multiplications that

$$\mathbf{A}^r \mathbf{u}_i = \lambda_i^r \mathbf{u}_i. \tag{2.19}$$

Let the ith eigenvector be represented as

$$\mathbf{u}_i = \begin{bmatrix} u_{1i} \\ u_{2i} \\ \vdots \\ u_{ni} \end{bmatrix}. \tag{2.20}$$

Grouping the eigenvector columns now forms a square matrix $\mathbf{U}$:

$$\mathbf{U} = \begin{bmatrix} u_{11} & u_{12} & u_{13} & \cdots & u_{1n} \\ u_{21} & u_{22} & u_{23} & \cdots & u_{2,n} \\ \vdots & \vdots & \vdots & \cdots & \vdots \\ u_{n1} & u_{n2} & u_{n3} & \cdots & u_{nn} \end{bmatrix}. \tag{2.21}$$

If $\mathbf{D}$ is the square diagonal matrix of n eigenvalues in the form

$$\mathbf{D} = \begin{bmatrix} \lambda_1 & 0 & 0 & \cdots & 0 \\ 0 & \lambda_2 & 0 & \cdots & 0 \\ \vdots & \vdots & \vdots & \cdots & \vdots \\ 0 & 0 & 0 & \cdots & \lambda_n \end{bmatrix}, \tag{2.22}$$

then the n equations of (2.17) can be written in the form

$$\mathbf{UD} = \mathbf{AU}. \tag{2.23}$$

It can be shown that matrix $\mathbf{U}$ is nonsingular and possesses an inverse $\mathbf{U}^{-1}$ if n eigenvalues are distinct. Consequently, (2.23) can be written in the form

$$\mathbf{D} = \mathbf{U}^{-1}\mathbf{AU}, \tag{2.24}$$

or

$$\mathbf{A} = \mathbf{UDU}^{-1}. \tag{2.25}$$

Thus, a matrix $\mathbf{U}$ (and its inverse) that diagonalizes $\mathbf{A}$ may be formed by grouping the eigenvectors into a square matrix. This process is termed the diagonal transformation of $\mathbf{A}$, which is possible if the eigenvalues of $\mathbf{A}$ are all different. When two or more of the eigenvalues of $\mathbf{A}$ are equal to each other, such transformation into diagonal form is not always possible.

Similarity. Two matrices are said to be similar if they have the same eigenvalues. If a matrix $\mathbf{A}$ is pre- and postmultiplied by $\mathbf{W}$ and $\mathbf{W}^{-1}$, where $\mathbf{W}$ is a nonsingular matrix, then $\mathbf{WAW}^{-1}$ is similar to $\mathbf{A}$ because if λ is an eigenvalue of $\mathbf{A}$, we have

$$\mathbf{Ax} = \lambda\mathbf{x}.$$

Using this relation in the expression $\mathbf{WAW}^{-1}(\mathbf{Wx})$, we get

$$\mathbf{WAW}^{-1}(\mathbf{Wx}) = \lambda(\mathbf{Wx}).$$

Therefore the eigenvalues of $\mathbf{WAW}^{-1}$ are the same as that of $\mathbf{A}$, with corresponding eigenvector $\mathbf{Wx}$.

A real $(n \times n)$ matrix has n eigenvalues, all of which are necessarily distinct. Consequently, such a matrix has at most n independent eigenvectors. Generally, the eigenvalues of a real matrix are complex numbers (roots of the characteristic equation). However, it can be proved that the eigenvalues of a symmetric matrix (with the property $\mathbf{A} = \mathbf{A}^T$) are all distinct. Also as a consequence, all n eigenvectors are distinct. Fortunately, the formulation of most physical problems, including those addressed in this book, result in symmetric matrices.

If a matrix $\mathbf{A}$ is nonsingular, all its eigenvalues are nonzero. Then $\mathbf{A}^{-1}$ has eigenvalues that are reciprocals of the eigenvalues of $\mathbf{A}$.

The *spectral radius* of a matrix $\mathbf{A}$ is defined as the maximum absolute value of the eigenvalue; that is, $\rho(\mathbf{A}) = \max |\lambda_j[\mathbf{A}]|$.

The eigenvalues are bounded by the following upper and lower limits:

$$\lambda_{\max}[\mathbf{A}] = \max_{x \neq 0} \frac{\mathbf{x}^T\mathbf{Ax}}{\mathbf{x}^T\mathbf{x}}$$

and

$$\lambda_{\min}[\mathbf{A}] = \min_{x \neq 0} \frac{\mathbf{x}^T\mathbf{Ax}}{\mathbf{x}^T\mathbf{x}}.$$

The n distinct eigenvectors can be made to form an orthonormal set. That is,

$$\mathbf{U}_j^T\mathbf{U}_i = 0, \quad j \neq i; \qquad \mathbf{U}_j^T\mathbf{U}_j = 1.$$

Therefore, an orthonormal basis for R^n can be formed from the eigenvectors using this procedure.

2.6.1 Definiteness

If a symmetric matrix has all eigenvalues that are positive, the matrix is said to be *positive definite*. A positive definite matrix has the property

$$\mathbf{x}^T \mathbf{A} \mathbf{x} > 0.$$

Corresponding definition holds for a negative definite matrix. If a symmetric matrix has positive and negative eigenvalues, it is said to be *indefinite*. If a matrix has some eigenvalues that are positive and some that are zero (i.e., all eigenvalues are nonnegative), the matrix is said to be *semidefinite*. Test for definiteness based on the determinants of principal minors is discussed further in Section 5.4.

2.7 LINEAR EQUATIONS

A fundamental problem in linear algebra is that of solving a system of linear equations. The problem is stated as follows: Given an $m \times n$ matrix $\mathbf{A}$, find a n-vector $\mathbf{x}$ in order to satisfy the relationship

$$\mathbf{A}\mathbf{x} = \mathbf{b}, \tag{2.26}$$

where $\mathbf{b}$ is a m vector. In (2.26), $\mathbf{x}$ is the vector of unknowns that is transformed by $\mathbf{A}$ to $\mathbf{b}$. Also, the vector $\mathbf{x}$ is the set of coefficients in a linear combination of the columns of matrix $\mathbf{A}$. Equation (2.26) has a solution if $\mathbf{b}$ lies in the space spanned by the columns of $\mathbf{A}$. Then the system of equations (2.26) is said to be *compatible*. For example, the system of equations

$$\begin{bmatrix} 1 & 4 \\ 2 & 1 \end{bmatrix} \begin{bmatrix} x_1 \\ x_2 \end{bmatrix} = \begin{bmatrix} b_1 \\ b_2 \end{bmatrix}$$

is compatible for any $\mathbf{b}$ since the columns of $\mathbf{A}$ are linearly independent (the columns span R^2). Even if $\mathbf{A}$ is not a square matrix, one can have a compatible system of equation with a unique solution. For example, for the system

$$\begin{bmatrix} 1 & 0 \\ 2 & 1 \\ 3 & 1 \end{bmatrix} \begin{bmatrix} x_1 \\ x_2 \end{bmatrix} = \begin{bmatrix} 2 \\ 5 \\ 7 \end{bmatrix}$$

the solution is $x_1 = 2$, $x_2 = 1$. Since the columns of the matrix are independent, and the right-hand side can be expressed as a linear combination of the columns of the matrix with the coefficients equal to the values of x_1 and x_2.

In an *incompatible* system of equations, the system of (2.26) will have no solution; that is, a vector of **x** cannot be found that satisfies the relation (2.26). For example,

$$\begin{bmatrix} 1 & 0 \\ 2 & 1 \\ 1 & 0 \end{bmatrix} \begin{bmatrix} x_1 \\ x_2 \end{bmatrix} = \begin{bmatrix} -2 \\ 4 \\ 2 \end{bmatrix}$$

is incompatible because, despite the linear independence of the columns of **A**, no linear combination of these columns can produce the right-hand-side vector.

However, if **b** is compatible with **A**, the system of (2.26) has a unique solution if and only if the columns of **A** are linearly independent.

If columns in **A** are linearly dependent, we have a nonzero vector **y** such that $\mathbf{Ay} = \mathbf{0}$. If **x** is the solution of (2.26), then for an arbitrary scalar ε, we have

$$\mathbf{A}(\mathbf{x} + \varepsilon\mathbf{y}) = \mathbf{Ax} + \varepsilon\mathbf{Ay} = \mathbf{Ax} = \mathbf{b}.$$

Consequently, $\mathbf{x} + \varepsilon\mathbf{y}$ also is a solution of (2.26). This means that there are an infinite number of solutions to (2.26) if the columns of **A** are linearly dependent.

For example,

$$\begin{bmatrix} 1 & 2 \\ 2 & 4 \\ 3 & 6 \end{bmatrix} \begin{bmatrix} x_1 \\ x_2 \end{bmatrix} = \begin{bmatrix} -1 \\ -2 \\ -3 \end{bmatrix} \tag{2.27}$$

is a compatible system with a solution of $(1, -1)^T$. The matrix has linearly dependent columns since

$$\begin{bmatrix} 1 & 2 \\ 2 & 4 \\ 3 & 6 \end{bmatrix} \begin{bmatrix} -2 \\ 1 \end{bmatrix} = \begin{bmatrix} 0 \\ 0 \\ 0 \end{bmatrix}. \tag{2.28}$$

Therefore, (2.27) has an infinite number of solutions of the form

$$\begin{bmatrix} 1 \\ -1 \end{bmatrix} + \varepsilon \begin{bmatrix} -2 \\ 1 \end{bmatrix}.$$

2.7.1 Solution of Linear Algebraic Equations

The rank of a matrix is equal to the order of the largest nonvanishing determinant contained in the matrix. For example, the ranks of

$$\begin{bmatrix} 2 & -1 & 0 \\ 2 & 2 & -2 \end{bmatrix}, \quad \begin{bmatrix} 0 & 0 & 0 \\ 0 & 0 & 0 \\ 0 & 0 & 0 \end{bmatrix}, \quad \begin{bmatrix} 1 & 0 \\ 0 & 1 \end{bmatrix}, \quad \begin{bmatrix} 3 & 0 & 0 \\ 0 & 4 & 0 \\ 0 & 0 & 3 \end{bmatrix}, \tag{2.29}$$

and

$$\begin{bmatrix} 1 & 0 & 2 \\ -2 & 3 & -4 \\ 1 & 2 & 2 \end{bmatrix} \tag{2.30}$$

are 2, 0, 2, 3, and 2, respectively. Observe that the concept of rank also applies to nonsquare matrices as shown in the above example. It is easily seen that for a square matrix, if its rank is equal to the order of the matrix, it necessarily has an inverse.

Consider the system of equations

$$\mathbf{Ax} = \mathbf{b}, \tag{2.31}$$

where $\mathbf{A}$ is an $m \times n$ coefficient matrix, $\mathbf{x}$ is a $n \times 1$ vector of n unknowns, and $\mathbf{b}$ is an $m \times 1$ vector of constants. Define an $m \times (n+1)$ augmented matrix as

$$[\mathbf{Aug}] = [\mathbf{A}, \mathbf{b}]$$

formed by attaching an additional column of the vector of known constants to the coefficient matrix.

The system of equations in (2.31) is said to be consistent if a solution vector $\mathbf{x}$ exists which satisfies the system. If no solution exists, the system is said to be inconsistent.

For m algebraic equations in n unknowns described by (2.31), we now state the following three conditions without proof.

1. Rank of $[\mathbf{A}] = \text{rank}[\mathbf{A}, \mathbf{b}] = r \leq m$ (rank of coefficient matrix is less than the number of equations): A solution for any r unknowns in terms of $n - r$ remaining can be obtained.
2. Rank of $[\mathbf{A}] = \text{rank}[\mathbf{A}, \mathbf{b}] = r = n$ (rank of coefficient matrix is equal to number of unknowns): A unique solution for all n unknowns exists.
3. Rank of $[\mathbf{A}] = \text{rank}[\mathbf{A}, \mathbf{b}] = r < n$ (rank of coefficient matrix is less than the number of unknowns): A solution always exists.

Note that the above conditions require the ranks of the coefficient matrix $[\mathbf{A}]$ and the augmented matrix $[\mathbf{A}, \mathbf{b}]$ to be equal. This requirement is basic and the *only* requirement for the existence of a solution to (2.31). In other words, this requirement ensures that a system of equations is consistent and compatible.

2.8 VECTOR AND MATRIX NORMS

A norm is means of "measuring" vectors and matrices in order to assess their relative "magnitudes." A norm is a nonnegative scalar associated with a vector or a matrix.

- A vector norm is denoted by $\|\cdot\|$. It satisfies the following properties.

  1. $\|\mathbf{x}\| \geq 0$ for all vectors $\mathbf{x}$, and $\|\mathbf{x}\| = 0$ if and only if $\mathbf{x}$ is a null vector.
  2. $\|\delta\mathbf{x}\| = |\delta|\,\|\mathbf{x}\|$, where δ is any real number.
  3. Triangle inequality: For any two vectors $\mathbf{x}$ and $\mathbf{y}$, $\|\mathbf{x}+\mathbf{y}\| \leq \|\mathbf{x}\| + \|\mathbf{y}\|$.

- The p norm of a vector is defined as

$$\|\mathbf{x}\|_p = \left(\sum_{i=1}^{n} |x_i|^p\right)^{1/p}.$$

  The p norm satisfies the above properties (1) to (3).
- The *one, two,* and *infinity* norms are derived from the p norm with $p = 1, 2,$ and ∞. The two norm is also termed the *Euclidean* norm. Note that

$$\|\mathbf{x}\|_2^2 = x_1^2 + x_2^2 + \cdots + x_n^2 = \mathbf{x}^T\mathbf{x}.$$

  The infinity norm represents the absolute value of the maximum element of the vector. For example, if $\mathbf{x}^T = (4, -6, -1)$, then, $\|\mathbf{x}\|_1 = 11$, $\|\mathbf{x}\|_2 = \sqrt{53}$, and $\|\mathbf{x}\|_\infty = 6$. The above concepts lead to *Schwartz inequality* that states

$$\left|\mathbf{x}^T\mathbf{y}\right| \leq \|\mathbf{x}\|_2\|\mathbf{y}\|_2.$$

  The proof of this inequality is left to the reader.

As for the vector, a *matrix norm* denoted by $\|.\|$ is a nonnegative scalar that satisfies three properties analogous to those of the vector:

1. $\|\mathbf{A}\| \geq 0$ for all $\mathbf{A}$, and $\|\mathbf{A}\| = 0$ if and only if $\mathbf{A}$ is a null matrix.
2. $\|\delta\mathbf{A}\| = |\delta|\,\|\mathbf{A}\|$, where δ is any real number.
3. $\|\mathbf{A}+\mathbf{B}\| \leq \|\mathbf{A}\| + \|\mathbf{B}\|$.

Additionally, a matrix norm satisfies the property $\|\mathbf{A}+\mathbf{B}\| \leq \|\mathbf{A}\|\|\mathbf{B}\|$.

Matrix norms can be defined in terms of vector norms or *induced* by vector norms. Consider $\|\mathbf{Ax}\|$ for all vectors $\mathbf{x}$ such that the vector has a unit norm—that is, $\|\mathbf{x}\| = 1$. The corresponding matrix norm *induced* by, or *subordinate* to, the vector norm is given by

$$\|\mathbf{A}\| = \max_{\|\mathbf{x}\|=1} \|\mathbf{Ax}\|.$$

In the above, the maximum is obtained for some vector $\mathbf{x}$ that has a unit norm.

Three earlier defined vector norms induce three norms corresponding to them. For an $m \times n$ matrix they are

$$\|\mathbf{A}\|_1 = \max_{1 \le j \le n} \left(\sum_{i=1}^{m} |a_{ij}| \right),$$

the maximum absolute column sum;

$$\|\mathbf{A}\|_2 = \left(\lambda_{\max} \left[\mathbf{A}^T \mathbf{A} \right] \right)^{1/2},$$

the square root of the largest eigenvalue of $\mathbf{A}^T\mathbf{A}$; and

$$\|\mathbf{A}\|_\infty = \max_{1 \le j \le n} \left(\sum_{i=1}^{m} |a_{ij}| \right),$$

the maximum absolute row sum.

Another norm not induced by a vector norm is the *Forbenius norm*, denoted as $\|.\|_F$. It is computed as the Euclidean norm of a vector that contains all the elements of an $m \times n$ matrix:

$$\|.\|_F = \left(\sum_{i=1}^{m} \sum_{j=1}^{n} a_{ij}^2 \right)^{\frac{1}{2}}$$

2.9 CONDITION NUMBER

The condition number of a matrix $\mathbf{A}$, denoted by Cond($\mathbf{A}$), indicates how the solution of a system of equations as in (2.26) is affected by small changes in the right-hand side, or by small changes to the matrix elements of $\mathbf{A}$. The condition number is given by a product of the matrix norm and the norm of its inverse as

$$\text{Cond}(\mathbf{A}) = \|\mathbf{A}\| \|\mathbf{A}^{-1}\| \ge 1.$$

For a proof of this see Gill et al. (1981). As an illustration, consider

$$\mathbf{Ax} = \mathbf{b},$$

where

$$\mathbf{A} = \begin{bmatrix} 0.05 & 0.04 \\ 0.44 & 0.35 \end{bmatrix}, \quad \mathbf{x} = \begin{bmatrix} x_1 \\ x_2 \end{bmatrix}, \quad \text{and} \quad \mathbf{b} = \begin{bmatrix} 0.09 \\ 0.79 \end{bmatrix},$$

which has a solution $\mathbf{x} = (1.0, -1.0)^T$. Using infinity norms, we have $\|\mathbf{A}\|_\infty = 0.79$, $\|\mathbf{x}\|_\infty = 1.0$, and $\|\mathbf{b}\|_\infty = 0.79$. We obtain (check arithmetic either by using Excel or MATLAB)

$$\mathbf{A}^{-1} = \begin{bmatrix} -3500 & 400 \\ 4400 & -500 \end{bmatrix}$$

such that $\|\mathbf{A}^{-1}\|_\infty = 4900$, giving $\mathrm{Cond}(\mathbf{A}) = \|\mathbf{A}\|_\infty \|\mathbf{A}^{-1}\|_\infty = 0.79 \times 4900 = 3871$.

This indicates that because of the "large" condition number, "small" changes in the data (**b** or the entries of **A**) will produce "substantial" changes to the solution. To this end, consider a perturbation of $(10^{-4}, 10^{-4})^T = \Delta\mathbf{b}$ to the **b** vector such that the input **b** vector is $(0.0901, 0.7901)^T$. For **x** we now obtain a solution $\mathbf{x} = (0.69, 1.39)^T$. The change in solution $\Delta\mathbf{x}$ is given by $\Delta\mathbf{X} = (0.31, -0.39)^T$. Consequently, we get $\|\Delta\mathbf{b}\|/\|\mathbf{b}\| = 10^{-4}/0.79 = 0.000126$, whereas $\|\Delta\mathbf{x}\|/\|\mathbf{x}\| = 0.39/1.0 = 0.39$. Clearly, this shows that the relative change in the solution for **x** is larger than the changes to the right-hand-side vector **b**. A similar analysis perturbing the elements of **A** derives the same conclusion.

2.10 SOLVED PROBLEMS

2.10.1 Inverse of a Complex Matrix

This example deals with the computation of an inverse matrix to a matrix whose entries are complex numbers of the type $x + jy$, where $j = \sqrt{-1}$. Complex.Xls

Let a complex matrix be represented by $\mathbf{X} + j\mathbf{Y}$, whose inverse is $\mathbf{U} + j\mathbf{V}$. Then, by definition of the inverse, for the real and imaginary parts of the product of the matrix and its inverse, we have

$$\mathbf{XU} - \mathbf{YV} = \mathbf{I} \tag{2.32}$$

and

$$\mathbf{YU} + \mathbf{XV} = 0. \tag{2.33}$$

From (2.33), one obtains

$$\mathbf{YU} = -\mathbf{XV},$$

or

$$\mathbf{U} = -\mathbf{Y}^{-1}\mathbf{XV}. \tag{2.34}$$

A substitution of **U** obtained from (2.34) into (2.32) gives

$$(-\mathbf{XY}^{-1}\mathbf{X} - \mathbf{Y})\mathbf{V} = \mathbf{I}.$$

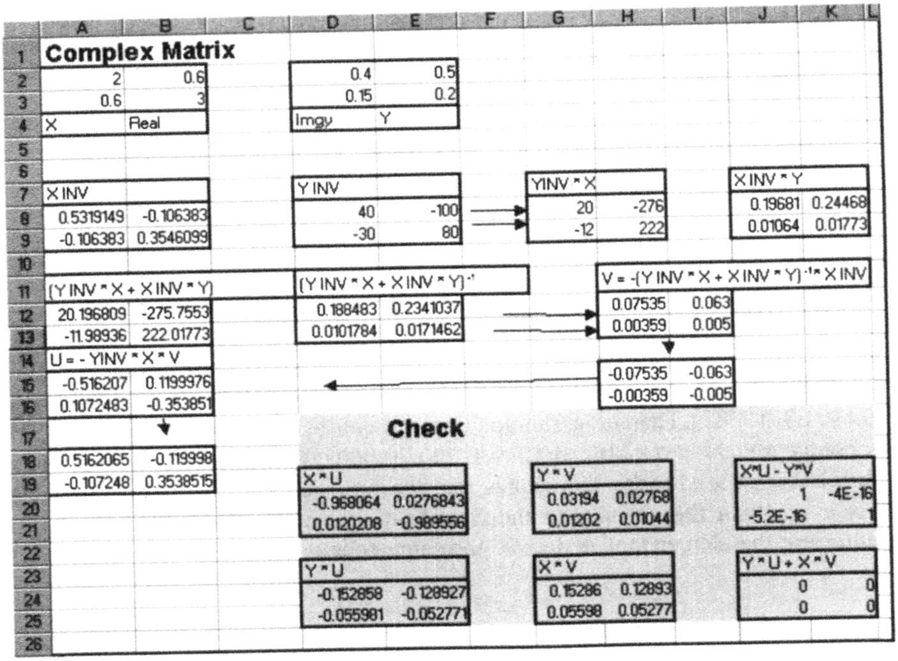

Figure 2.1. Inverse of a complex matrix.

A premultiplication of the above by $\mathbf{X}^{-1}$ gives

$$(-\mathbf{I}\mathbf{Y}^{-1}\mathbf{X} - \mathbf{X}^{-1}\mathbf{Y})\mathbf{V} = \mathbf{X}^{-1},$$

or

$$\mathbf{V} = (-\mathbf{Y}^{-1}\mathbf{X} - \mathbf{X}^{-1}\mathbf{Y})^{-1}\mathbf{X}^{-1}. \tag{2.35}$$

Figure 2.1 shows the spreadsheet of "Complex.Xls," in which the formulae of (2.34) and (2.35) are used to compute the inverse. Arbitrarily, a 2×2 matrix has been chosen. Its real components are shown in cells A2:B3, and imaginary components are shown in cells D2:E3. The above equations have been incorporated to obtain the real and imaginary parts, **U** and **V** of the inverse matrix, in cells A15:B16, and H15:I16. A check to confirm that the product of the inverse and the original matrix gives $\mathbf{I} + j\mathbf{0}$ is also made.

2.10.2 Computation of Eigenvalue

Find the eigenvalues of

$$[\mathbf{A}] = \begin{bmatrix} 4 & 2 & -2.5 \\ 2 & 3 & 3 \\ -2.5 & 3 & 4 \end{bmatrix}. \tag{2.36}$$

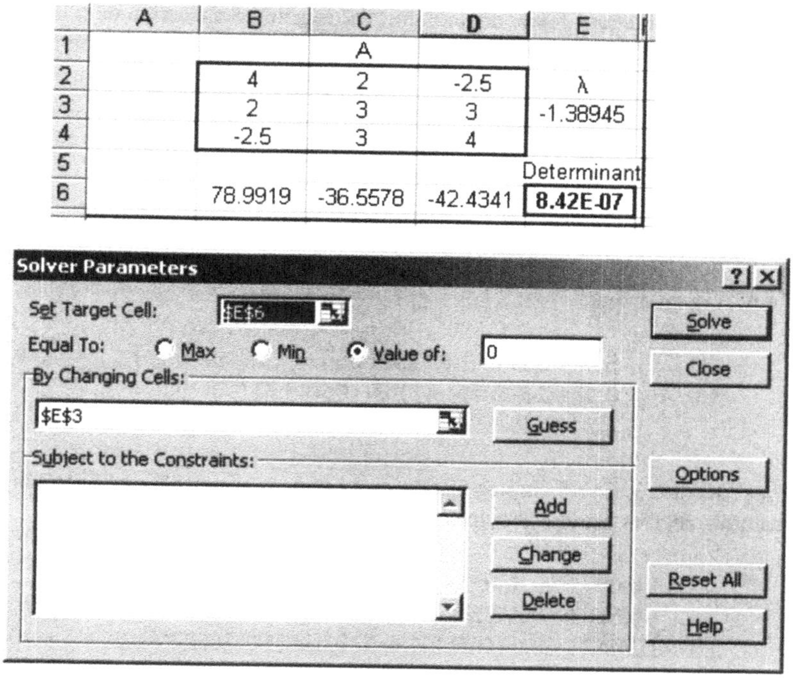

Figure 2.2. Computation of eigenvalues.

Since [**A**] is symmetric, all the three eigenvalues are distinct.[2] According to (2.15), we require the determinant of

$$(\mathbf{A} - \lambda\mathbf{I}) = \begin{bmatrix} (4-\lambda) & 2 & -2.5 \\ 2 & (3-\lambda) & 3 \\ -2.5 & 3 & (4-\lambda) \end{bmatrix}$$

to be zero. The determinant is given by

$$\begin{aligned} \mathrm{Det}(\mathbf{A}) = {} & (4-\lambda)\,[(3-\lambda)(4-\lambda) - 9] - 2\,[(4-\lambda)2 + 7.5] \\ & +2.5\,[6 + (3-\lambda)2.5]\,. \end{aligned} \tag{2.37}$$

The above represents the characteristic polynomial of third degree in λ. The solution of $\mathrm{Det}(\mathbf{A}) = 0$ for three roots gives eigenvalues. **eigen-value.Xls**

Figure 2.2 shows the spreadsheet "eigenvalue.Xls." Cell E6 is the determinant per (2.37) which is required by the Solver routine to be equal to zero. The choice variable λ is in cell E3. The spreadsheet shows a simplistic way of searching for the roots of the polynomial.

[2]From MATLAB we know that the eigenvalues are -1.3895, 5.554, and 6.83.

The solution obtained for λ depends on the starting value assumed in cell E3. If we start with an assumed value of zero, a solution of $\lambda = -1.3895$ is obtained. Similarly, assume different starting values to obtain the other two solutions for λ, namely 5.554, and 6.83. Unfortunately, there can be no *a priori* guidance as to what the starting values should be. One may be better off using a procedure to find the roots of a polynomial to compute eigenvalues.

2.10.3 Confirmation of Eigenvector

Problem: Given a matrix

$$\mathbf{A} = \begin{bmatrix} -5.509882 & 1.870086 & 0.422908 \\ 0.287865 & -11.81165 & 5.71190 \\ 0.49099 & 4.308033 & -12.970687 \end{bmatrix},$$

setup a procedure using Excel Solver to test and confirm which of the following four column vectors are the eigenvectors to matrix **A**.

0.094129	−0.860553	0.8760	2.804268
−0.766896	0.813572	−0.127869	0.275404
0.7452438	0.644055	0.521354	0.172627

Solution: Figure 2.3 is a copy of the spreadsheet. We resort to minimize the square error of differences[3] until convergence. Therefore, to solve this problem we use the nonlinear option of the Solver.

Recall the relationship

$$\mathbf{Au} = \lambda\mathbf{u}. \tag{2.38}$$

Eigen-vector.xls

In Figure 2.3, first the eigenvector corresponding to the first column vector is entered in cells F3:F5 for confirmation. The choice variable is λ in cell D7. Then **Au** is computed in cells B10:12, and $\lambda\mathbf{u}$ is computed in cells D10:D12. Next $(\mathbf{Au} - \lambda\mathbf{u})^2$, the square of difference between the two sides of (2.38), is computed in cells F8:F10. The sum of these differences in cell F11 becomes the objective function.

The Solver formulation is to choose an eigenvalue λ to minimize the error.

In this case, the problem has converged to a value of $\lambda = -17.3978$, making this error almost zero. Hence the first column is an eigenvector.

Exercises: Try other column vectors to examine if they are eigenvectors.

If the eigenvector is not given, how can you obtain all the eigenvalues of a matrix?

[3]The concept of minimizing the sum of squares of differences is discussed in later chapters starting in Sections 3.3.1 and 3.3.3.

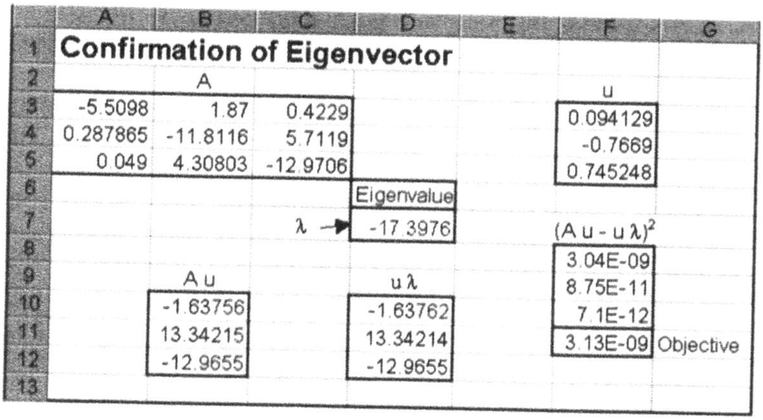

Figure 2.3. Confirmation of eigenvector.

2.10.4 Eigenvector Computation

Problem: From the previous example, the solution in Figure 2.3 shows that −17.3976 is an eigenvalue. By a similar exercise, it can be confirmed that −5.30012 is an eigenvalue for the fourth column vector of section 2.10.3. Suppose that we did not know that the fourth column vector was an eigenvector and that we were told that −5.30012 was an eigenvalue. How do we compute the eigenvector? Will we obtain the fourth column vector as the eigenvector?

Solution: Figure 2.4 shows the spreadsheet "e.vector.restructure.Xls."

`e.vector.restructure.Xls`

Equation (2.15) is entered in cells A2:C4. The values represent the **A** matrix with the eigenvalue of −5.3002 subtracted from the diagonal terms. Cells E2:E4 represent the eigenvector **u** that is to be computed, and G2:G4 represents (**A** − λ**I**)**u**. Since there are infinite solutions for **u**, we specify the needed magnitude in cell E6; in this case a unit magnitude is specified for the sum of squares of cells E2:E4. In the Solver routine, we specify that column G should be zero and the sum of squares of the three components of **u** in column of eigenvectors E should be unity.

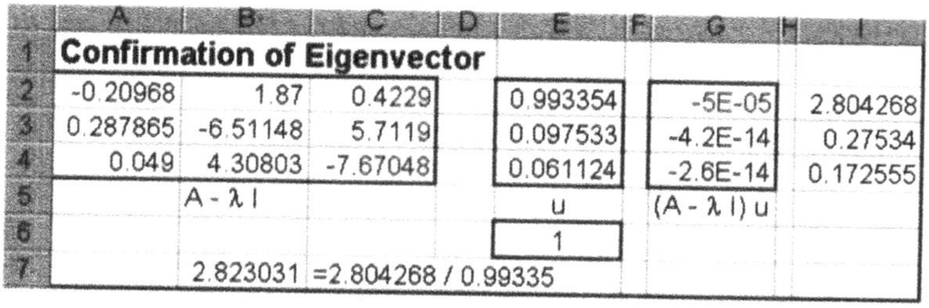

Figure 2.4. Reconstitution of eigenvector.

Choice variables for the Solver are values of three components of the eigenvector. The target cell is set as E6, requiring the Solver to set its value at 1.0 against a constraint that cells G2:G4 are equal to zero.

Because of rounding errors, the Solver comes up with a statement that it could not find a solution. In column G the resulting product is not quite zero. Therefore a smaller tolerance has to be chosen.[4] Despite the nonconvergence, a solution for the eigenvector has been obtained. Is the resulting solution in cells E2:E4 the same as the fourth column vector? Indeed so, because multiplying cells by the scaling factor of 2.823 (cell B7) gives values of column I, which are identical to the fourth column vector.

2.11 CONCLUSIONS

The above has outlined some fundamental features of matrix algebra required to understand linear and nonlinear optimization techniques covered in the remaining chapters of this book. It is needless to say that this is just a cursory sketch of matrix algebra. For those who want to delve deeper into the subject of matrix algebra as it impinges on the development of optimization algorithms, several good reference books Pipes and Harvill (1970), Bellman (1970), and Gantmacher (1950) are listed in the Bibliography.

2.12 EXERCISE PROBLEMS

1. Select arbitrary matrices of dimension at least 3×3 on an Excel spreadsheet. Practice (if not already confident) to add, multiply, and invert matrices. Check the answer obtained for the inverse by testing if $\mathbf{AA}^{-1} = \mathbf{I}$.

2. Find the condition number for the matrices

   (a)

$$\begin{bmatrix} 1 & 3 \\ 1.0001 & 2 \end{bmatrix},$$

   (b)

$$\begin{bmatrix} 3.9 & 1.6 \\ 6.9 & 2.885 \end{bmatrix}.$$

[4]This can be done by requiring cells G2:G4 to be less than a very small value, say 10^{-5}, and simultaneously requiring the cells to be greater than a small negative number, say 10^{-5}. Alternatively, the Solver should be asked to minimize the sums of squares of the three components in column G.

3. The matrix

$$A = \begin{bmatrix} 4.0 & -2.0 & -2.5 \\ -2.0 & 3.0 & 3.0 \\ -2.5 & 3.0 & 4.0 \end{bmatrix}$$

has eigenvalues of 0.4543, 1.8314, and 8.7143 with corresponding eigenvectors $[-0.0482, 0.7455, -0.6647]^T$, $[0.8259, 0.4040, 0.3932]^T$, and $[-0.5617, 0.5301, 0.6352]^T$.

Using the procedures of Sections 2.10.3 and 2.10.4, confirm using an Excel spreadsheet that for known eigenvalues, one obtains the above eigenvectors. Correspondingly, knowing the eigenvectors, confirm that one obtains the above eigenvalues.

Note: The exercise using Excel is intended to reinforce the mathematics of eigenvalues and eigenvectors. However, the alternative of using MATLAB to obtain eigenvalues and eigenvectors is a trivial exercise for advanced students.

4. The eigenvalues of matrix **A** of (2.36) are -1.3895, 5,554, and 6.83.

(a) Compute three normalized (unit magnitude) eigenvectors $\mathbf{v}_1$, $\mathbf{v}_2$, $\mathbf{v}_3$ corresponding to these eigenvalues using the procedure of Section 2.10.4.

(b) Group the three eigenvectors as a matrix $\mathbf{U} = [\mathbf{v}_1\mathbf{v}_2\mathbf{v}_3]$. Compute $D = \mathbf{U}^{-1}\mathbf{A}\mathbf{U}$ and $\mathbf{X} = \mathbf{U}\mathbf{D}\mathbf{U}^{-1}$. What are your observations?

PART II

LINEAR OPTIMIZATION

CHAPTER 3

SOLUTION OF EQUATIONS, INEQUALITIES, AND LINEAR PROGRAMS

3.1 EXTREME VALUES, RELATIVE MAXIMUM AND MINIMUM

As a background to inequalities and linear programs, we first recapitulate some fundamentals of calculus as it pertains to unconstrained maxima and minima of functions. We start by examining simple functions in two dimensions and examine some properties related to their extreme values. Using these properties, we extend our observations to functions in multidimensions.

3.1.1 Necessary and Sufficient Conditions

The process of optimization implies finding the maximum or the minimum value of a function, either subject to certain constraints, or unconstrained. The collective mathematical term used for the maximum or the minimum value of a function is *extreme value*. The first and second derivatives of a function have the following properties:[1]

$$\left.\begin{array}{l} f'(x_i) > 0 \\ f'(x_i) < 0 \end{array}\right\} \begin{array}{l} \text{means that the value of} \\ \text{function } f(x) \text{ tends to} \end{array} \left\{\begin{array}{l} \text{increase} \\ \text{decrease,} \end{array}\right.$$

$$\left.\begin{array}{l} f''(x_i) > 0 \\ f''(x_i) < 0 \end{array}\right\} \begin{array}{l} \text{means that the slope of} \\ \text{the curve } f(x) \text{ tends to} \end{array} \left\{\begin{array}{l} \text{increase} \\ \text{decrease.} \end{array}\right.$$

[1]For completeness, recall that: slope of tangent is $f'(x)$, slope of Normal is $\frac{-1}{f'(x)}$, Radius of curvature ρ is $\frac{f''(x)}{[1+f'(x)]^{1.5}}$.

Optimization Principles, by Narayan S. Rau
ISBN 0-471-45130-4

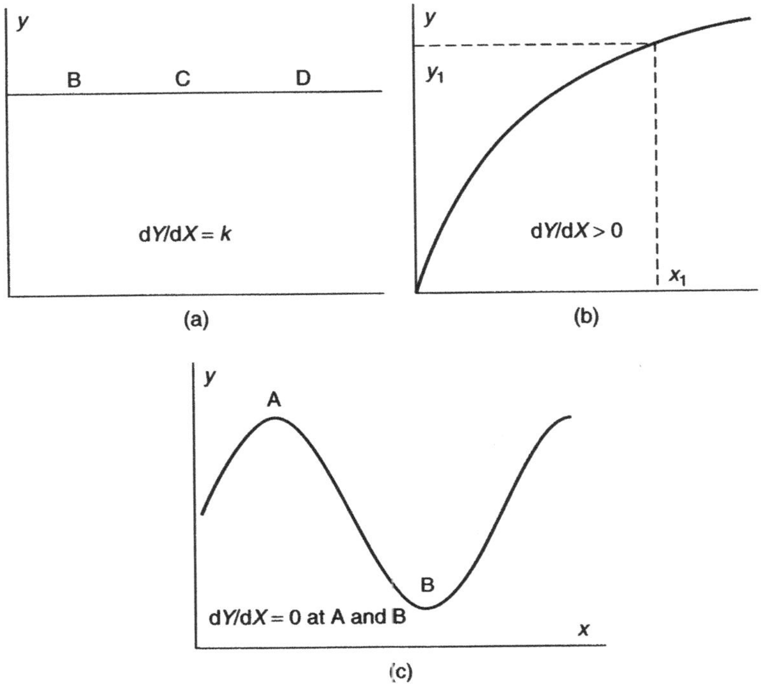

Figure 3.1. Functions and derivatives.

Any point in space R^n at which the derivative of a function is zero is called a *stationary* point. The stationary point is situated on that part of the curve with zero slope and the function value is stationary—neither increasing or decreasing.

Consider functions shown in Figure 3.1. The first two functions have no extreme values—the function in 3.1a is constant whereas the function in Figure 3.1b is constantly increasing. Consequently, the former has a zero (constant) derivative, and the latter has a positive derivative. The former can be considered to have an infinite number of stationary points while the later has no stationary points in the region of x shown. In Figure 3.1b, we can say that in the interval $[0, x_1]$, the function has two extreme values: a relative minimum value of zero at $x = 0$, and a relative maximum of y at $x = x_1$. These points represent *local* or *relative* extrema as they are extrema in the immediate neighborhood of the points.[2] However, the derivative of the function is not zero at these extreme points. Therefore, if the maximum or a minimum value of a function is determined in a space confined by some constraints (in this case $0 \leq x \leq x_1$), the derivative of the function at extreme values may not be equal to zero.[3]

[2]The term relative is used to signify that these points represent maximum or minimum in the immediate neighborhood. There is no guarantee that either one of them represent a global extremum in the space R^2.
[3]This case is discussed in detail under nonlinear optimization outlining Kuhn–Tucker conditions in Section 7.5.

From a visual observation, we see that the third function in Figure 3.1c, however, has two extrema at points A and B. These points represent local or relative extrema in their immediate neighborhood.

From the above discussion, we can assert that at an extreme point, $f'(x) = 0$, and hence it is a stationary point. This condition for extrema is called the first order condition. While this is a *necessary* condition for an extremum, it is not a *sufficient* condition because of the following.

Consider the Function in Figure 3.2a that has a zero derivative at three indicated points A, B, and C. Since the first derivative is zero at these points, they are stationary points, but they are extreme points as well. In contrast, point J in Figure 3.2b is a *point of inflection.* It is a stationary point, but it is not an extreme point. Consequently, we confirm that all points where the derivative is zero are stationary points, but not all stationary points are necessarily extreme points.

In Figure 3.2a, points A and B and C represent *local,* or *relative* extremum. These points represent an extremum in their immediate neighborhood. There is no guarantee that they represent *global* minimum in the whole range of the x-axis, although that might be the case. A local minimum or a maximum in the immediate neighborhood is called a relative extremum. However, point C in Figure 3.2a, can be considered a *global minima* only in the range of the one dimensional space X addressed in the figure (since the value of the function at C is less than the value at A).

First-Derivative Condition for Relative Extrema. The above has shown that,

1. A point $\overline{x}$ is a relative minimum if $f'(\overline{x})$ changes sign from negative value at the immediate left of $\overline{x}$ to a positive value to the immediate right of $\overline{x}$.
2. A point $\overline{x}$ is a relative maximum if $f'(\overline{x})$ changes sign from positive value at the immediate left of $\overline{x}$ to a negative value to the immediate right of $\overline{x}$.
3. A point $\overline{x}$ is neither a relative minimum nor a relative maximum if $f'(\overline{x})$ has the same sign on the immediate left and immediate right side of $\overline{x}$.

Consequently, in addition to the necessary first order condition, we need to determine a sufficient condition to identify extreme points.

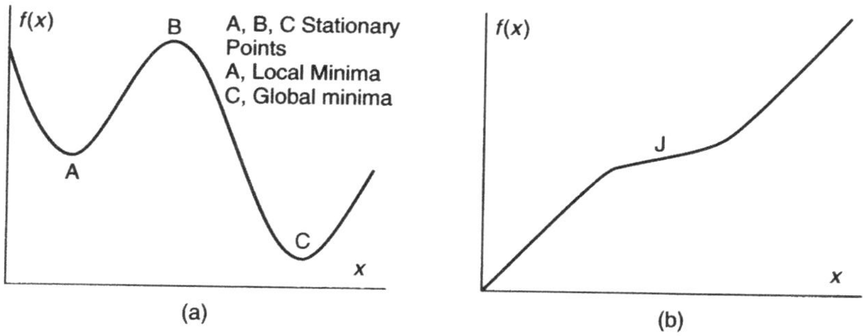

Figure 3.2. Local minima, global minima, and inflection point.

Change of Sign of First Derivatives. The derivative of a function changes sign at relative extreme points. We examine if this could suggest a test for a sufficient condition.

Example: Consider the function

$$y = f(x) = x^3 - 12x^2 + 36x + 4.$$

This function is shown in Figure 3.3.

We have: $f'(x) = 3x^2 - 24x + 36$. Hence, the extremum is given by the solution of $3x^2 - 24x + 36 = 0$ which gives $\overline{x}_1 = 2$, $f(2) = 36$; and $\overline{x}_2 = 6$, $f(6) = 4$.

In Figure 3.3, $f'(x) > 0$ for $x < 2$, and $f'(x) < 0$ for $2 \leq x \leq 6$. Hence $f'(x)$ changes sign at extrema. A similar reasoning will indicate that $f'(x)$ changes sign at $x = 6$. It is clear that the extreme point at $x = 6$ is a global minimum in R^2. However, the other extreme point at $x = 2$ with $f(x) = 36$ is a local maximum in the immediate neighborhood of $x = 2$, but it is not a global maximum in the region $0 \leq x \leq \infty$.

While the test for change of sign of first derivative works for the above function, it fails for functions with inflection points. Examine Figures 3.4a and 3.4b in which two types of inflection points are shown. In Figure 3.4a, the first derivatives is positive at the inflection point and on either side of it; in 3.4b it is zero at the point of inflection and positive on either side of it! Hence, we note that the first derivative

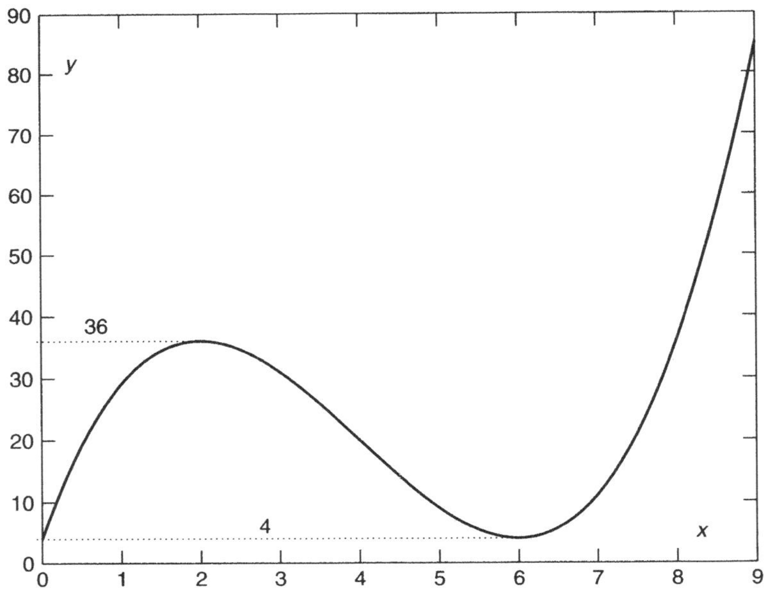

Figure 3.3. Graph of example problem.

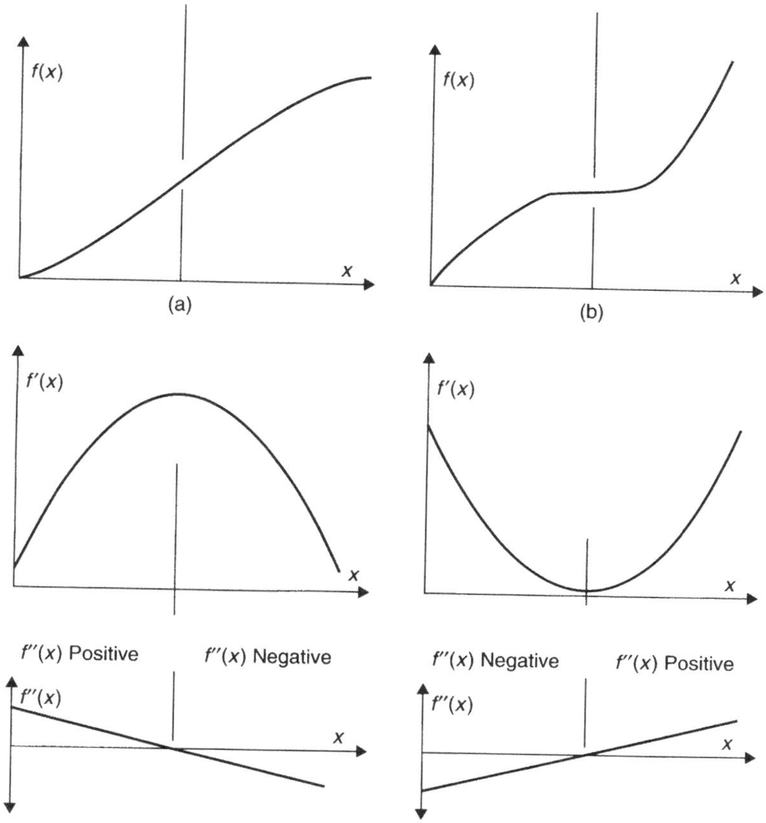

Figure 3.4. Change of sign of second derivative at point of inflection.

does not change sign at the point of inflection in either of the functions. The fact that the function of Figure 3.4b has $f'(x) = 0$ at the inflection point identifies it as a stationary point. That is not the case in 3.4a because $f'(x)$ is positive at the point of inflection.

Because of the foregoing, a test for the change in sign of the first derivative cannot always serve as a sufficient condition. However, we observe in Figures 3.4a and 3.4b that the sign of the second derivative changes at the point of inflection. Hence, we now examine a test related to the property of second derivative that can be used as a sufficient condition for extrema.

Second derivative Test. Since the property of second derivative is

$$\left.\begin{matrix} f''(x_i) > 0 \\ f''(x_i) < 0 \end{matrix}\right\} \begin{matrix} \text{means that the slope of} \\ \text{the curve } f(x) \text{ tends to} \end{matrix} \left\{\begin{matrix} \text{increase} \\ \text{decrease,} \end{matrix}\right.$$

the second-derivative test for relative extremum is as follows.

If the necessary condition $f'(\overline{x}) = 0$ at $\overline{x}$ is satisfied, then the value of the function will be

- a relative maximum if the second derivative $f''(\overline{x}) < 0$ at $\overline{x}$,
- a relative minimum if the second derivative $f''(\overline{x}) > 0$ at $\overline{x}$.

This test for sufficiency rules out the inflection point of the type shown in Figure 3.4a from consideration because $f' \neq 0$. However, the test is not decisive for the point of inflection in Figure 3.4b because both the first and second derivatives are zero.

The second derivative test can be indecisive for some functions without an inflection point. Consider the function $f(x) = 3 + x^4$ of Figure 3.6. It has $f'(x) = 4x^3$ and $f''(x) = 12x^2$. At the stationary point $\overline{x} = 0$, which is an extreme global minimum point, $f''(0) = 0$. Consequently, we cannot assert by the above criterion of second derivative alone that $\overline{x} = 0$ is an extreme minimum. To confirm that $\overline{x} = 0$ is indeed a minimum, tests on higher order derivatives using Taylor's expansion have to be conducted (see section and references [5], [27]).

Summary of Necessary and Sufficient Conditions. A relative extremum must be at a stationary point ($f' = 0$), but it may be either a relative extremum or a point of inflection. To find if the point is a relative maximum or minimum, the second derivative test can be applied. Therefore, $f' = 0$ is a *necessary* condition, called the first order condition, for extrema, but this is not *sufficient* because it does not rule out the possibility of inflection points. A nonzero positive definite (or negative definite) sign of $f''(x)$ is sufficient to determine if the stationary point under examination is a relative minimum (or maximum). This latter test is called the *second order condition*. If $f''(x) = 0$, higher order derivatives have to be checked to ascertain if it is an extreme point or a point of inflection (saddle point in n dimensions). Fortunately, for most practical problems, the second derivative test gives definite answers.

3.2 CONCAVE AND CONVEX FUNCTIONS

The definition of concave and convex functions is as follows.[4]

For a concave function, a line joining any two points P and Q on the curve is

$$\left\{\begin{array}{l}\text{at or below the function} \rightarrow \\ \text{strictly below the function} \rightarrow\end{array}\right\} \text{if the function is} \left\{\begin{array}{c}\leftarrow \text{concave} \\ \leftarrow \text{strictly concave}\end{array}\right\}.$$

[4]Some use the terminology concave down and concave up instead of concave and convex functions; see Stewart James (1999) and Finney and Thomas (1993).

Similarly, for a convex function, a line joining any two points P and Q on the curve is

$$\left\{\begin{array}{l}\text{at or above the function} \rightarrow \\ \text{strictly above the function} \rightarrow\end{array}\right\} \text{if the function is} \left\{\begin{array}{l}\leftarrow \text{convex} \\ \leftarrow \text{strictly convex}\end{array}\right\}.$$

Another way of describing above properties is by observing tangents. A concave function is along (concave) or below (strictly concave) the tangent at any point. The opposite is true for convex functions.

The following observations arise from Figure 3.5.

Figure 3.5a: Concave			Figure 3.5b: Convex		
x	Derivative	Point	x	Derivative	Point
$x = x_1$	$f'(x_1) > 0,\ f''(x_1) < 0$	A	$x = x_4$	$f'(x_4) < 0,\ f''(x_4) > 0$	D
$x = x_2$	$f'(x_2) = 0,\ f''(x_2) < 0$	B	$x = x_5$	$f'(x_5) = 0,\ f''(x_5) > 0$	E
$x = x_3$	$f'(x_3) < 0,\ f''(x_3) < 0$	C	$x = x_6$	$f'(x_6) > 0,\ f''(x_6) > 0$	F

In Figure 3.5a, irrespective of the sign of $f'(x)$, $f''(x) < 0$. Correspondingly, we see that in Figure 3.5b $f''(x) > 0$. From these observations, we define concavity and convexity of functions as follows.

Function is			
Concave	Strictly Concave	Convex	Strictly Convex
$f''(x) \leq 0$	$f''(x) < 0$	$f''(x) \geq 0$	$f''(x) > 0$

The above second-derivative test can be ambiguous or insufficient under some circumstances. An example is the function of Figure 3.6 used earlier, which is a strictly convex curve. The second derivative is zero at the stationary point. Hence, it is not possible to say if the function is convex or concave based only on the above second-derivative criterion.

Figure 3.4 shows $f(x)$ and its derivative $f'(x)$ for two types of functions that contain inflection points. At points of inflection, the function changes curvature. As indicated in the figure, the sign of the second derivative is different on either side of the point of inflection. This change in sign of the second derivative indicates concavity and convexity (and vice versa) on either side, which is the characteristic of the inflection point.[5]

[5] In multidimensional functions, a point corresponding to the point of inflection is the saddle point which will be discussed later.

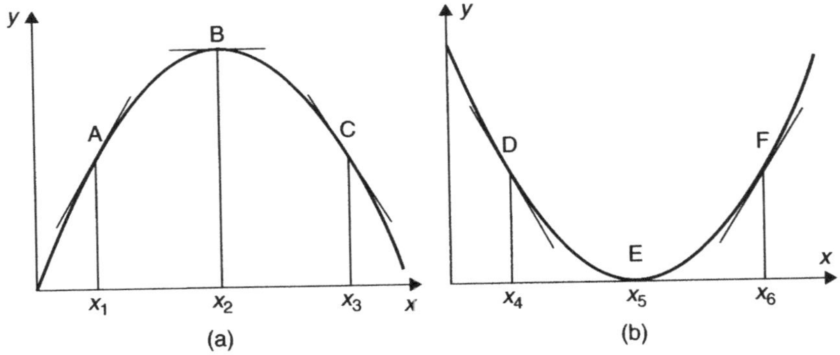

Figure 3.5. Concave and convex functions.

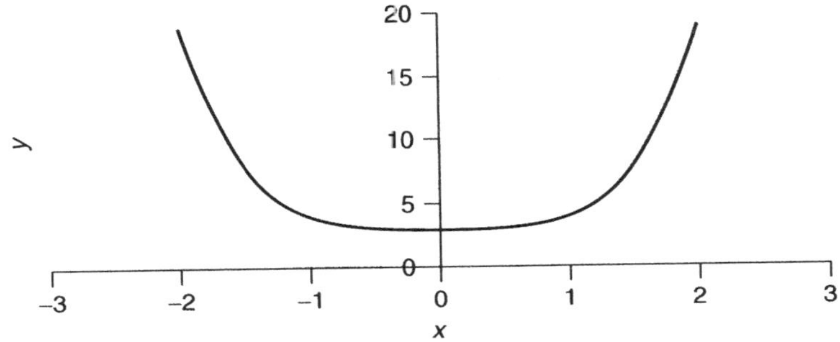

Figure 3.6. Function $y = 3 + x^4$.

In consideration of the above matters, we are now in a position to define the convexity and concavity of multidimensional functions in a rigorous manner as follows:

A differentiable function $f(\mathbf{x})$ is $\left\{\begin{array}{c}\text{concave}\\ \text{convex}\end{array}\right\}$ if and only if for any given point $\boldsymbol{u} = (x_1, x_2, \ldots, x_n)$ and any other point $\boldsymbol{v} = (x_1', x_2', \ldots, x_n')$ when

$$f(\boldsymbol{v}) \left\{\begin{array}{c}\leq\\ \geq\end{array}\right\} f(\boldsymbol{u}) + f'(\boldsymbol{u})(\boldsymbol{v} - \boldsymbol{u}). \tag{3.1}$$

For multidimensional functions, the above equation should be valid along any of the orthogonal coordinate axes that form the basis of the function. For example, if the partial derivative f' is taken for all $x = x_1, x_2, \ldots, x_n$, and the points U and V are measured along each of these directions, (3.1) will be valid for each direction. This relationship is illustrated in Figure 3.7 for a function in two dimensions.

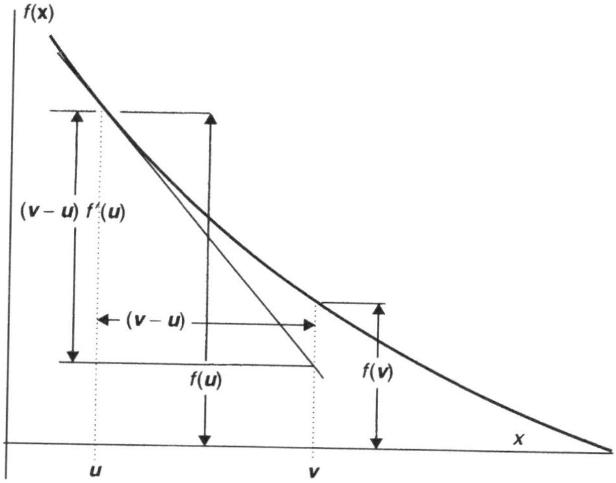

Figure 3.7. Definition of convex function.

Concavity and convexity will be strict if weak inequalities in the above expression are replaced by strict inequalities.

The following theorems can be deduced easily from the foregoing:

- If a function $f(x)$ is concave, then $-f(x)$ is convex, and vice versa. Correspondingly, the negative of a strictly concave function is strictly convex, and vice versa.
- If $f(x)$ and $g(x)$ are both convex (concave) functions, then $f(x) + g(x)$ is also a convex (concave) function. If either of the functions $f(x)$ or $g(x)$ is strictly convex (concave), or if both the functions are convex (concave), then $f(x) + g(x)$ is strictly convex (concave).
- A linear function $f(x)$ is a concave function as well as a convex function, but not strictly so.

3.2.1 Convexity and Concavity of Multivariate Functions

The above explanation of convexity and concavity has addressed a function of one variable. In practice, optimization involves the maximization and minimization of an multivariate objective function that is not necessarily a linear function. Such a function containing several variables $x_1, x_2, \ldots, x_n$ can be expressed as

$$f(\mathbf{x}) = f(x_1, x_2, \ldots, x_n).$$

We state without proof [see Rao (1996)] that a multivariate function is convex if the Hessian[6] matrix (symmetric) $\mathbf{H(x)}$ given by

[6]Hessian matrix will be discussed in Section 5.3.1.

$$\mathbf{H} = \begin{bmatrix} \dfrac{\partial^2 f(\mathbf{x})}{\partial x_1^2} & \dfrac{\partial^2 f(\mathbf{x})}{\partial x_1 \partial x_2} & \cdots & \dfrac{\partial^2 f(\mathbf{x})}{\partial x_1 \partial x_n} \\ \dfrac{\partial^2 f(\mathbf{x})}{\partial x_2 \partial x_1} & \dfrac{\partial^2 f(\mathbf{x})}{\partial x_2^2} & \cdots & \dfrac{\partial^2 f(\mathbf{x})}{\partial x_2 \partial x_n} \\ \vdots & \vdots & \vdots & \vdots \\ \dfrac{\partial^2 f(\mathbf{x})}{\partial x_n \partial x_1} & \dfrac{\partial^2 f(\mathbf{x})}{\partial x_n \partial x_2} & \cdots & \dfrac{\partial^2 f(\mathbf{x})}{\partial x_n^2} \end{bmatrix} \tag{3.2}$$

is positive semidefinite.[7] Correspondingly, it is concave if the Hessian matrix is negative semidefinite. Test for positive definiteness is discussed later in Section 5.4, indicating that the determinants of principal minors of the Hessian matrix should be positive.

The following examples illustrate tests for concavity and convexity of functions.

Examples on Convexity. Examine if the following functions are concave, convex, or neither:

1. $f(x) = -3x^2 - 4x + 4$.
   We have $f' = -6x - 4$ and $f'' = -6 < 0$ for all x. Hence, the function is strictly concave from the conditions indicated in Section 3.2.
2. $f(x) = e^{x^2}$.
   We get $f'(x) = 2xe^{x^2}$ and $f''(x) = 4x^2e^{x^2} + 2e^{x^2}$. Since $f''(x) > 0$ for all values of x, the function is strictly convex.
3. $f(x_1, x_2, x_3) = 6x_1^2 + 4x_2^2 + 2x_3^2 + 6x_1x_2 - 3x_1 - 4x_2 + 20$.
   The Hessian matrix **H**according to (3.2) is

$$\mathbf{H} = \begin{bmatrix} 12 & 6 & 0 \\ 6 & 8 & 0 \\ 0 & 0 & 4 \end{bmatrix}.$$

For the determinants of principal minors we have

$$|12| > 0, \quad \begin{vmatrix} 12 & 6 \\ 6 & 8 \end{vmatrix} = 60 > 0, \quad \text{and} \quad \begin{vmatrix} 12 & 6 & 0 \\ 6 & 8 & 0 \\ 0 & 0 & 4 \end{vmatrix} = 240 > 0,$$

indicating a positive definite matrix. Therefore the function is strictly convex.

[7] The concept of definiteness has been introduced in Section 2.6.1. A practical method to test for definiteness is shown in Section 5.4.

3.2.2 Convex Sets

We now extend the concept of convexity of functions to convex sets.

Let S be a set of points in two-dimensional or three-dimensional space. If a line segment connecting any two points in this set S lies entirely in S, then the set S is said to be a convex set. By convention, a set consisting of a single point as well as a null set are considered convex sets. For example, a straight line satisfies this definition of convex sets; the set consists of all points on the line. To reinforce the concept of convex sets further, observe in Figure 3.8 that a circle containing all points within it is a convex set because a line joining any two points within this set is in this set. Lines XY show this. The same applies to a closed geometric shape such as the triangle or the hexagon. However, this is not true for either the pallet or the nonconcentric circles illustrated in the figure.

We now seek to extend the above definition applicable to two to an algebraic definition applicable to n-dimensions. Toward this end, consider a vector $\mathbf{r}$ which is a linear combination of two vectors $\mathbf{x}$ and $\mathbf{y}$ as

$$\mathbf{r} = a\mathbf{x} + b\mathbf{y}.$$

When the scalars a and b are both ≤ 1, if a and b sum to one, such a linear combination is said to be a convex combination, which can be expressed as

$$\theta\mathbf{X} + (1-\theta)\mathbf{Y}, \quad (0 \leq \theta \leq 1).$$

For example, the linear combination of two vectors $0.4\begin{bmatrix} 2 \\ 5 \end{bmatrix} + 0.6\begin{bmatrix} 4 \\ 2 \end{bmatrix}$ is a convex combination. It can be confirmed easily that this resulting vector lies on

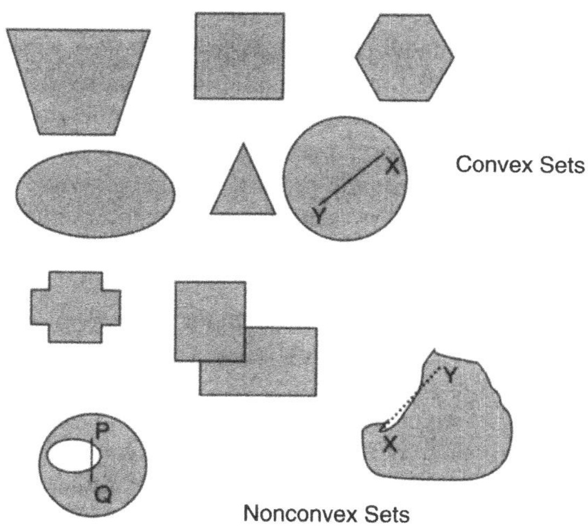

Figure 3.8. Concave and convex sets.

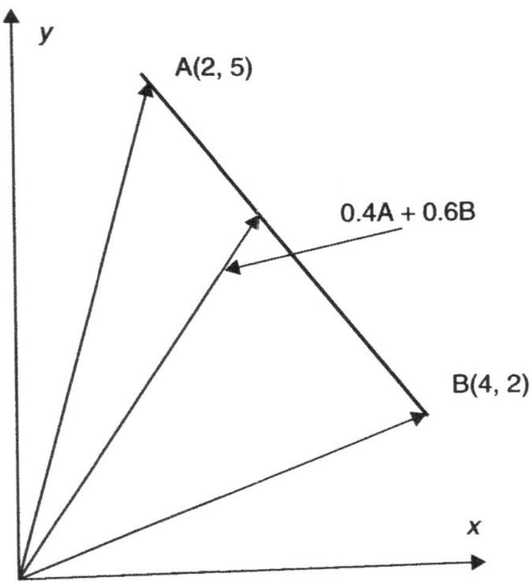

Figure 3.9. Convex combination of vectors.

the line joining the tips of the two vectors $[2, 5]^T$ and $[4, 2]^T$. The position of the resulting vector on the line depends on the value of θ. Figure 3.9 portrays this.

In light of the above, a convex set can be redefined algebraically as follows: A set S is convex if and only if, for any two points X and Y contained in the set S, the convex combination $Z = \theta X + (1 - \theta)Y$ is also in S. This algebraic definition is valid in n-dimensional space as well. The concept of the convex set defines how the sets are packed together—without indentations in their boundary or containing holes. In contrast, the concept of convex functions defines the curvature of functions.

The points in a set can be classified as boundary points, interior points, and extreme points. The distinction between boundary and interior points is intuitively obvious. Referring to Figure 3.8, for example, the points lying on the six sides of the hexagon are boundary points of the set S, and the points not on the boundary are interior points. Extreme points are also boundary points, but they do not lie on a line segment joining any two points in the set. In other words, an extreme point is a point that cannot be obtained by a convex combination of any other two points in a set.

As the above underscores the distinction between the concepts of convex functions and sets, there is a connection between them as well. Recall that in (3.1) the implicit assumption was that the domain of $(x_1, x_2, \ldots, x_n)$ was R^n. Equation (3.1) also implies that for any two points U and V in the domain, all convex combinations of U and V given by this equation must also be in the domain. This requires the domain to be a convex set. Consequently, we can be precise in saying that the domain of a function is a convex subset of R^n rather than the entire R^n.

Some Definitions. In light of the above, we now outline some concepts applicable to n-dimensional space R^n and define them without mathematical rigor.

1. Point in R^n: A point Y is characterized by a set of n coordinates ($y_1, y_2, \ldots, y_n$). The coordinates are also termed components in n directions.
2. Line Segment in R^n: Let the coordinates of two points P and Q be given by $y_i^{(1)}$ and $y_i^{(2)}$. The line segment (L) joining the two points P and Q is the collection of points $Y(\lambda)$ whose coordinates are given by $y_j = \lambda y_j^{(1)} + (1 - \lambda)y_j^{(2)}$, $j = 1, 2, \ldots, n$, where $0 \leq \lambda \leq 1$.
   Thus

$$L = \left\{Y | Y = \lambda y^{(1)} + (1 - \lambda) y^{(2)}\right\}.$$

   In one dimension, as Figure 3.10 illustrates, we have

$$y(\lambda) = \lambda y^{(1)} + (1 - \lambda)y^{(2)}, \qquad 0 \leq \lambda \leq 1.$$

   If $X^{(1)} = (2, 3)$ and $X^{(2)} = (6, 7)$, then, for any $0 \leq \lambda \leq 1$, say $\lambda = 0.6$, we have $0.6 \times 2 + 0.4 \times 6 = 3.6$ and $0.6 \times 3 + 0.4 \times 7 = 4.6$. Thus, the point (3.6, 4.6) is on a line joining $X^{(1)}$ and $X^{(2)}$. Consequently, a collection of such points on the line joining the two points represents the line.
3. Hyperplane: In a two-dimensional x_1–x_2 plane, a set of points which satisfy the linear equation $a_1x_1 + a_2x_2 = k$ represents a straight line with a slope of $-a_1/a_2$ and an x_1 intercept of k/a_1. An extension of this in n-dimensional space to represent a hyperplane is a set of points whose n coordinates satisfy the linear equation

$$a_1x_1 + a_2x_2 + \cdots + a_nx_n = a^T\mathbf{x} = k.$$

   A set of points whose coordinates satisfy a linear inequality of the type $a_1x_1 + a_2x_2 + \cdots + a_nx_n = a^T\mathbf{x} \leq k$ is called a *closed half-space*. Because of the inclusion of equality sign in this inequality, the half-space is called closed. A hyperplane partitions an n-dimensional space into two closed half-spaces, so that

$$H^+ = \left\{X | a^T\mathbf{x} \geq k\right\}, \tag{3.3}$$

$$H^- = \left\{X | a^T\mathbf{x} \leq k\right\}. \tag{3.4}$$

   Figure 3.11 illustrates this concept for the case of the two-dimensional space.

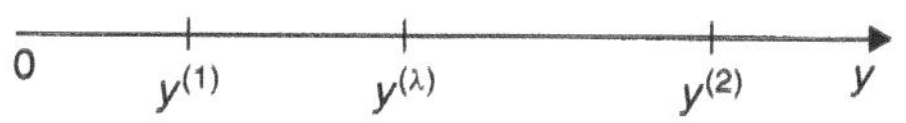

Figure 3.10. Line segment in two dimensions.

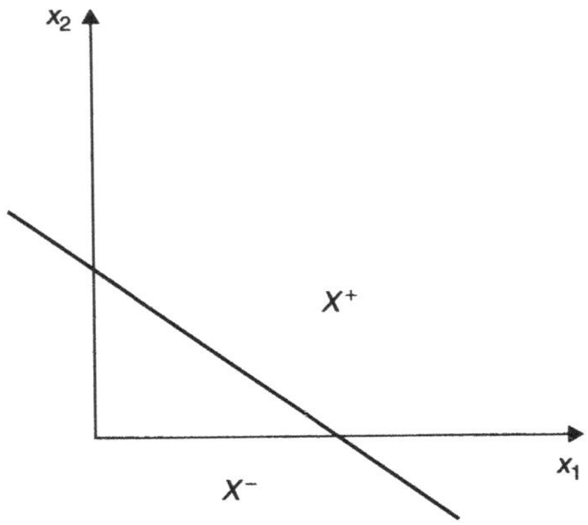

Figure 3.11. Illustration of a hyperplane in two dimensions.

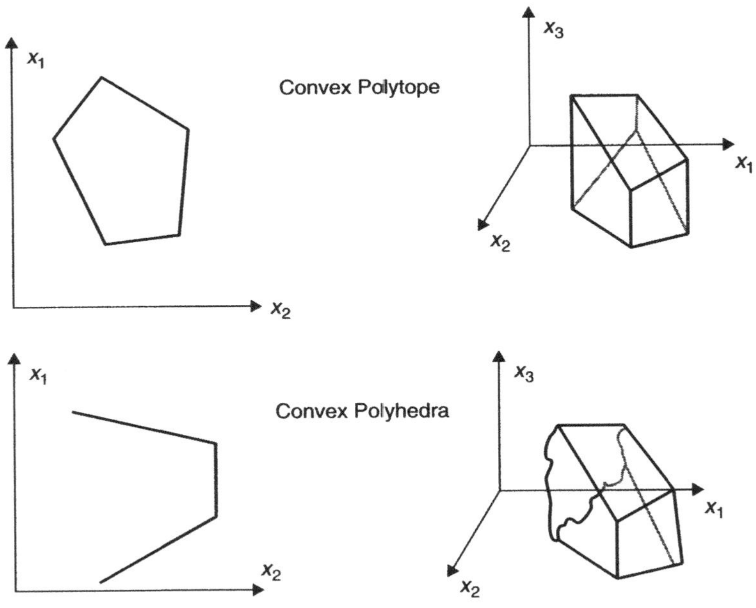

Figure 3.12. Polytopes and polyhedra in two and three dimensions.

4. Convex Set: A convex set is a collection of points such that for any two points in the collection, the line joining the two points is also in the collection. Earlier, it was defined mathematically as follows: If $Y^{(1)}, Y^{(2)} \in S$, then $Y \in S$, where

$$Y = \lambda Y^{(1)} + (1 - \lambda) Y^{(2)}, \qquad 0 \leq \lambda \leq 1.$$

Note from the definition of the line segment earlier that Y is on the line joining $Y^{(1)}$ and $Y^{(2)}$, and therefore is in the convex set.

5. Convex Polyhedron and Polytope: A convex polyhedron is a set of points common to one or more half-spaces. A convex polyhedron, which is bounded, is called a convex polytope.

   Convex polytopes and convex polyhedra in two and three dimensions are represented in Figure 3.12.
6. Extreme Point—Vertex: An extreme point is a point in a convex set that does not lie on a line segment joining any two other points of the set. For example, every corner point of a polytope is an extreme point or a vertex. Similarly, every point on the circumference of a circle is an extreme point.

The applicability of above concepts of convex sets and convex functions to the solution of linear optimization problems will become clear later.

3.3 SOLUTION OF LINEAR EQUATION SYSTEMS

We first examine the nature of systems of linear equations and their solution. Subsequently, we show that the solution of linear optimization problems is related to and draws from the solution method of linear equation systems.

3.3.1 Existence of Solution for Systems

Consider the system of linear equations

$$\mathbf{A}\mathbf{x} = \mathbf{b}, \tag{3.5}$$

in which $\mathbf{A}$ is a $m \times n$ matrix, $\mathbf{x}$ is an n element vector to be determined (or estimated), and $\mathbf{b}$ is an m element vector of constants. The following indicates under what conditions one can obtain a solution for $\mathbf{x}$, and it also lists the characteristics of solution.

1. If $m = n$ and $rank(\mathbf{A}) = n$, $\mathbf{x}$ has a unique solution. This is called a *fully specified* or *fully determined* case.
2. If when $m < n$ and $rank(\mathbf{A}) = m$, there are an infinite number of solutions to $\mathbf{x}$ which exactly satisfy $\mathbf{b} - \mathbf{A}\mathbf{x} = 0$. This is called a *underspecified* or *underdetermined* system of equations. In order to find a unique solution, it is often useful to find a solution which minimizes the norm $\|\mathbf{x}\|^2$. This is called the *minimum-norm solution* to an underspecified system of linear equations.
3. When $m > n$ and $rank(\mathbf{A}) = n$, the solution to the problem is unique; the problem is referred to as finding the *least-squares solution*. The least-squares problem is to find $\overline{\mathbf{x}}$ in order to obtain

$$\min_{\mathbf{x}} (\mathbf{b} - \mathbf{A}\overline{\mathbf{x}})^T (\mathbf{b} - \mathbf{A}\overline{\mathbf{x}}), \tag{3.6}$$

which minimizes the sum of the squares of the residuals since the above equation is the same as minimizing R, where

$$R = \|(\mathbf{b} - \mathbf{A\bar{x}})\|^2. \tag{3.7}$$

3.3.2 Solution of Fully Specified System Using Canonical Form

The solution of a fully specified system is in the curriculum of high schools. Its solution using canonical form—meaning reduced to the simplest or clearest scheme—is familiar to all the readers. Despite this, we sketch the solution procedure, because the extension of this procedure (shown subsequently) to an underspecified system of equation is the foundation of the simplex method to solve linear programs.

Consider the following system[8]:

$$3x_1 + 2x_2 - x_3 = 1 \quad (I) \tag{3.8}$$

$$2x_1 + x_2 + x_3 = 6 \quad (II) \tag{3.9}$$

$$x_1 - x_2 + x_3 = 4. \quad (III) \tag{3.10}$$

This system is expressed in matrix form

$$\mathbf{Ax} = \mathbf{b} \tag{3.11}$$

with the coefficient matrix **A**given by

$$\mathbf{A} = \begin{bmatrix} 3 & 2 & -1 \\ 2 & 1 & 1 \\ 1 & -1 & 1 \end{bmatrix}. \tag{3.12}$$

The method of solution is by reducing it to canonical form. In this system of equations, there are three unknowns ($nu = 3$) defined by three equations ($ne = 3$), thus giving a coefficient matrix of full rank ($R = 3$). Therefore, the system has a unique solution for the unknowns, which can be obtained by the traditional method of *canonical form* as follows.

[8] It is assumed that the reader is familiar with various forms of writing these equations such as:

- The detailed form as in (3.8) to (3.10).
- The compact form $\mathbf{Ax} = \mathbf{b}$, where $\mathbf{A} = \begin{bmatrix} 3 & 2 & -1 \\ 2 & 1 & 1 \\ 1 & -1 & 1 \end{bmatrix}$, $\mathbf{x} = [x_1, x_2, x_3]^T$, and $\mathbf{b} = [1, 6, 4]^T$.

Pivoting[9] first on $a_{11} = 3$, the first element of the coefficient matrix, we get

$$x_1 + \frac{2}{3}x_2 - \frac{1}{3}x_3 = \frac{1}{3} \qquad (I_1), \tag{3.13}$$

$$0 - \frac{1}{3}x_2 + \frac{5}{3}x_3 = \frac{16}{3} \qquad (II_1 = II - 2I_1), \tag{3.14}$$

$$0 - \frac{5}{3}x_2 + \frac{4}{3}x_3 = \frac{11}{3} \qquad (III_1 = III - I_1). \tag{3.15}$$

Pivoting next on $a_{22} = -1/3$, we get

$$x_1 + 0x_2 - \frac{9}{3}x_3 = \frac{33}{3} \qquad \left(I_2 = I_1 + \frac{2}{3}II_2\right), \tag{3.16}$$

$$0 - x_2 + 5x_3 = 16 \qquad (II_2 = 3II_1), \tag{3.17}$$

$$0 - x_2 + \frac{-21}{3}x_3 = \frac{-69}{3} \qquad \left(III_2 = III_1 - \frac{5}{3}II_2\right). \tag{3.18}$$

The solution for x_3 is readily apparent by the last equation as $69/21 = 3.2857$. From successive back-substitution,from (3.17) we obtain $x_2 = (345/21) - 16 = 0.42857$, and from (3.16) $x_1 = 1.142857$.

3.3.3 Overspecified or Underspecified Systems

We illustrate the solution of such systems of equations in the following.

- In case of the overspecified system, we use the concept of a *least-square error fit* to be described later. Also, a solution using the *pseudo-inverse* is exactly equivalent to that of a least-square error fit as proved in Appendix C.

  The least-squares problem of finding a solution to the overdetermined system of (3.5) could be shown to be the same as that of solving a system of equations

$$\begin{pmatrix} \mathbf{I} & \mathbf{A} \\ \mathbf{A}^T & \mathbf{0} \end{pmatrix} \begin{pmatrix} \mathbf{r} \\ \mathbf{x} \end{pmatrix} = \begin{pmatrix} \mathbf{b} \\ \mathbf{0} \end{pmatrix}, \tag{3.19}$$

  in which the vector $\mathbf{r}$ represents the residuals.

Example: Consider

$$\begin{bmatrix} 1 & 2 \\ 4 & 5 \\ 6 & 7 \end{bmatrix} \begin{bmatrix} x_1 \\ x_2 \end{bmatrix} = \begin{bmatrix} 1 \\ 3 \\ 4 \end{bmatrix},$$

in which $m = 3$ and $n = 2$, and the definitions of $\mathbf{A}$, $\mathbf{x}$, and $\mathbf{b}$ are obvious.

[9]The definition of pivoting element is evident from this as the element whose coefficient is made equal to unity.

Since

$$\begin{pmatrix} \mathbf{I} & \mathbf{A} \\ \mathbf{A}^T & \mathbf{0} \end{pmatrix} = \begin{bmatrix} 1 & 0 & 0 & 1 & 2 \\ 0 & 1 & 0 & 4 & 5 \\ 0 & 0 & 1 & 6 & 7 \\ 1 & 4 & 6 & 0 & 0 \\ 2 & 5 & 7 & 0 & 0 \end{bmatrix},$$

we get **W** as its inverse (check using Excel or other program)

$$\mathbf{W} = \begin{bmatrix} 0.105 & -0.263 & 0.157 & -1.315 & 1.105 \\ -0.263 & 0.657 & -0.394 & -0.210 & 0.236 \\ 0.157 & -0.394 & 0.236 & 0.526 & -0.342 \\ -1.315 & -0.210 & 0.526 & -2.052 & 1.684 \\ 1.105 & 0.236 & -0.342 & 1.684 & -1.394 \end{bmatrix},$$

and hence the left-hand vector in (3.19) amounts to

$$\begin{pmatrix} \mathbf{r} \\ \mathbf{x} \end{pmatrix} = \mathbf{W}^{-1} \begin{pmatrix} \mathbf{b} \\ \mathbf{0} \end{pmatrix} = [0.052,\ -0.131,\ 0.078,\ -1.157,\ 1.552]^T.$$

In the above, the last two elements represent $\mathbf{x} = [-1.157, 1.552]^T$ and the first three elements represent the residuals, for

$$\begin{bmatrix} 1 & 2 \\ 4 & 5 \\ 6 & 7 \end{bmatrix} \begin{bmatrix} -1.157 \\ 1.552 \end{bmatrix} + \begin{bmatrix} 0.052 \\ -0.131 \\ 0.078 \end{bmatrix} = \begin{bmatrix} 1 \\ 3 \\ 4 \end{bmatrix}.$$

Pseudo-Inverse. A solution using the *pseudo-inverse* is exactly equivalent to that of a least-square error fit as proved in Appendix C.

The procedure to obtain a pseudo-inverse is as follows.

Let the system of equations for a vector **x** of dimension *nu* be given by

$$\mathbf{A}\mathbf{x} = \mathbf{b}. \tag{3.20}$$

Premultiplying (3.20) by $\mathbf{A}^T$, we get

$$\mathbf{A}^T\mathbf{A}\mathbf{x} = \mathbf{A}^T\mathbf{b}. \tag{3.21}$$

Although **A** is nonsquare, $\mathbf{A}^T\mathbf{A}$ is a square matrix. Let **D** denote $\mathbf{A}^T\mathbf{A}$. Premultiplication of (3.21) by $\mathbf{D}^{-1}$ gives

$$\mathbf{D}^{-1}\mathbf{D}\mathbf{x} = \mathbf{D}^{-1}\mathbf{A}^T\mathbf{b}, \tag{3.22}$$

or

$$\mathbf{Ix} = \mathbf{D}^{-1}\mathbf{A}^T\mathbf{b}, \tag{3.23}$$

which gives values for $\mathbf{x}$.

Consequently the matrix $\mathbf{D}^{-1}\mathbf{A}^T = (\mathbf{A}^T\mathbf{A})^{-1}\mathbf{A}^T$ is called the pseudo-inverse of $\mathbf{A}$.

The computation of a pseudo-inverse matrix is possible only when the columns of $\mathbf{A}$ are linearly independent.

The pseudo-inverse represented by the symbol $\mathbf{A}^+$ has the following properties:

$$\mathbf{A}\mathbf{A}^+\mathbf{A} = \mathbf{A}, \tag{3.24}$$

$$\mathbf{A}^+\mathbf{A}\mathbf{A}^+ = \mathbf{A}, \tag{3.25}$$

$$\mathbf{A}^+\mathbf{A} = (\mathbf{A}^+\mathbf{A})^T, \tag{3.26}$$

and

$$\mathbf{A}\mathbf{A}^+ = (\mathbf{A}\mathbf{A}^+)^T. \tag{3.27}$$

- In an underspecified system, $ne < nu$, one can solve for ne unknowns in terms of the $nu - ne$ unknowns. For instances, if six unknowns are described by four equations, we can solve for any four unknowns in terms of the remaining two unknowns. The following examples reinforce the above discussion of over and underspecified systems.

Illustrative Problems

Overspecified System; Pseudo-inverse Matrix. Consider the above canonical system example of (3.8) to (3.10) with a solution of $x_1 = 1.142857, x_2 = 0.4287$, and $x_3 = 3.2857$. We now add two more equations to (3.8), (3.9), and (3.10) as follows:

$$\begin{aligned} 3x_1 + 2x_2 - x_3 &= 1, \\ 2x_1 + x_2 + x_3 &= 6, \\ x_1 - x_2 + x_3 &= 4, \\ x_1 + x_2 + x_3 &= 4.8, \\ 2x_1 + 3x_2 - 2x_3 &= 3.1, \end{aligned}$$

making this an overspecified system.

From the earlier solution of the first three equations, the actual sum of x_1, x_2, and x_3 is 4.857, but it has been represented now as 4.8 in the newly added fourth equation. Similarly, the right-hand side of the last equation on the basis of the previously obtained solution should be 3. However, a value of 3.1 has been used.

These arbitrarily chosen differences represent errors in observations or measurements of a physical phenomenon that result when several independent observations greater than the number of variables are represented mathematically by a system of equations. It is now of interest to see how these additional (redundant) equations influence the solution.

In order to obtain the pseudo-inverse on an Excel spreadsheet, represent the coefficient matrix **A** and the right-hand vector **b** as

$$\mathbf{A} = \begin{bmatrix} 3 & 2 & -1 \\ 2 & 1 & 1 \\ 1 & -1 & 1 \\ 1 & 1 & 1 \\ 2 & 3 & -2 \end{bmatrix}, \qquad \mathbf{b} = \begin{bmatrix} 1 \\ 6 \\ 4 \\ 4.8 \\ 3.1 \end{bmatrix}. \tag{3.28}$$

On the spreadsheet, form a transpose of matrix **A** and obtain the product 3×3 matrix $\mathbf{D} = \mathbf{A}^T\mathbf{A}$ as

$$\mathbf{D} = \begin{bmatrix} 19 & 14 & -3 \\ 14 & 16 & -7 \\ -3 & -7 & 8 \end{bmatrix},$$

and its inverse

$$\mathbf{D}^{-1} = \begin{bmatrix} 0.2095 & -0.2413 & -0.1326 \\ -0.2413 & 0.3793 & 0.2413 \\ -0.1326 & 0.2413 & 0.2864 \end{bmatrix}.$$

Next, form the product

$$\mathbf{D}^{-1}\mathbf{A}^{\mathrm{T}} = \begin{bmatrix} 0.2785 & 0.0450 & 0.3183 & -0.1644 & -0.0397 \\ -0.2069 & 0.1379 & -0.3793 & 0.3793 & 0.1724 \\ -0.2015 & 0.2625 & -0.0875 & 0.3952 & -0.1140 \end{bmatrix},$$

which is the pseudo-inverse of matrix **A**.

The solution vector is obtained as a product of the pseudo-inverse matrix and the **b** vector of (3.28) as $[1.1562, 0.3896, 3.2725]^T$. This solution is somewhat different from the earlier solution obtained by three equations using a matrix of full rank = 3. It is shown in the Appendix C that this solution has a minimum value for the sum of squares of deviation from the previous solution.

Underspecified System. Consider the system of (3.11) to include an additional unknown x_4 as

$$3x_1 + 2x_2 - x_3 - 7x_4 = 1 \qquad \text{(I)},$$

$$2x_1 + x_2 + x_3 + 3x_4 = 6 \qquad \text{(II)},$$

$$x_1 - x_2 + x_3 + 5x_4 = 4 \qquad \text{(III)}.$$

A reduction of this into canonical form by the same procedure as indicated earlier will result in the following which is identical to (3.18) with the addition of coefficients for x_4.

$$x_1 + 0x_2 - \frac{9}{3}x_3 - \frac{39}{3}x_4 = \frac{33}{3},$$
$$0 - x_2 + 5x_3 + 23x_4 = 16,$$
$$0 - 0x_2 - \frac{21}{3}x_3 - \frac{93}{3}x_4 = \frac{-69}{3}.$$

From the last equation of this system we get

$$x_3 = \frac{69}{21} - \frac{93}{21}x_4,$$

and corresponding solutions for x_2 and x_1 in terms of x_4 can be easily obtained by successive back-substitution.

A general way of stating the solution procedure for underspecified system of equations is as follows.

A system of equations

$$\mathbf{Ax} = \mathbf{b} \tag{3.29}$$

where $\mathbf{x}$ is a vector of n unknowns, $\mathbf{b}$ is a m vector of constants, and $\mathbf{A}$ is a $m \times n$ matrix with $m < n$ can be written as

$$\mathbf{A_1x_1} + \mathbf{A_2x_2} = \mathbf{b}, \tag{3.30}$$

in which $\mathbf{A_1}$ is an $m \times m$ matrix, $\mathbf{x_1}$ is an m vector, $\mathbf{A_2}$ is a matrix of dimensions $m \times (n - m)$, and $\mathbf{x_2}$ is an $n - m$ vector. For example,

$$\begin{bmatrix} 3 & 2 & -1 & -7 & 6 \\ 2 & 1 & 1 & 3 & -1 \\ 1 & -1 & 1 & 5 & 3 \end{bmatrix} \begin{bmatrix} x_1 \\ x_2 \\ x_3 \\ x_4 \\ x_5 \end{bmatrix} = \begin{bmatrix} 1 \\ 6 \\ 4 \end{bmatrix} \tag{3.31}$$

can be written as

$$\begin{bmatrix} 3 & 2 & -1 \\ 2 & 1 & 1 \\ 1 & -1 & 1 \end{bmatrix} \begin{bmatrix} x_1 \\ x_2 \\ x_3 \end{bmatrix} + \begin{bmatrix} -7 & -6 \\ 3 & -1 \\ 5 & 3 \end{bmatrix} \begin{bmatrix} x_4 \\ x_5 \end{bmatrix} = \begin{bmatrix} 1 \\ 6 \\ 4 \end{bmatrix}. \tag{3.32}$$

Consequently, for any m unknowns ($m = 3$ in this case), we can write a solution in terms of the $(n - m)$ unknowns as

$$\mathbf{Iu}_1 + \mathbf{A}_1^{-1}\mathbf{A}_2\mathbf{u}_2 = \mathbf{A}_1^{-1}\mathbf{b}, \tag{3.33}$$

where

$$\mathbf{u}_1 = \begin{bmatrix} x_1 \\ x_2 \\ x_3 \end{bmatrix}$$

and

$$\mathbf{u}_2 = \begin{bmatrix} x_4 \\ x_5 \end{bmatrix}.$$

From (3.31) we get

$$\mathbf{I}\begin{bmatrix} x_1 \\ x_2 \\ x_3 \end{bmatrix} + \begin{bmatrix} 3 & 2 & -1 \\ 2 & 1 & 1 \\ 1 & -1 & 1 \end{bmatrix}^{-1} \begin{bmatrix} -7 & -6 \\ 3 & -1 \\ 5 & 3 \end{bmatrix} \begin{bmatrix} x_4 \\ x_5 \end{bmatrix} = \begin{bmatrix} 3 & 2 & -1 \\ 2 & 1 & 1 \\ 1 & -1 & 1 \end{bmatrix}^{-1} \begin{bmatrix} 1 \\ 6 \\ 4 \end{bmatrix}, \tag{3.34}$$

which simplifies to (confirm the values for inverse matrix on the PC)

$$\begin{bmatrix} x_1 \\ x_2 \\ x_3 \end{bmatrix} + \begin{bmatrix} -0.2857 & 3.1428 \\ -0.8571 & -3.5714 \\ 4.4285 & -3.7142 \end{bmatrix} \begin{bmatrix} x_4 \\ x_5 \end{bmatrix} = \begin{bmatrix} 1.1428 \\ 0.4285 \\ 3.2857 \end{bmatrix}. \tag{3.35}$$

In the above, we have arbitrarily chosen x_1, x_2, and x_3 to be expressed in terms of x_4, and x_5. One may obtain a solution to any m unknowns in terms of the other unknowns. For instance, in the above example, a solution to x_1, x_2, and x_4 can be obtained in terms of x_3 and x_5 as

$$\mathbf{I}\begin{bmatrix} x_1 \\ x_2 \\ x_4 \end{bmatrix} + \begin{bmatrix} 3 & 2 & -7 \\ 2 & 1 & 3 \\ 1 & -1 & 5 \end{bmatrix}^{-1} \begin{bmatrix} -1 & -6 \\ 1 & -1 \\ 1 & 3 \end{bmatrix} \begin{bmatrix} x_3 \\ x_5 \end{bmatrix} = \begin{bmatrix} 3 & 2 & -7 \\ 2 & 1 & 3 \\ 1 & -1 & 5 \end{bmatrix}^{-1} \begin{bmatrix} 1 \\ 6 \\ 4 \end{bmatrix}. \tag{3.36}$$

If zero values for the $n - m$ unknowns are assumed, we can readily obtain a solution for the m unknowns. For instance, if x_4, and x_5 are assumed to be zero, from (3.35), we have $x_1 = 1.1428$, $x_2 = 0.4285$, and $x_3 = 3.2857$. Of course, one obtains the very same solution for the equation system using the matrix with a full rank of 3 by ignoring the coefficients of x_4 and x_5 from the outset.

Subsequently, we shall show that a solution of linear optimization problems using linear programs entails the solution of an underspecified system of equations as shown above.

3.4 LINEAR OPTIMIZATION PROBLEMS

Problems in engineering, economics, and physical and social sciences entail the maximization or the minimization of an objective function (also called by other names such as performance index or goal) subject to certain restrictions and tradeoffs. In electric power engineering, a typical example is that of dispatching generation to minimize the total cost of production, subject to the equality constraint that the sum of generator outputs is equal to the demand plus losses. An additional constraint may be that the flows in lines are below certain limits. In manufacturing, a typical example is that of maximizing production or profit, subject to constraints of machining times by different processes, and the number of hours different machines are available. One can go on with a myriad of examples in several disciplines. Importantly, however, all such problems have a common structural characteristic that we will soon discuss. Such problems, called linear programs, can be solved by very efficient commercially available software. Almost all softwares use the simplex method, the revised simplex method or its derivatives, originally due to George Dantzig (1951). Several further advancements have been made to LP solution algorithms such as the interior point method suggested by Karmarker (1984).

In the following, our initial agenda is to examine an optimization problem for two choice variables so that a graphical interpretation of the solution space is possible. Subsequently, we study the general nature of the Linear Program (LP) followed by a cursory understanding of the simplex method using an Excel spreadsheet. Our purpose is not to attain proficiency in the development of algorithms, but to become familiar with problem formulation and solution of linear programs using the Solver tool of Microsoft Excel or other commercial software. Toward this end, in subsequent sections we formulate and solve certain illustrative examples with particular emphasis on problems related to the electric power industry.

3.4.1 Graphical Solution of a Simple Linear Programming Problem

General Structure of Linear Programs

We select choice variables

$$x_1, x_2, \ldots, x_n$$

in order to minimize or maximize the objective function

$$\pi = c_1x_1 + c_2x_2 + \cdots + c_nx_n, \tag{3.37}$$

where $c_1, c_2, \ldots, c_n$ are constant coefficients, subject to constraints of the form

$$\begin{aligned} a_{11}x_1 + a_{12}x_2 + \cdots + a_{1n}x_n &\leq k_1 \\ a_{21}x_1 + a_{22}x_2 + \cdots + a_{2n}x_n &\leq k_2 \\ \vdots \qquad\qquad &\leq \vdots \\ a_{m1}x_1 + a_{m2}x_2 + \cdots + a_{mn}x_n &\leq k_m \end{aligned} \tag{3.38}$$

and the nonnegativity restrictions

$$x_1, x_2, \ldots, x_n \geq 0, \tag{3.39}$$

where a_{ij}, $i = 1, m$; $j = 1, n$ are constant coefficients.

In compact form, this can be written as follows:

Choose a vector $\mathbf{x}$, in order to maximize

$$\pi = \mathbf{c}^{\mathrm{T}}\mathbf{x},$$

subject to

$$\mathbf{A}\mathbf{x} \leq \mathbf{k}.$$

Graphical Interpretation. Consider a simple formulation of the above in two dimensions amenable to graphical interpretation.

Example: In a hypothetical drug treatment plan for a patient in a hospital, two drugs x_1 and x_2 are available at per dose costs of \$0.60 and \$1, respectively. The physician can choose between these two drugs and has to decide on an optimal mix of these two drugs to administer. Obviously, the total cost of the drugs should be minimized (objective function). The quantities of the two drugs chosen are choice variables.

Both drugs contain a particular antibiotic, vitamin E, and a pain suppressor in the following percentages. Drug x_1 contains 10 units, 5 units, and 2 units of these ingredients per dose. The corresponding composition of drug x_2 is 4 units, 5 units, and 6 units, respectively. The physician wants to ensure that the treatment dispenses at least 20 units of antibiotic, 20 units of vitamin E, and 12 units of pain suppressor per dose.

Problem Formulation: If x_1 and x_2 are the quantities of the two drugs chosen, this problem of optimum combination of the two drugs per dose in order to minimize cost can be written as follows:

Minimize

$$\pi = 0.6x_1 + x_2, \tag{3.40}$$

subject to

$$10x_1 + 4x_2 \geq 20 \quad \text{[antibiotic constraint]}, \tag{3.41}$$

$$5x_1 + 5x_2 \geq 20 \quad \text{[vitamin E constraint]}, \tag{3.42}$$

$$2x_1 + 6x_2 \geq 12 \quad \text{[pain supressor constraint]} \tag{3.43}$$

and the simple nonnegativity constraints

$$x_1, x_2 \geq 0. \tag{3.44}$$

Figure 3.13 shows the three lines in the x_1–x_2 plane corresponding to the above three inequality constraints. Since the administered quantities of three ingredients should be equal to or greater than indicated quantities, the solution for x_1 and x_2 can be anywhere above the three constraint lines. This region is called the feasible region of solution indicated in Figure 3.13. Notice that the boundary of the feasible region comprises constraint borders and the axes. The intersection of the borderlines produces corner points, which is referred to as *extreme points*. Extreme points occur at the intersection of two border lines (for example, at (3, 1) and $(\frac{2}{3}, 3\frac{1}{3})$). As we shall see, the extreme points are of vital significance for the solution to the problem.

Solution: In order to minimize π, consider the objective function written in the form

$$x_2 = \pi - 0.6x_1.$$

Considering π to be a parameter, this equation can be plotted as a family of lines with slope = −0.6. Figure 3.13 shows these *isocost* lines for $\pi = 6$ and $\pi = 2.8$.

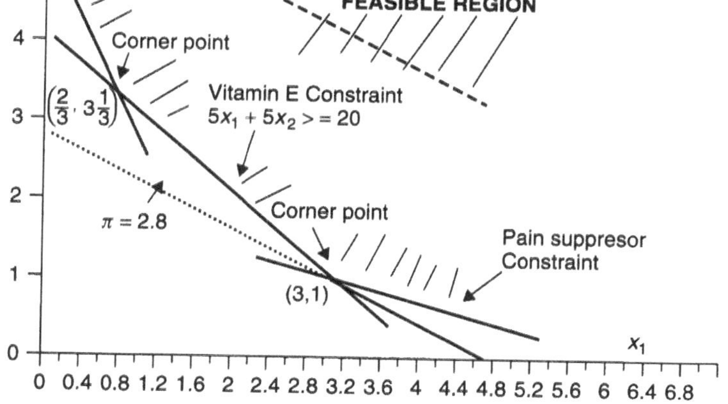

Figure 3.13. Graphical interpretation.

Any solution for x_1 and x_2 on any isocost line that lies within the shaded feasible region is feasible. Such a solution, although feasible (satisfies all constraints), is not necessarily optimal. For that matter, all extreme points are feasible solutions. Since our objective is to minimize cost, the feasible lowest cost isocost is the one tangential to the feasible region at the extreme point (3,1). Therefore, the optimal solution to the problem is $\bar{x}_1 = 3$ and $\bar{x}_2 = 1$. It turns out that the lowest total cost of the drug therapy at the indicated individual drug prices is $\pi = \$2.80$.[10] Since all corner points are feasible solutions, one corner point gives the optimal solution. Therefore, an efficient method of attaining the best corner point through a minimum number of corner point searches is the characteristic of a good algorithm. As will be seen soon, the simplex method achieves this.

The solution to this problem, while exactly satisfying the vitamin E and pain suppressor quantities, overfulfills the antibiotic requirement. This is evident from the solution of $\bar{x}_1 = 3$ and $\bar{x}_2 = 1$, which upon substitution in three constraint equations (3.41) to (3.43) give values of 34, 20, and 12 for the right-hand side of constraints.

Another important observation is in respect of price changes. What will happen if the respective price of the two drugs change? Price changes effect the slope of isocost lines of Figure 3.13. The slope of the isocost lines may change from the present -0.6 to a different value. If the two prices change in exactly the same proportion, the isocosts' slope will remain unchanged. The above optimal solution will still be valid although the isocost lines may shift up or down.

If the two prices change by different proportions, but by a "minor" amount, the earlier optimal solution may still hold. For example, if the prices change so that the isocosts have a slope of -0.7, our earlier solution still holds. This can be confirmed by seeing that a line of slope -0.7 passing through (3,1) is still tangential to the feasible region at this extreme point.

In Figure 3.13, the vitamin E constraint line has a slope of -1.0. Therefore, if the prices of drugs change such that their ratios give a slope of -1.0 to the isocost line, it will then be parallel to the vitamin constraint line. Consequently, the optimal solution can then be anywhere along the vitamin line including the points (3,1) and $(\frac{2}{3}, 3\frac{1}{3})$. Then, any value of x_1 and x_2 falling along this line in the interval between the two extreme points will result in the same value for the objective. This is called a degenerate solution because it gives an infinite possible solutions.

3.5 LINEAR PROGRAMS AND CONVEX SETS — RELATION

Earlier, it was said that a convex set has the property that if X and Y are contained in a set S, then a set $W = \theta X + (1 - \theta)Y$ is also contained in S. Now take the objective function $\pi = c_1x_1 + c_2x_2 + \cdots, c_nx_n$ of (3.37), for example. This

[10]This problem and its solution are identical to an example in Chiang (1984) derived from a discipline of dietary science. The choice of examples with identical solution is deliberate in order to show the universality of optimization principles in different disciplines.

represents a n-dimensional hyperplane H in n-space. We now demonstrate that the hyperplane is a convex set.

Let points $\mathbf{u}(u_1, u_2, \ldots, u_n)$ and $\mathbf{v}(v_1, v_2, \ldots, v_n)$ belong to the set H. Since $\mathbf{u}$ and $\mathbf{v}$ are possible solutions, we have

$$\pi = \mathbf{c}^T \mathbf{u}$$

and

$$\pi = \mathbf{c}^T \mathbf{v}.$$

In order to prove that set H is convex, it is necessary to show that any convex combination, say $\mathbf{w}$, of $\mathbf{u}$ and $\mathbf{v}$ satisfies (3.37). Then for $\mathbf{c}^T \mathbf{w}$, since $\mathbf{w}$ is a convex combination of $\mathbf{u}$ and $\mathbf{v}$, we have

$$\begin{aligned}\mathbf{c}^T \mathbf{w} &= \mathbf{c}^T [\theta \mathbf{u} + (1-\theta)\mathbf{v}] = \mathbf{c}^T \theta \mathbf{u} + \mathbf{c}^T (1-\theta)\mathbf{v} \\ &= \theta\pi + (1-\theta)\pi = \pi,\end{aligned}$$

which completes the proof.

A hyperplane divides the n-dimensional space into two *half-spaces*. For example, in two dimensions, a straight line divides the coordinate plane to two spaces, one on either side of it. Depending on whether the dividing hyperplane is considered a part of a half-space in question, we will have either an *open* or a *closed* half-space. For instance, consider two inequalities $a^T x < k$ and $a^T x \le k$. The former defines an *open* half-space on one side of the hyperplane $a^T x = k$ while the latter defines the *closed* half-space containing the points on one side of the hyperplane $a^T x = k$ as well as the points lying on the hyperplane itself.

In view of the above, the constraint equations (3.45) of a linear program,

$$\begin{aligned}a_{11}x_1 + a_{12}x_2 + \cdots + a_{1n}x_n &\le k_1 \\ a_{21}x_1 + a_{22}x_2 + \cdots + a_{2n}x_n &\le k_2 \\ \vdots \qquad\qquad &\le \vdots \\ a_{m1}x_1 + a_{m2}x_2 + \cdots + a_{mn}x_n &\le k_m\end{aligned} \tag{3.45}$$

define closed half-spaces. Clearly, these closed half-spaces again are convex sets. This argument applies to nonnegativity restrictions as well, since nonnegativity constraints are nothing but a special case of the constraint equations.

We can summarize the above argument as follows. The solution to a linear programming problem is within a feasible region if it satisfies all the inequalities. Such a feasible region is a closed convex set. Such a feasible region represents the intersection of $m + n$ convex sets satisfying $m + n$ weak inequalities comprising of m constraints and n nonnegativity conditions [see (3.38) and (3.39)]. Thus the feasible region is in the intersection of $m + n$ closed convex sets since it is simultaneously a member of $m + n$ closed half-spaces.

Optimal Solution and Extreme Points. Interior and extreme points have been defined earlier in Section 3.4.1. Let F represent the feasible region of solution. The object of the linear program solution is to push the objective hyperplane to the highest (or the lowest) point by varying the value of π while staying in F. Clearly, when pushed to the highest position (lowest), the hyperplane can contain no interior points of the set F, but it can contain only the boundary points. Thus evolves the concept of *supporting hyperplane* $(\overline{H})$ as one with one or more of its points common with the convex set F, and situated such that the set F lies to one side of $(\overline{H})$ exclusively. Figure 3.14 illustrates a supporting hyperplane in three dimensions.

We now state two theorems without proof:

1. For a given boundary point $\mathbf{x}$ of a closed convex set, there is at least one supporting hyperplane at $\mathbf{x}$.
2. There is at least one extreme point of a closed convex set bounded from below in every supporting hyperplane.

The implication of the above is penetratingly obvious. At the optimal solution, the objective hyperplane will be a supporting hyperplane. Then from the first statement above, we deduce that every point of the closed convex set (defining the problem) is a possible candidate for optimal solution. The second statement, however, narrows our focus only to the extreme points. Even though there may be nonextreme boundary points of F in the same supporting hyperplane, since a hyperplane represents a particular value of the objective π, such nonextreme boundary points do not offer any better solution. Consequently, such points can be discarded without loss.

It is this fact that is exploited in the simplex method by using a procedure to systematically search only the extreme points till the optimum is attained. This

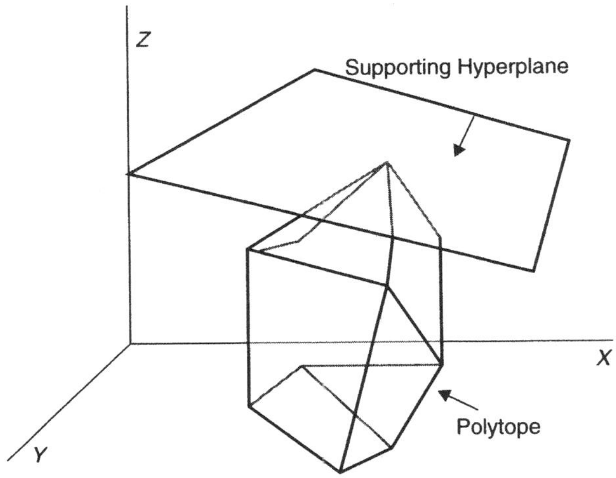

Figure 3.14. Supporting hyperplane.

feature of narrowing the search only to extreme points is a special feature of linear programming, distinguishing it from classical optimization problems.

3.6 LINEAR PROGRAMS

The first step in the solution of linear programming problems is to create an auxiliary problem by the introduction of *artificial variables* into the problem formulation. The intent of this artifice is to bring the auxiliary problem into canonical form from which a basic feasible solution (BFS) can be obtained immediately, if one exists. Starting from the BFS, the optimal solution to the original problem is obtained by moving from one BFS to another along the vertices. This process is eloquently discussed in Rao (1996), Gill et al. (1981), and Bazaraa et al. (1979).

In the following, since our intent is only to gain a general understanding of the solution procedure, the process is described using an Excel spreadsheet without the full rigor of mathematics. Only the mathematical underpinnings required to understand the elements of the process are included. Toward this end, the problem is first formulated by using slack and surplus variables, and subsequently it is solved in an Excel spreadsheet.

Problem Formulation. In general, the inequalities in LP problems may be of the less than or equal to kind.[11] Let there be p inequalities of the less than or equal to type, r inequalities of greater than or equal to kind, and s equality relations. The problem can be written as follows:

Minimize

$$\pi = f(x_1, x_2, \ldots, x_n), \tag{3.46}$$

subject to:

p inequalities the kind

$$\begin{array}{ccccccccc} a_{11}x_1 & + & a_{12}x_2 & + & \cdots & + & a_{1n}x_n & \leq & b_1 \\ \vdots & & \vdots & & \vdots & & \vdots & \vdots & \vdots \\ a_{p1}x_1 & + & a_{p2}x_2 & + & \cdots & + & a_{pn}x_n & \leq & b_p, \end{array}$$

r inequalities of the kind

$$\begin{array}{ccccccccc} c_{11}x_1 & + & c_{12}x_2 & + & \cdots & + & c_{1n}x_n & \geq & d_1 \\ \vdots & & \vdots & & \vdots & & \vdots & \vdots & \vdots \\ c_{r1}x_1 & + & c_{r2}x_2 & + & \cdots & + & c_{rn}x_n & \geq & d_r, \end{array}$$

[11] The convention for inequalities is that of an algebraic expression being on the left-hand side. In other words, we always write the algebraic expression of the constraint on the left-hand side; that is, we write the inequality as $ax_1 + bx_2 \leq 10$, but not as $10 \geq ax_1 + bx_2$.

s equality equations of the kind

$$\begin{array}{ccccccccc} e_{11}x_1 & + & e_{12}x_2 & + & \cdots & + & e_{1n}x_n & = & f_1 \\ \vdots & & \vdots & & \vdots & & \vdots & = & \vdots \\ e_{s1}x_1 & + & e_{s2}x_2 & + & \cdots & + & e_{sn}x_n & = & f_s, \end{array}$$

and the nonnegativity restrictions

$$(x_1, x_2, \ldots, x_n) \geq 0.$$

If we now add slack variables $s_1, \ldots, s_n$ to the first set of less than or equal to inequalities, and surplus variables $r_1, \ldots, r_n$ to greater than or equal to inequalities, the above constraints can be written as

p equations of the kind

$$\begin{array}{ccccccccccc} a_{11}x_1 & + & a_{12}x_2 & + & \cdots & + & a_{1n}x_n & + & s_1 & = & b_1 \\ \vdots & & \vdots & & \vdots & & \vdots & & \vdots & = & \vdots \\ a_{p1}x_1 & + & a_{p2}x_2 & + & \cdots & + & a_{pn}x_n & + & s_p & = & b_p, \end{array} \tag{3.47}$$

r equations of the kind

$$\begin{array}{ccccccccccc} c_{11}x_1 & + & c_{12}x_2 & + & \cdots & + & c_{1n}x_n & - & r_1 & = & d_1 \\ \vdots & & \vdots & & \vdots & & \vdots & & \vdots & = & \vdots \\ c_{r1}x_1 & + & c_{r2}x_2 & + & \cdots & + & c_{rn}x_n & - & r_r & = & d_r, \end{array} \tag{3.48}$$

s equality equations of the kind

$$\begin{array}{ccccccccc} e_{11}x_1 & + & e_{12}x_2 & + & \cdots & + & e_{1n}x_n & = & f_1 \\ \vdots & & \vdots & & \vdots & & \vdots & = & \vdots \\ e_{s1}x_1 & + & e_{s2}x_2 & + & \cdots & + & e_{sn}x_n & = & f_s, \end{array} \tag{3.49}$$

and the nonnegativity restrictions

$$(x_1, x_2, \ldots, x_n), (s_1, \ldots, s_p), (r_1, \ldots, r_r) \geq 0.$$

The above is a general formulation of linear optimization problems. This formulation can also accommodate the maximization of an objective function. To maximize an objective, the above formulation for minimization is written with a negative sign for the objective. Clearly, minimizing the negative of an objective function is tantamount to maximizing the objective function. In the above, all constraints are expressed as equalities in terms of slack and surplus variables.

The problem formulation has $(p + r + s)$ equations, $(n + p + r)$ unknowns comprising n choice variables, p slack variables, and r surplus variables. It turns out that in most practical problems, $p < n$; that is, the number of equality constraints is less than the number of decision variables.[12] Consequently, solving this problem is akin to solving a system of under specified equations since $(p + r + s) < (n + p + r)$. As explained earlier, one can obtain a solution to $(p + r + s)$ unknowns in terms of $(n - s)$ variables. For example, if any slack or surplus variable is not zero, the above equations show that the corresponding specified inequality is more than satisfied. However, if a slack or a surplus variable is zero, the equality of the "less than or equal to" or "greater than or equal to" holds, indicating a corresponding "*binding constraint*."

The question now is, Which $(n - s)$ unknowns are to be assumed zero? To answer this, one can try all combinations in a "brute force" way examining which trial gives the least (highest) value for the objective function. Such a procedure, in essence, examines the value of the objective function at each extreme point in the solution hyperspace. The number of solutions to be inspected is equal to the number of ways in which m variables can be selected from a set of n choice variables$(m < n)$. This is given by

$$\binom{n}{m} = \frac{n!}{(n-m)!m!}.$$

If $n = 12$ and $m = 6$, we have 924 combinations, which increases to 12,850 for $m = 16$ and $n = 8$. Clearly, the number of solutions to be examined by a trial-and-error method becomes formidably large as n increases and as the difference between n and m increases. Hence we need a method to organize the search efficiently to obtain the minimum point in some logical fashion. The basis of simplex method is to search an optimal extreme point, starting from a given feasible solution, efficiently.

3.6.1 Definitions Pertaining to Linear Optimization

Before a discussion of rudiments of the simplex method, certain further definitions of terms in linear optimization are in order.

1. *Basic Solution:* This is the solution for m unknowns in an underspecified set of linear equations in which $n - m$ variables are set equal to zero. As said earlier, $n > m$.
2. *Basic Variables:* Variables not set equal to zero to obtain the basic solution is called the basis.

[12]If this is not the case, the linear program becomes a trivial exercise. For example, if there are as many equality constraints as the number of choice variables, the problem is fully specified, and its solution is obtained by a solution of a system of equations. If there are more equality constraints than choice variables, the problem is one of least-square estimation via a pseudo-inverse matrix.

3. *Feasible Solution:* Any solution that satisfies constraints of the linear programming problem

$$\mathbf{Ax} = \mathbf{b},$$

$$\mathbf{x} \geq \mathbf{0}$$

is called a *feasible solution*. All constraints have been expressed above as equalities incorporating appropriate slack and surplus variables.

4. *Basic Feasible Solution:* A basic solution that satisfies the nonnegativity conditions of the linear program is called basic feasible solution.
5. *Optimal Solution:* This is a feasible solution that optimizes the objective function.

3.7 ILLUSTRATION OF THE SIMPLEX METHOD

Readers interested expressly in practical techniques and commercially available software may omit this section that illustrates the simplex method. Such readers may directly proceed to Chapter 4, which addresses solutions to practical problems.

The simplex method of obtaining the optimal solution is illustrated through the following simple example.

Consider the optimization problem:

Minimize

$$\pi = -x_1 - 2x_2 - x_3, \tag{3.50}$$

subject to

$$2x_1 + x_2 - x_3 \leq 2, \tag{3.51}$$

$$2x_1 - x_2 + 5x_3 \leq 6, \tag{3.52}$$

$$4x_1 + 1.5x_2 + 1.5x_3 \leq 6, \tag{3.53}$$

and nonnegativity constraints

$$x_1, x_2, x_3 \geq 0, \tag{3.54}$$

and slack variables to make inequalities into equalities as in (3.47):

$$s_1, s_2, s_3 \geq 0. \tag{3.55}$$

We shall illustrate the procedure used by the simplex method through a tableau (which sets up these equations in a table) on an Excel spreadsheet "Simplex.Xls" (Open the program "Simplex.Xls" on the PC). Subsequently, the solution obtained

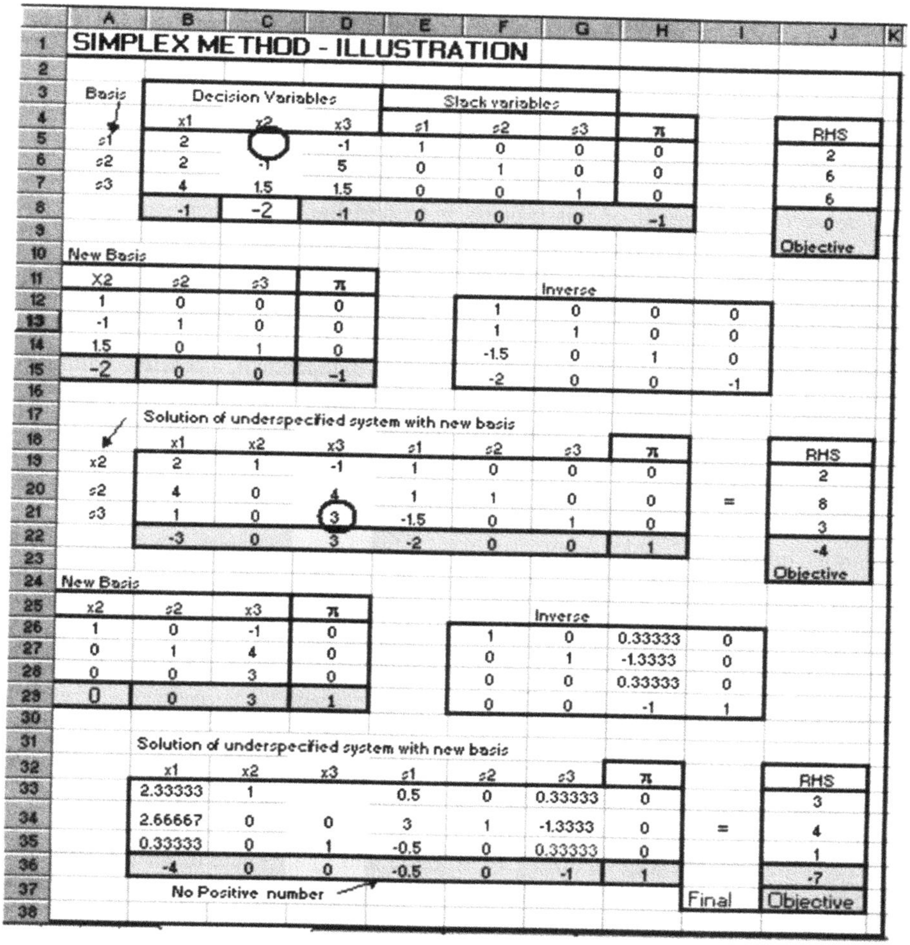

Figure 3.15. Principle of simplex method.

by using the Excel Solver routine is compared to the results of the simplex tableau. As expected, the two solutions will be identical.

Simplex.Xls A print of the "Simplex.Xls" spreadsheet is shown in Figure 3.15. Cells B5 to G7 show the coefficients of the constraint equations (3.51), (3.52), (3.53) and (3.55) expressed as equalities by adding slack variables. For example, row 5 represents

$$2x_1 + x_2 - x_3 + s_1 = 2.$$

Cells J5:J7 indicate the right-hand side of constraint equations. Row 8 shows the objective function (3.50) as $-x_1 - 2x_2 - x_3 - \pi = 0$.

By converting the above equations to equalities by the addition of slack variables, we have six choice variables, x_1, x_2, x_3, s_1, s_2, and s_3. This represents the auxiliary problem.

In the optimization step, since there are three constraint equations, there can only be three linearly independent vectors, which form the *basis* for a three-dimensional space. Therefore, we can solve for any three of these choice variables with the other three as zero. The pursuit of this process is to find the three basic variables and their values that obtain the minimum of the objective function.

We first select s_1, s_2, and s_3 as the basis to obtain a solution making x_1, x_2, and x_3 equal to zero. Such an assumption does not violate any nonnegativity constraints ($x_1, x_2, x_3 \geq 0$). Since the three column vectors $(1, 0, 0)$, $(0, 1, 0)$, and $(0, 0, 1)$ in columns E, F, and G of the spreadsheet are linearly independent, a ***feasible*** solution can be obtained. Therefore, these three vectors form a *basic feasible set* (BFS).[13] From the earlier explanation of solution of linear equations, it is clear that s_1, s_2, and s_3 can be expressed in terms of x_1, x_2, and x_3, which are assumed to be zero. The value of the objective function ($-\pi$ shown in cell J8) for this assumption that x_1, x_2, and x_3 are equal to zero is in cell J8.

Can we now reject any one of s_1, s_2, or s_3, from this initial BFS and substitute it with one from x_1, x_2, or x_3 to reduce π further? An examination of the coefficients of the objective in row 8 shows that the highest coefficient belongs to x_2, a value of -2. Since the objective function is most affected by this, bringing in x_2 into BFS should result in a maximum amount of reduction of the objective. Since only three variables can form the vectors of a BFS, one of the present variables (s_1, s_2, s_3) has to be jettisoned in favor of x_2. The following procedure is used to decide which should be such outgoing variable. To recapitulate our argument so far, by bringing in x_2, we will have $x_2 \neq 0$, $x_1 = 0$, $x_3 = 0$, and one of the variables to be jettisoned (s_1, s_2, or s_3) becomes zero.

To examine the effect of incoming x_2, since x_1 and x_2 are equal to zero, equations including the objective function expressed in cells B5:J8 of Figure 3.15 can be written as

$$\begin{aligned} s_1 + 1x_2 &= 2, \\ s_2 - 1x_2 &= 6, \\ s_3 + 1.5x_2 &= 6, \\ -\pi - 2x_2 &= 0, \end{aligned} \tag{3.56}$$

in which one of s_1, s_2, or s_3 will be zero.

The last of these equations in (3.56) suggests that x_2 should be made as large as possible in order to achieve a maximum reduction of π. However, by doing so, we see from other equations of (3.56) that some of the s_1, s_2, or s_3 may become

[13] For some minimization problems with greater than type of equalities, it may not be possible to obtain a BFS by assuming x_1, x_2, and x_3 to be zero. Under such circumstances, *artificial variables* are introduced in order to obtain an initial BFS. The reader will find the details of such a procedure in Chiang (1984).

negative in violation of (3.55). For instance, if $x_2 > 2$, the first equation of (3.56) indicates that s_1 becomes negative. However, if all the coefficients of the incoming variable x_2 in (3.56) (cells C5:C7) are negative, x_2 can be made infinitely large and the objective function of the minimization problem will have a value of minus infinity. Such a linear program problem is said to have an unbounded solution.

If at least one or more coefficients of x_2 in (3.56) is positive, the limit for the value of x_2 that does not make any of old basis vectors s_1, s_2, and s_3 negative can be found as the minimum of the ratio of the right-hand side of (3.56) and the nonpositive coefficients of the incoming variable. In the above example, in order to keep s_1 positive, x_2 cannot exceed $\frac{2}{1}$. Similarly, for the other nonpositive coefficient 1.5 of x_2, it cannot exceed $\frac{6}{1.5}$ in order to keep s_3 positive. The lower of these[14] ratios is the former pertaining to s_1, thus limiting the value of $x_2 = 2$.

Substituting this value of $x_2 = 2$ in (3.56), and transposing terms, we get

$$\begin{aligned} x_2 &= 2 \qquad\;\; = 2, \\ s_1 &= 2 \quad -2 = 0, \\ s_2 &= 6 \quad +2 = 8, \\ s_3 &= 6 \quad -3 = 3, \\ \pi &= 0 \quad -4 = -4. \end{aligned} \tag{3.57}$$

In (3.57) observe that the value of the objective function is reduced further to a value of −4 from its previous value of zero with new basic variables x_2, s_2, and s_3.

The following is a summary of the above process of moving from one set of choice basic variables to a better set.

1. Select the variable with the highest negative coefficient in the objective function as the incoming variable (x_2 in this example).
2. Examine the ratio of the right-hand side of the constraint and the nonnegative coefficients of possible incoming variables in each constraint equation (these ratios are equal to 2 and 4 in this example). Select the lowest value of this ratio as the pivot element (shown circled in Figure 3.15). Pivotal element means that the variable corresponding to this row is the outgoing (s_1 in this example).
3. Having made this determination, solve the underspecified system of equations with the new BFS in a manner similar to that of solving (3.29) and (3.30).

When we continue this process of seeking better objective function in the Excel spreadsheet of Figure 3.15, the columns corresponding to the new basis, x_2, s_2, s_3, and π are entered as a matrix in cells A12:D15. The inverse of this matrix (cells F12:I15) premultiplies the original matrix in cells B5:H8, and the previous RHS in cells J5:J8. The resulting matrix and the RHS are in cells B19:H22 and J19:J22.

[14]In problems where this ratio is the same for two or more variables resulting in a tie, the choice of any one of them will give the same objective function. Such a solution is called a degenerate solution. Rao (1996) and Chiang (1984) discuss details of such problems.

From these equations, we note that

$$\begin{aligned} x_2 &= 2, \\ s_2 &= 8, \\ s_3 &= 3, \\ x_1 &= 0, \\ x_3 &= 0, \\ s_1 &= 0. \end{aligned} \tag{3.58}$$

The last row of equations (row 22) states:

$$-3x_1 + 3x_3 - 2s_1 + \pi = -4. \tag{3.59}$$

This suggests that π can be further reduced by bringing in x_3 since it has a positive coefficient of 3. Bringing in x_1, or s_1 would increase π because of the negative coefficients associated with these variables. Consequently. such an action would negate the goal of reducing π.

With x_3 as the new incoming variable, one of the variables x_2, s_2, or s_3 has to be rejected. As before, we examine the ratio of the right-hand side to the positive coefficient entries for x_3 in cells D19:D21. These ratios being $\frac{8}{4}$ and $\frac{3}{3}$, the latter is the minimum of the two. Hence, s_3 is chosen as the pivot (outgoing variable) giving the new basis as x_2, s_2, and x_3. The columns corresponding to these and π are copied onto cells A26:D29 to form a matrix. The inverse of this matrix (cells F26:I29) premultiplies equations in cells B19:J22 to give a new set of relations for the basic variables x_2, x_3, and s_2 in cells B33:J36. These relations in expanded form are

$$\begin{aligned} x_2 &= 3 \\ s_2 &= 4 \\ x_3 &= 1 \\ x_1 &= 0 \\ s_1 &= 0 \\ s_3 &= 0. \end{aligned} \tag{3.60}$$

The last row of the relation (row 36) states

$$-4x_1 - 0.5s_1 - 1s_3 + \pi = -7. \tag{3.61}$$

At this stage, there are no positive coefficients for x_1, s_1, or s_3 in (3.61). Consequently, the solution cannot be improved any further by bringing in any new basic variable. Hence, cell J38 gives the final solution to the minimization problem with an objective of -7.

3.7.1 Sensitivity of Optimal Solution

Very importantly, the above procedure obtains coefficients for other possible candidates for the BFS shown in row 36. A rearrangement of (3.61) gives

$$\pi = 4x_1 + 0.5s_1 + s_3 - 7. \tag{3.62}$$

The above equation describes how π changes as x_1, s_1, or s_3 is changed. Therefore, these are sensitivity factors that contain very useful information.

From (3.62), we get

$$\frac{d\pi}{dx_1} = 4, \quad \frac{d\pi}{ds_1} = 0.5, \quad \text{and} \quad \frac{d\pi}{ds_3} = 1.$$

Therefore, we know that increasing x_1 by one will increase the objective function by four, and so on. These sensitivity characteristics provide very important information for decision-making. The Excel Solver routine (as any other LP software) provides sensitivity information along with the solution. In subsequent sections we address this matter further. However, note that the solution procedure of the simplex method provides the sensitivities as a byproduct.

3.8 SOLUTION OF LINEAR PROGRAMS USING EXCEL SOLVER

The following demonstrates how the Excel Solver option can be used to solve the problem of (3.50) subject to (3.51) to (3.53).

The file "LP Demo put on Solver.Xls" on Wiley's ftp site at ftp://ftp.wiley.com/public/sci_tech_med/electric_power/ shows the procedure. Figure 3.16 is the solution copied from the spreadsheet, which is shown along with the sensitivity sheet obtained from Solver.

3.8.1 Setup of Spreadsheet to Use Solver

It is assumed that most readers know how to set up Excel spreadsheets to solve such problems.[15] Nevertheless, since this is the first encounter, steps in the solution of this problem are briefly explained below.

LP Demo put on Solver.Xls

In reference to Figure 3.16, cells C2:C4 are allocated as choice variables. To start with, some arbitrary value, say zero, is entered in these cells. The formula for the objective function is then entered into cell E2.[16] Similarly, three constraint equations are entered in cell G5, G6, and G7. The right-hand-side bound for constraints is entered in cells E5, E6, and E7, respectively. Next by clicking on the Tools button, open the Solver frame. Then select the minimization of cell E2 (objective function). For constraints, enter the requirement that cells G5 to G7 have to

[15] Readers unfamiliar with Excel spreadsheet procedures are urged to click on each cell entry and understand how formulae are entered and related to other cells.

[16] The formulae in cells can be confirmed by clicking on the appropriate cell and examining the formula bar of the Excel sheet.

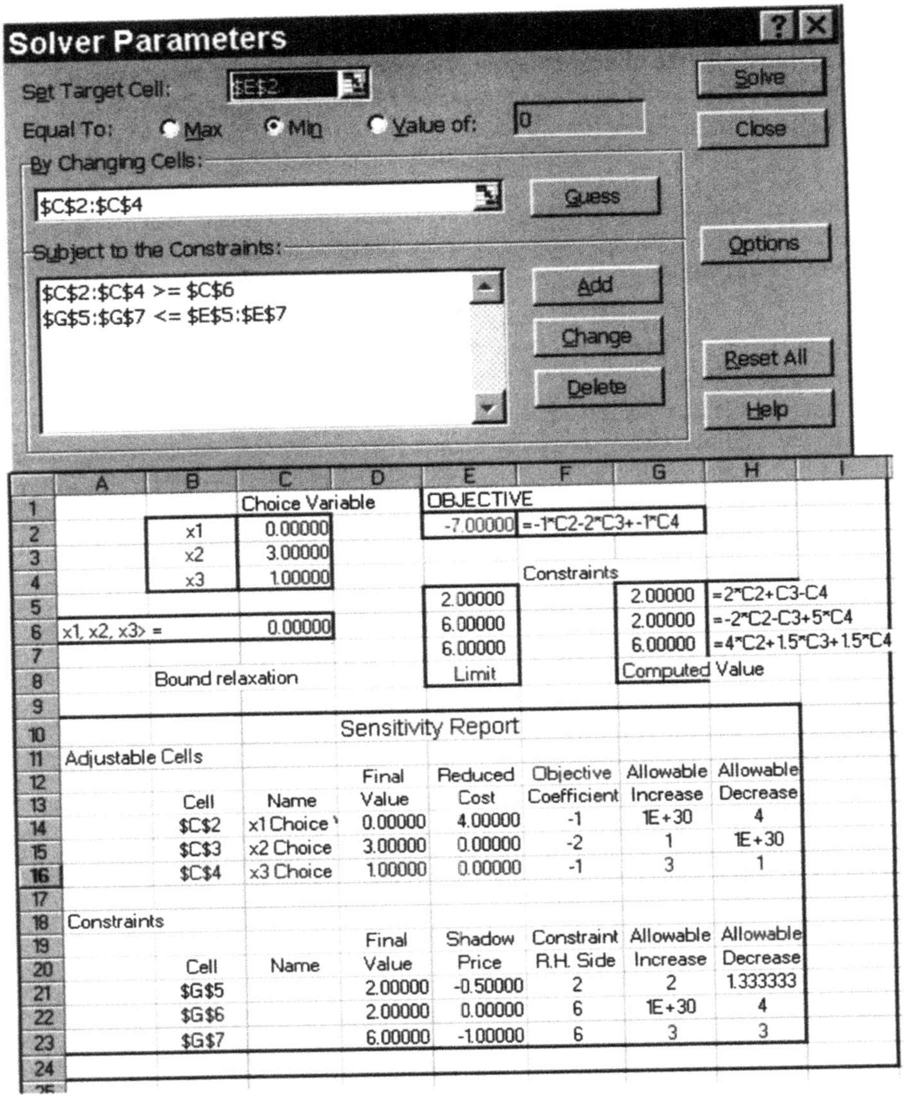

Figure 3.16. Solution using Solver.

be less than or equal to cells E5 to E7. Also, enter another requirement that cells C2 to C4 have to be greater than or equal to zero (cell C6). Next by choosing the options button, choose a linear model and appropriate entries for accuracy and time limit. Then obtain the solution to the problem by clicking the "solve" button. When the solution is obtained, the program alerts the user asking if a sensitivity report is required. The resulting solution and the sensitivity report are shown in Figure 3.16.

The result obtained from the Solver is identical to the one obtained earlier using a tableau.

3.8.2 Sensitivity Report

We now embark on a more detailed examination of the sensitivity report produced by Solver. In a later section, we show that from the information contained in such sensitivity reports we can determine the locational marginal price (LMP) for energy delivery in electrical networks.

The following is a description of the sensitivity report shown in Figure 3.16.

Shadow Prices. These are rates of change of objective function for an increase in the right-hand side of the constraint. Thus, it is the ratio of the increase in the objective function to the increase in the right-hand side of the constraint. Therefore, a positive shadow price indicates that the objective will increase as the right-hand side of the constraint is increased. Correspondingly, a negative shadow price indicates that the objective will decrease. For a nonbinding constraint, the shadow price is zero since the right-hand side of the constraint is nonbinding—that is, it is not constraining the optimal solution. In this example, the second constraint is nonbinding. Consequently, it has a zero shadow price. The first and the third constraints have shadow prices of 0.5 and 1. Relaxing the right-hand side of these constraints by one will reduce the value of objective by 0.5 and 1, respectively. It is clear from the above that the shadow price indicates the marginal value of a constraint to the reduction (or increase) of the objective function.

A reduction in the objective function in minimization problems and the increase in maximization problems are intuitively considered to be improvements. Consequently, relaxing constraints is equivalent to increasing the size of the feasible region. Hence, a "$\leq$" constraint is weakened by *increasing* the right-hand side, and a "$\geq$" constraint is weakened by *decreasing* the right-hand side. Consequently, it follows that the signs of shadow prices for "$\leq$" and "$\geq$" are of opposite sign.

Equality constraints can be programmed into the Excel Solver routine as they are. An equality constraint can be deemed as two separate constraints. For example, $x = 6.5$ can be written as two separate constraints: $x \leq 6.5$ and $x \geq 6.5$. In such cases, at most one of these will have a nonzero shadow price. The sign of the shadow price indicates the nature of the binding constraint. In a minimization problem, a negative shadow price for an equality constraint indicates that the objective will decrease (improve) if the right-hand side of the equality is increased. The inference then is that it is the "$\leq$" constraint and not the "$\geq$" constraint that is binding. The corresponding argument for maximization problems is obvious.

Reduced Cost. This is the marginal value of a decision variable. It is equal to the rate of change of the objective function as the bound (upper or lower) on the variable is increased. The reduced cost of a basic variable is zero since its bounds are nonbinding and do not constrain the optimal solution. If a nonbasic variable has a positive reduced cost, the objective function will increase when the binding bound is relaxed. For example, if the binding bound is a lower bound ("$\geq$" inequality), relaxation of the bound implies reducing the right-hand side. If the binding bound is an upper bound, relaxation of the bound means an increase to the right-hand side. Clearly, these changes increase the size of the feasible region.

In the sensitivity sheet of Figure 3.16, we see that x_2 and x_3 are basic variables with solution values of 3 and 1 (cells C3,C4). They are nonbinding, and, therefore, they have zero reduced costs. The variable x_1 is a nonbasic variable; consequently, it is at its lower bound of zero. It has no upper bound in the problem formulation. It has a positive reduced cost of 4. This means that if the lower limit is increased from zero to one, the objective function will increase by four; it will change from the present -7 to -3. The reader is encouraged to make this change in the spreadsheet and confirm that this indeed is the case.

The value of the objective function is a summation of each choice variable multiplied by its objective cost coefficient. Clearly, by adjusting the objective cost coefficients of a nonbasic variable, it can be made a basic variable and, therefore, becomes active. The reduced costs of nonbasic variables in the sensitivity report indicates the amount by which the cost coefficient of the variable has to be ***reduced*** (for minimization problems) in order to make it active. A variable with a positive reduced cost will become active if its cost coefficient is lowered, while the cost coefficient of a variable with a negative reduced cost must be increased.

In the above example, x_1 is a nonbasic variable with a reduced cost of 4. The coefficient of x_1 in the objective function is -1. Therefore, in order to make x_1 a basic variable, its cost coefficient has to be reduced by 4; its coefficient in the objective function should be changed from its present value of -1 to -5 to make it a basic variable.

Importantly, the reduced costs and the shadow prices are all derived during the optimization procedure as a byproduct. Observe in Figure 3.15 that the values shown in row 36 are identical to reduced costs and shadow prices in the sensitivity sheet of Figure 3.16.

As a necessary exercise, the reader is asked to make the above changes to the objective coefficients and bound values in order to examine the corresponding solutions and sensitivity reports. Such an exercise will not only help in the retention of the above concepts but will make the reader more confident about formulating and tackling complex problems arising in physical systems.

3.9 TRANSPORTATION PROBLEMS

In the early 1940s, Hitchcock (1941) and Koopmans (1947) proposed a classic problem amenable to efficient solution by using linear programming techniques. The formulation addressed movement of materials from several sources to locations of demand. Because efficient transportation of material or troops during World War II were of great concern, transportation algorithms found a great following.

The characteristic of the problem can be explained as follows. Suppose there are n resource locations—for example, warehouses—and k destinations for delivery (factories or shopping centers). Let the amount of supply of a particular commodity at the origin be a_i, $i = 1, 2, \ldots, n$, and let the commodity required at destinations be b_j, $j = 1, 2, \ldots, k$. Let the cost of transporting the commodity from origin i to destination j be t_{ij}. The objective is to minimize transportation cost by determining the optimum amount of commodity c_{ij} transported from origin i to destination j. The mathematical formulation is as follows:

Choose

$$c_{ij}, \qquad i = 1, n; \quad j = 1, k \tag{3.63}$$

in order to minimize

$$\pi = \sum_{i=1}^{n} \sum_{j=1}^{k} c_{ij} t_{ij}, \tag{3.64}$$

subject to

$$\sum_{j=1}^{k} c_{ij} = a_i, \qquad i = 1, 2, \ldots, n, \tag{3.65}$$

$$\sum_{i=1}^{n} c_{ij} = b_j, \qquad j = 1, 2, \ldots, k, \tag{3.66}$$

$$c_{ij} \geq 0, \qquad i = 1, 2, \ldots, n; \quad j = 1, 2, \ldots, k. \tag{3.67}$$

This is an LP formulation in $n \times k$ variables and $n + k$ constraints (equality, for now). Equation (3.64) specifies the minimization of transportation cost, while (3.65) defines that the total quantity supplied from a warehouse should be equal to the quantity available,[17] and (3.66) specifies that the quantity delivered to each of the destinations is equal to the demand at that location. In this formulation with equality constraints, in addition to equations (3.64) to (3.67), we require that the total supply is equal to the total demand, that is,

$$\sum_{i=1}^{n} a_i = \sum_{j=1}^{k} b_j. \tag{3.68}$$

The formulation of this problem has a special structure from which a transportation matrix can be written. Such a transportation array can be solved very easily. We refer the reader to Rao (1996) for details regarding the special structure.

[17]Instead of equality, this can be an inequality as well. The quantity shipped from a warehouse can be less than the quantity available. A solved problem later for supply of gas for energy generation uses inequalities.

3.10 DUALITY IN LINEAR PROGRAMMING

In the previous sections, we have formulated the LP program in a general way as in (3.46) to (3.49) to include maximization and minimization of objective functions. Associated with every LP problem, called the *primal*, there is a corresponding LP problem called its *dual*. These two problems possess interesting relationships and properties. Corresponding to every primal minimization problem, the corresponding dual is a maximization program, and vice versa. Clearly, the dual of a dual is the primal problem itself. Consequently, it is immaterial which problem is designated the primal problem and which is designated the dual problem.

3.10.1 Primal Dual Relations

The following illustrates the relationships between the primal and the dual problems for a set of inequalities.

Primal Problem

Minimize

$$\pi = c_1 x_1 + c_2 x_2 + \cdots + c_n x_n, \tag{3.69}$$

subject to m inequality constraints

$$\begin{aligned} a_{11}x_1 + a_{12}x_2 + \cdots + a_{1n}x_n &\le k_1 \\ a_{21}x_1 + a_{22}x_2 + \cdots + a_{2n}x_n &\le k_2 \\ \vdots \qquad\qquad &\le \vdots \\ a_{m1}x_1 + a_{m2}x_2 + \cdots + a_{mn}x_n &\le k_m, \end{aligned} \tag{3.70}$$

where a_{ij}, $i = 1, m$; $j = 1, n$ are constant coefficients, and subject to the nonnegativity restrictions

$$[x_1, x_2, \ldots, x_n] \ge 0.$$

Dual Problem

Maximize

$$\pi' = k_1 y_1 + k_2 y_2 + \cdots + k_m y_m, \tag{3.71}$$

subject to inequality constraints

$$\begin{aligned} a_{11}y_1 + a_{21}y_2 + \cdots + a_{m1}y_m &\ge c_1 \\ a_{12}y_1 + a_{22}y_2 + \cdots + a_{m2}y_m &\ge c_2 \\ \vdots \qquad\qquad &\ge \vdots \\ a_{1n}x_1 + a_{2n}x_2 + \cdots + a_{mn}x_n &\ge c_n, \end{aligned} \tag{3.72}$$

and to the nonnegativity restrictions

$$\left[y_1, y_2, \ldots, y_m\right] \geq 0.$$

Briefly, the rules of transformation of the primal problem to the dual problem are as follows:

1. Corresponding to m inequalities in the primal problem, there are m dual variables.
2. Corresponding to n choice variables in the primal, there are n constraints in the dual problem.
3. The transpose of the column vector of constants in the constraints of the primal problem becomes the row vector of coefficients of the dual objective functions.
4. Change "minimize" to "maximize."
5. Reverse the inequalities signs of the primal constraints in the dual.
6. Take the transpose of the coefficients of the primal constraint matrix into the dual.
7. The transpose of the row vector of coefficients in the objective function of the primal problem becomes the column vector of constants in the dual constraints.

Examples. Two examples reinforce the above concepts.
Example 1: Primal Problem
Maximize

$$\pi = 3x_1 + 4x_2 + 6x_3,$$

subject to

$$\begin{bmatrix} 19 & 14 & 3 \\ 14 & 16 & 7 \end{bmatrix} \begin{bmatrix} x_1 \\ x_2 \\ x_3 \end{bmatrix} \leq \begin{bmatrix} 12 \\ 15 \end{bmatrix},$$

and $x_1, x_2, x_3 \geq 0$.

Dual Problem
Minimize

$$\pi' = 12y_1 + 15y_2,$$

subject to

$$\begin{bmatrix} 19 & 14 \\ 14 & 16 \\ 3 & 7 \end{bmatrix} \begin{bmatrix} y_1 \\ y_2 \end{bmatrix} \geq \begin{bmatrix} 3 \\ 4 \\ 6 \end{bmatrix},$$

and $y_1, y_2 \geq 0$.

Example 2: Primal Problem
Maximize

$$\pi = 4x_1 + 5x_2 + 8x_3,$$

subject to

$$\begin{bmatrix} 1 & 1 & 5 \\ 2 & 4 & 1 \end{bmatrix} \begin{bmatrix} x_1 \\ x_2 \\ x_3 \end{bmatrix} \leq \begin{bmatrix} 12 \\ 40 \end{bmatrix},$$

and $x_1, x_2, x_3 \geq 0$.

Dual Problem
Minimize

$$\pi' = 12y_1 + 40y_2,$$

subject to

$$\begin{bmatrix} 1 & 2 \\ 1 & 4 \\ 5 & 1 \end{bmatrix} \begin{bmatrix} y_1 \\ y_2 \end{bmatrix} \geq \begin{bmatrix} 4 \\ 5 \\ 8 \end{bmatrix},$$

and $y_1, y_2 \geq 0$.

Primal–Dual.Xls

The spreadsheet "Primal–Dual.Xls" incorporates these problems in sheets 1 and 2. The results are shown in Figures 3.17 and 3.18. Furthermore, the objective function of both the primal and dual problems converges to the same value of 52. It will be clear from the duality theorem in Section 3.10.2 and from solved problems in Section 4.10 that the solution for the primal can be obtained directly

	A	B	C	D	E	F
1	Primal Problem					
2						
3	x1	4		1	1	5
4	x2	8		2	4	1
5	x3	0				
6	Choice	Variables		Constraints		
7				12	12	
8	Objective	52		40	40	

Figure 3.17. Solution of primal problem.

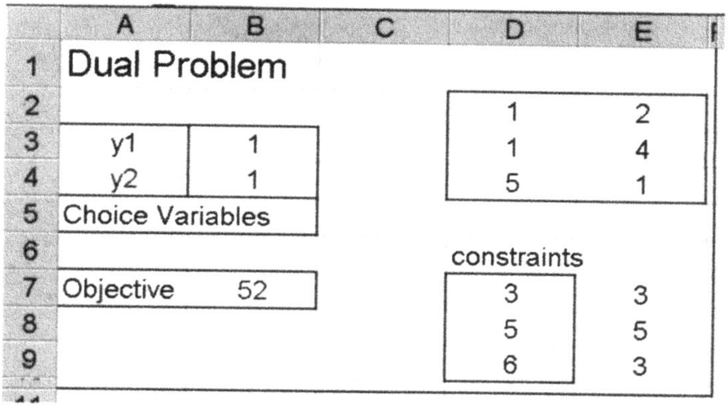

Figure 3.18. Solution of dual problem.

from a solution of the dual. It turns out that the shadow prices of the primal problem become the optimum choice variables for the dual, and vice versa.

For some problems, therefore, solution of the dual problem rather than the primal problem can be of advantage. The computational effort of solving linear programs increases with increasing number of constraint equations. Under such circumstances, it may be advantageous to solve the dual of the original problem with a larger number of dual variables, but with a smaller number of constraint equations. The reader will find details of such possibilities in Gill et al. (1981), Chiang (1984), and Rao (1996).

3.10.2 Duality Theorems

The following theorems indicate the relationship between the primal and dual solutions.

1. The dual of the dual is the primal.
2. If both the primal and the dual have feasible solutions, the optimum solutions of both are the same; that is, maximum $\pi =$ minimumπ'.
3. If either the primal or the dual has an unbounded solution, the other problem is infeasible.
4. If a certain choice variable in a linear program (primal or dual) is optimally *nonzero*, the corresponding *dummy* variable[18] in the counterpart program (dual or primal) is optimally zero. To elaborate this, if x_i and s_i are the choice and dummy variables in the primal and if y_j and r_j are the corresponding choice and dummy variables of the dual, then

$$\overline{y}_i > 0 \Rightarrow \overline{s}_i = 0,$$

[18]Dummy variable is a collective term used for slack and surplus variables of equations (3.47) and (3.48).

and

$$\bar{x}_j > 0 \Rightarrow \bar{r}_j = 0.$$

In the above, bars on variables indicate their values at the optimum solution. From this theorem, we also have

$$\bar{s}_i > 0 \Rightarrow \bar{y}_i = 0,$$

and

$$\bar{r}_j > 0 \Rightarrow \bar{x}_j = 0.$$

3.10.3 General Relations Between Primal and Dual Programs

In above illustrations of primal and dual programs, we have only shown inequalities. For the primal problem,

Choose

$$x_i, \qquad i = 1, \ldots, n,$$

in order to minimize

$$\mathbf{c}^T\mathbf{x},$$

subject to

$$\mathbf{Ax} = \text{or} \leq \mathbf{b},$$

the relation between primal and dual solutions is shown in Table 3.1 considering inequalities and equalities. These relations will be used in a solved problem in Section 4.10 to formulate and solve the dual problem.

Table 3.1. Primal–Dual Relations

Primal	Corresponding Dual
Objective, Minimize $\mathbf{c}^T\mathbf{x}$	Maximize $\mathbf{y}^T\mathbf{b}$
ith variable $x_i \geq 0$	Inequality for ith constraint, $\mathbf{y}^T\mathbf{A_i} \leq c_i$
ith variable x_i unrestricted in sign	Equality in ith constraint, $\mathbf{y}^T\mathbf{A_i} = c_i$
Equality in kth constraint, $\mathbf{A_k x} = b_k$	Unrestricted sign for kth variable
Inequality in kth constraint, $\mathbf{A_k x} \geq b_k$	kth variable $y^k \geq 0$
Coefficient matrix $\mathbf{A} = \begin{bmatrix} \mathbf{A_1} \\ \vdots \\ \mathbf{A_m} \end{bmatrix}$	Coefficient matrix $\mathbf{A^T} = [\mathbf{A_1} \cdots \mathbf{A_n}]$
Right-hand vector $\mathbf{b}$	Right-hand vector $\mathbf{c}$
Cost coefficients $\mathbf{c}$	Cost coefficients $\mathbf{b}$

3.11 EXERCISE PROBLEMS

1. Consider an overspecified system of equations of Section 3.3.3 modified as follows:

$$\mathbf{Ax} = \mathbf{b},$$

where

$$\mathbf{A} = \begin{bmatrix} 3 & 2 & -1 \\ 2 & 1 & 1 \\ 1 & -1 & 1 \\ 1 & 1 & 1 \\ 2 & 3 & -2 \end{bmatrix}, \quad \text{and} \quad \mathbf{b} = \begin{bmatrix} 1 \\ 6 \\ 4 \\ 4.857 \\ 3.0 \end{bmatrix}.$$

(a) Obtain the solution for $\mathbf{x}$ by taking the pseudo-inverse matrix $A^+ = (A^T A)^{-1} A^T$.

(b) Check if the properties of the pseudo-inverse as required by (3.24) to (3.27) are satisfied.

(c) Obtain the solution using an Excel Solver procedure such that the estimated values for $\mathbf{x}$ result in the least value for the sum of squares of error.

2. Consider the underspecified system of equations in Section 3.3.3:

$$\begin{bmatrix} 3 & 2 & -1 & -7 \\ 2 & 1 & 1 & 3 \\ 1 & -1 & 1 & 5 \end{bmatrix} \begin{bmatrix} x_1 \\ x_2 \\ x_3 \\ x_4 \\ x_5 \end{bmatrix} = \begin{bmatrix} 1 \\ 6 \\ 4 \end{bmatrix}.$$

(a) Obtain the solution for $\mathbf{x}$ by taking the pseudo-inverse matrix $A^+ = (A^T A)^{-1} A^T$.

(b) Check if the properties of the pseudo-inverse as required by (3.24) to (3.27) are satisfied.

(c) Obtain the solution using an Excel Solver procedure such that the estimated values for $\mathbf{x}$ result in the least value for the sum of squares of error.

(d) What conclusions can you draw by comparing the results of this problem with those of problem 1 above?

CHAPTER 4

SOLVED LINEAR PROGRAM PROBLEMS

Agenda. In order to reinforce the theoretical development of previous chapters, solutions to problems using the Excel Solver are shown below. By and large, problems selected focus on the electrical power industry. This is not to say that problems from other disciplines are not of interest. In fact, it is hoped that the reader will eventually be able to conceive problems in many fields as optimization problems and formulate them in a manner amenable to solution with readily available software. Toward that end, we start our excursion into practical problems, first by addressing an illustration from a discipline other than electrical engineering.[1] Rao (1996) has a good collection of such problems, particularly in its exercise problems and review questions.

In terms of the structure of this chapter, as said earlier, the marginal notes refer to the appropriate Excel spreadsheet with embedded Solver solution included in the spreadsheet files offered as part of Wiley's ftp site. The reader is expected to use a personal computer to open those spreadsheets and examine the program input to the Solver tool. The following text should be studied in conjunction with the corresponding spreadsheet in order to derive maximum benefit.

4.1 PROBLEMS INVOLVING OPTIMAL MIXING

In many endeavors, particularly manufacturing processes, optimal mixing plays an important role. In chemical industries, the same product made at different plants may have different compositions. Then, it is important to mix such products in

[1] In subsequent examples we address problems relating to auctions, transportation, and reservoir management.

Optimization Principles, by Narayan S. Rau
ISBN 0-471-45130-4

some optimal fashion such that the blended product either has a uniform composition or meets regulations regarding minimum or maximum amounts of pollutants or toxins. In the petrochemical industry, the crude oil feed stock received from different sources may have different percentages of disparate hydrocarbons. It may be necessary to blend the feedstock to be refined in order that the yield of certain hydrocarbon compounds is maximized.

Another classic example in electric energy production is that of mixing fuels to be below certain targets of pollution. Problem 1 in Section 4.13 asks the reader to determine the optimum blending of coal types to meet certain restrictions.

The following problem addresses a typical merchandising problem. It is taken from the exercises in Rao (1996). Although it deals with mixing food items, its generality to commodities with varying composition should not be overlooked.

Problem: A snack food manufacturer markets two kinds of mixed nuts, labeled A and B. Mixed nuts A contain 20% almonds, 10% cashews, 15% walnuts, and 55% peanuts. Mixed nuts B contain 10% almonds, 20% cashews, 25% walnuts, and 45% peanuts. A customer wants to use mixed nuts A and B to prepare a mix that contains at least 4 lb of almonds, 5 lb of cashews, and 6 lb of walnuts, for a party. If mixed nuts A and B cost $2.50 and $3.00 per pound, respectively, determine the amount of mixed nuts A and B to be used to prepare the new mix at minimum cost.

Mixed Nuts.Xls

Solution: The Excel spreadsheet using the Solver routine is shown in Figure 4.1. Cells E3 and F3 are choice variables x_1, x_2. The objective function is $2.5x_1 + 3.0x_2$. The mathematical formulation of the problem is as follows: Choose x_1, x_2 to minimize $2.5x_1 + 3.0x_2$, subject to almond constraint

$$0.2x_1 + 0.1x_2 \geq 4,$$

cashew constraint

$$0.1x_1 + 0.2x_2 \geq 5,$$

and the walnut constraint

$$0.15x_1 + 0.25x_2 \geq 6,$$

and the nonnegativity constraints $x_1, x_2 \geq 0$.

Examine Figure 4.1 in which the Solver subject to the above constraints chooses choice variables. The optimum solution is 10 lb of mixed nuts A and 20 lb of mixed nuts B, for a total cost of $85.00.

	A	B	C	D	E	F	G	H
1	Mixed Nuts Problem							
2					Choice Variables			
3		A	B		10	20		
4	Almonds	0.2	0.1	Constraints				
5	Cashews	0.1	0.2	Almonds	0.2*E3 +0.1*F3 >= 4			4
6	Walnuts	0.15	0.25	Cashews	0.1*E3 +0.2*F3 >= 5			5
7	Peanuts	0.55	0.45	Walnuts	0.15*E3 +0.25*F3 >= 6			6.5
8								
9		Objective function		85				
10		2.5*E3+3.0*F3						

Figure 4.1. Solution to mixed nuts problem.

4.2 SIMPLE DISPATCH PROBLEM

One of the important tasks of the energy management system (EMS) software in the control room of an electrical dispatch center is to obtain energy from the least expensive set of generators to meet the demand at each instant. The following example is a simplified version of such a problem.

Formulation and Discussion of Problem: Consider a simple example of Figure 4.2 where there are 11 Generators at five locations 1, 2, 3, 4, and 5 offering to supply energy at prices indicated in the spreadsheet "Simple dispatch.Xls." There are two Generators in all locations except in location #3, where there are three Generators.[2] The quantity offered by each generator is shown in the spreadsheet column B and the associated offer prices in column C. If the total quantity to be procured is 250 (demand), what are the quantities to be procured from each generator in order to minimize the cost of procurement?

Simple-Dispatch.Xls

The solution to this trivial problem is readily obvious. The usual "stacking" provides the answer: Stacking implies procurement from the lowest source first progressing into higher and higher prices until the total quantity is procured. Figure 4.2 is a printout of the spreadsheet portraying the problem formulation. Choice variables are the quantities to be procured (column E). Column G shows the cost of procurement given by the summation of quantity times the price. The cost of procurement becomes the objective function to be minimized (cell G15).

There are three constraints: first that the quantity procured is greater than or equal to zero, second that the quantities procured cannot exceed the quantities offered, and finally that the total quantity procured must be equal to the demand.

[2]The location of Generators is irrelevant in this example. The location becomes important when the deliverability (of energy) is of concern. Such a situation arises if there is intervening transmission between supply and demand locations. Later examples that include the transmission system will consider the location of generators in resolving bids and offers.

	A	B	C	D	E	F	G
2	Offers			Simple Dispatch			
3	Location	Qty.	Price		Choice Variables		Cost
4	1	10	5		10		50
5	1	20	15		20		300
6	2	15	6		15		90
7	2	40	7		40		280
8	3	30	6		30		180
9	3	40	15		40		600
10	3	20	25		0		0
11	5	10	10		10		100
12	5	20	15		20		300
13	4	60	30		0		0
14	4	75	18		65		1170
15					250	sum	3070
16			Demand	250		Index	
17							

Figure 4.2. Generation dispatch based on offers.

This problem is similar to an auction to be discussed in the next section. However, in this example, there are no demand bids as in some future auction examples. Consequently, this can be considered as a one-sided auction.

Mathematical Statement of the Problem: Minimize

$$\sum_{i=1}^{11} Q_i P_i,$$

subject to

$$\sum_{i=1}^{11} Q_i = demand = 250,$$

and

$$Q_i \leq Q_i \text{ (offer)},$$

$$Q_i \geq 0,$$

where Q_i is the quantity procured from i, Q_i (offer) is the quantity offered, and P_i is the offer price from i.

Solution: See spreadsheet "Simple-Dispatch.Xls" and Figure 4.2. The solution obtained corresponds to that which can be easily obtained by stacking procedure. The minimum cost of procurement is 3070.

4.3 SIMPLE AUCTION

Auctions are commonplace in economic theory. Prior to deregulation of the electricity industry, the system operator had access to the information about the actual cost of generation from generators. In the deregulated industry, Generators offer prices to supply energy in an auction, and consumers, in some markets, may submit bid prices for the purchase of energy.[3] Although economic theory suggests that in the presence of competition, the bid and offer prices approach marginal cost in the longer run, they need not be related to actual cost in every hour, particularly in the shorter term. The system operator is the bid taker who chooses generators to supply energy and reconcile demand in a manner not dissimilar to that of an auction. The following problems related to the electricity industry illustrate the reason for our interest in auctions.

Formulation and Discussion of the Problem: The spreadsheet "Simple-Auction.Xls" shows the resolution of a simple auction in which there are offers to sell and bids to buy products. In the simple dispatch problem of the previous section, there were only offers to sell. In Figure 4.3, cells C6:D11 and C14:D19 show the price and quantity of offers and bids of different entities.

The offers and bids are first sorted in increasing order of price for offers and decreasing order of prices for bids.[4]

Simple-Auction.Xls

The sorted offers and bids represent an increasing and a decreasing curve. The search is for the point of intersection of the two curves. This is the goal in many other physical problems including that of finding the intersection of demand and supply curves in economics. If the demand curve (bids to buy) is denoted by $f_1(x)$ and the supply (offer to sell) curve is denoted by $f_2(x)$, then consider the expression

$$D = \int_{x=0}^{\overline{x}} f_1(x)\,dx - \int_{x=0}^{\overline{x}} f_2(x)\,dx,$$

which denotes the area between f_1 and f_2 from the origin to $\overline{x}$. Therefore, in the above expression, D is the difference between the two areas which is a maximum at the point of intersection of the two curves.[5] The discrete form of the above expression is used in simulating auctions in a linear program.

Therefore, the objective function of the problem is to maximize the area between the two curves of the sorted offers and bids. In order to do so, we set up choice variables in column H where we select quantities that are nonnegative and less than or equal to the offered (or bid) quantity. The Solver procedure is asked to select these choice variables subject to constraints. The product of the quantity selected and the price gives the area under the offer curve and the bid curve. These

[3] In some markets, consumers bid a price above which a bid quantity of consumption can be curtailed. Such loads are called price-sensitive or price-responsive loads.

[4] Sorting is not necessary to obtain a solution.

[5] The solution for the equilibrium point between demand and supply is also called welfare maximization in microeconomics (Samuelson, 1947); (Crew and Kleindorfer, 1986).

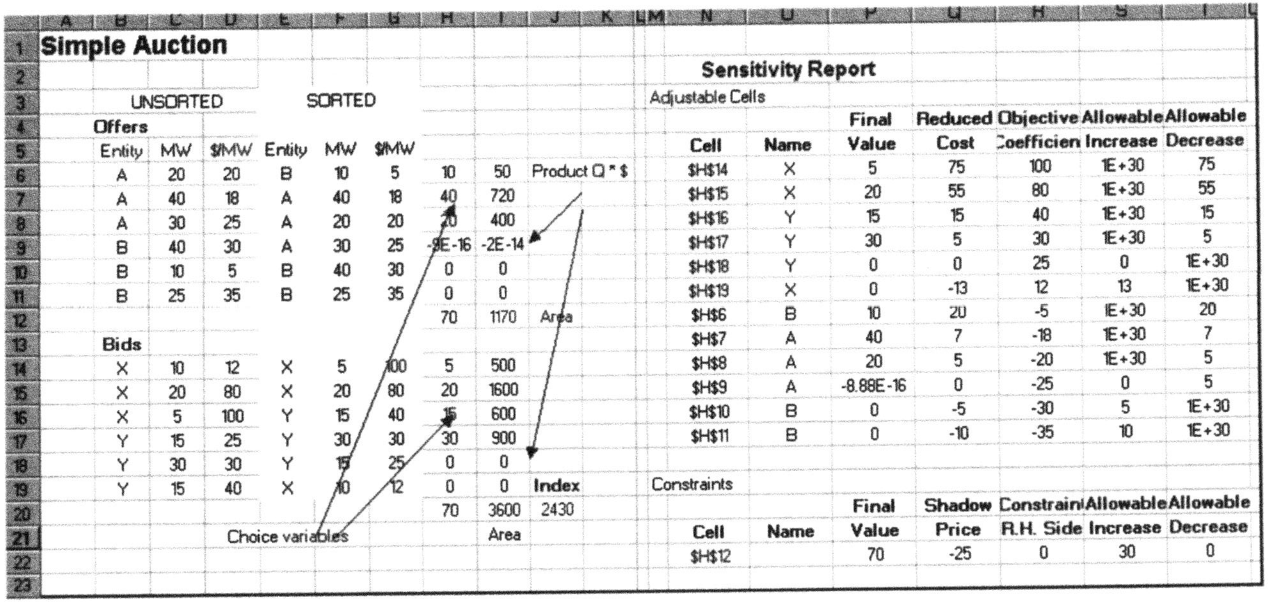

Simple Auction

UNSORTED			SORTED					
Offers								
Entity	MW	$/MW	Entity	MW	$/MW			
A	20	20	B	10	5	10	50	Product Q * $
A	40	18	A	40	18	40	720	
A	30	25	A	20	20	20	400	
B	40	30	A	30	25	-9E-16	-2E-14	
B	10	5	B	40	30	0	0	
B	25	35	B	25	35	0	0	
						70	1170	Area
Bids								
X	10	12	X	5	100	5	500	
X	20	80	X	20	80	20	1600	
X	5	100	Y	15	40	15	600	
Y	15	25	Y	30	30	30	900	
Y	30	30	Y	15	25	0	0	
Y	15	40	X	10	12	0	0	Index
						70	3600	2430
		Choice variables					Area	

Sensitivity Report

Adjustable Cells

Cell	Name	Final Value	Reduced Cost	Objective Coefficien	Allowable Increase	Allowable Decrease
$H$14	X	5	75	100	1E+30	75
$H$15	X	20	55	80	1E+30	55
$H$16	Y	15	15	40	1E+30	15
$H$17	Y	30	5	30	1E+30	5
$H$18	Y	0	0	25	0	1E+30
$H$19	X	0	-13	12	13	1E+30
$H$6	B	10	20	-5	1E+30	20
$H$7	A	40	7	-18	1E+30	7
$H$8	A	20	5	-20	1E+30	5
$H$9	A	-8.88E-16	0	-25	0	5
$H$10	B	0	-5	-30	5	1E+30
$H$11	B	0	-10	-35	10	1E+30

Constraints

Cell	Name	Final Value	Shadow Price	Constraint R.H. Side	Allowable Increase	Allowable Decrease
$H$12		70	-25	0	30	0

Figure 4.3. Example of an auction.

are shown in cells I12 and I20, and their difference in cell J20 is the objective to be maximized. There is one more constraint: The total quantity of offers selected must be equal to the quantity of bids awarded. Thus we want cell H12 = cell H20. The solution obtained by the Solver is shown in the spreadsheet.

Mathematical Statement of the Problem: Given s offers to supply and b bids to buy with quantity price relationships, choose $Q_i,\ i = 1, s$, and $Q_j,\ j = 1, b$ (where Q represents quantity and P its price), in order to Maximize

$$\pi = \sum_{i=1}^{i=s} Q_i P_i - \sum_{j=1}^{j=b} Q_j P_j,$$

subject to

$$Q_j, Q_i \geq 0, \qquad i = 1, b; \qquad j = 1, s,$$

$$\sum_{i=1}^{i=s} Q_i = \sum_{j=1}^{j=b} Q_j,$$

$$Q_i \leq Q_i \quad \text{(offer)}, \qquad i = 1, s,$$

and

$$Q_j \leq Q_j \quad \text{(sell)}, \qquad j = 1, b.$$

Solution: See Figure 4.3 and the spreadsheet Simple-Auction.Xls.

The sensitivity report is copied onto the spreadsheet, indicating a shadow price of 25 for cell H12. This is the point of intersection of the two curves indicating a clearing price of 25.

4.4 DISPATCH PROBLEM WITH DEMAND BIDS

Problem: This problem is in the spreadsheet "Dispatch-demand bid.Xls." Figure 4.4 is a printout of the spreadsheet. Here, in addition to the offer price for energy as in the problem of Section 4.3, there are seven bids to buy energy. As a consequence, dispatch with demand bids becomes identical (mathematically) to the previously discussed two sided auction.

Cells B4:C14 show the price and quantity bid for supply and cells A19:B25 show the demand. To resolve offers and bids, we have to decide how much to procure from whom and which demand bids to satisfy. **Dispatch-demand bid.Xls**

In Figure 4.5 the cumulative quantities and price are plotted in decreasing order of price for demand, and in increasing order for offers. The solution to this problem is tantamount to the finding the point of intersection of the two curves. At the point

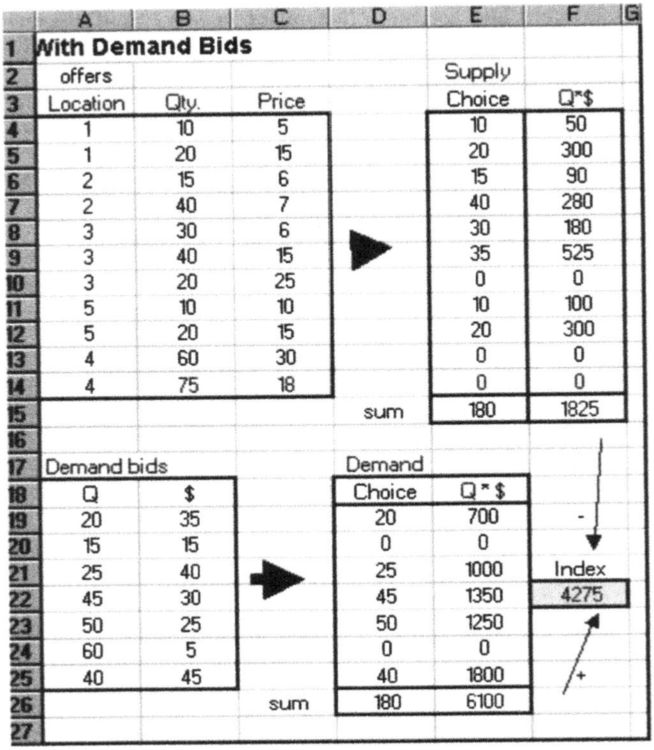

	A	B	C	D	E	F
1	With Demand Bids					
2	offers				Supply	
3	Location	Qty.	Price		Choice	Q*$
4	1	10	5		10	50
5	1	20	15		20	300
6	2	15	6		15	90
7	2	40	7		40	280
8	3	30	6		30	180
9	3	40	15		35	525
10	3	20	25		0	0
11	5	10	10		10	100
12	5	20	15		20	300
13	4	60	30		0	0
14	4	75	18		0	0
15				sum	180	1825
16						
17	Demand bids			Demand		
18	Q	$		Choice	Q * $	
19	20	35		20	700	-
20	15	15		0	0	
21	25	40		25	1000	Index
22	45	30		45	1350	4275
23	50	25		50	1250	
24	60	5		0	0	
25	40	45		40	1800	+
26			sum	180	6100	
27						

Figure 4.4. Generation dispatch based on offers and demand bids.

of intersection, the area between the two curves (welfare) is a maximum (Crew and Kleindorfer, 1986). The point of intersection is also called the equilibrium point.

Choice variables are in columns D and E of Figure 4.4. Cells D19:D25 indicate the accepted demand bids, and cells E4:E14 represent accepted supply offers. The areas under the demand and supply curves are given by the sums of quantity and price product shown in cells F4:F14, and F20:F25. The difference between the these sums shown in cell F22 represents the area between the demand and supply curves which is the objective function to be maximized.

As for constraints, of course, the quantity accepted has to be nonnegative and less than or equal to the offered or bid quantity. Additionally, the total of accepted offers must be equal to the total of demand bids satisfied.

Mathematical Statement of the Problem: The statement is identical to that of an auction.

Maximize

$$\sum_{i=1}^{11} Q_i P_i - \sum_{j=1}^{7} Q_j P_j,$$

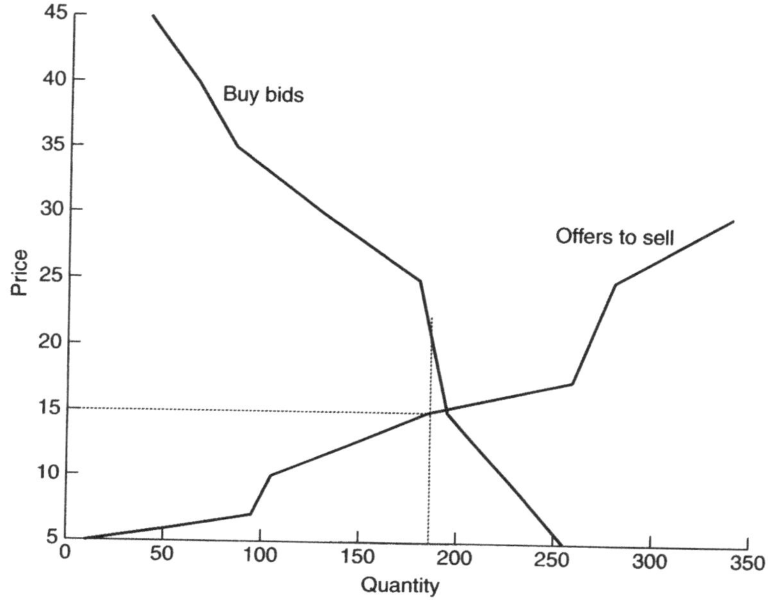

Figure 4.5. Sell offers and buy bids.

subject to

$$Q_i,\ Q_j \geq 0,$$

$$\sum_{i=1}^{11} Q_i = \sum_{j=1}^{7} Q_j,$$

$$Q_i \leq Q_i \quad \text{(offer)}, \qquad i = 1, \ldots, 11,$$

and

$$Q_j \leq Q_j \quad \text{(demand)}, \qquad j = 1, \ldots, 7$$

where Q_i are the quantities procured from offers i, Q_j are the quantities of demand supplied, P_i is the offer price from offers i, and P_j are the prices bid by demands.

Solution: In Figure 4.4, the solution shows that a quantity of 180 is to be procured giving 4275 as the maximum area between the two curves at the point of intersection.

4.4.1 Sensitivity Analysis

Figure 4.6 shows sensitivity figures associated with this optimization. At the outset, we see that a shadow price of 15 is associated with the total procured quantity of

Sensitivity Report

Adjustable Cells

Cell	Name	Final Value	Reduced Cost	Objective Coefficient	Allowable Increase	Allowable Decrease
$E$4	Choice	10	10	-5	1E+30	10
$E$5	Choice	20	0	-15	1E+30	0
$E$6	Choice	15	9	-6	1E+30	9
$E$7	Choice	40	8	-7	1E+30	8
$E$8	Choice	30	9	-6	1E+30	9
$E$9	Choice	35	0	-15	0	2.99999999
$E$10	Choice	0	-10	-25	10	1E+30
$E$11	Choice	10	5	-10	1E+30	5
$E$12	Choice	20	0	-15	1E+30	0
$E$13	Choice	0	-15	-30	15	1E+30
$E$14	Choice	0	-3	-18	3	1E+30
$D$19	Choice	20	20	35	1E+30	20
$D$20	Choice	0	0	15	0	1E+30
$D$21	Choice	25	25	40	1E+30	25
$D$22	Choice	45	15	30	1E+30	15
$D$23	Choice	50	10	25	1E+30	10
$D$24	Choice	0	-10	4.99999996	10	1E+30
$D$25	Choice	40	30	45	1E+30	30

Constraints

Cell	Name	Final Value	Shadow Price	Constraint R.H. Side	Allowable Increase	Allowable Decrease
$E$15	sum Choice	180	-15	0	5	35

Figure 4.6. Generation dispatch based on offers and demand bids—sensitivity analysis.

180. This is the auction clearing price indicating the price at the intersection of the demand and supply curves. Next, there are reduced costs associated with choice variables in cells E4:E14 and D19:D25 of Figure 4.4. In Figure 4.6 the reduced cost for the offers or bids not selected is ≤ 0. Recall that the reduced cost is the rate of change of the objective with the increase in the upper or lower bound of the variable. Since they are not accepted, unselected offers and bids are at their lower bound of zero, for example cell E14 in Figure 4.4. The reduced cost for this choice variable (see cell E14) is -3. This means that relaxing the bound on this variable—for example, from zero to plus 1 (since this is a maximization problem)—will result in a reduction of the objective by 3. Correspondingly, we note that the choice variable in cell E4 of Figure 4.4 is at its upper bound; if the quantity bid in cell B4 is relaxed from 10 to 11, the objective function becomes 4285. Therefore, the reduced cost associated with cell E4 is $+10$, as shown in Figure 4.6.

The reader is urged to try these and other changes in the spreadsheet and study corresponding sensitivity sheets. Such a self-learning process will aid in establishing an indelible understanding of the process of optimization.

4.5 DISPATCH PROBLEM WITH DEMAND BIDS AND TRANSMISSION NETWORK

Disp-demand-trans.Xls

In the above examples, we did not consider the effect of intervening transmission between supply and demand points. Consequently, locations of the offers and bids were irrelevant because it was implicit that all quantities were deliverable. In this example "Disp-demand bid-transmission.Xls," the simulation of a hypothetical transmission network with five nodes and six branches is shown in Figure 4.7. The shift factor matrix for this network and the locations of offers and demands in the network are also shown in the spreadsheet. The product of shift factor matrix and the vector of injections in five nodes give the power flow in branches. The line flows in cells J29:J34 are computed by this procedure. Appendix B.3.1 describes this procedure and the method of computing the shift factor matrix.

Formulation of Problem: Choice variables and the objective function remain the same as in the previous example. In addition, it is now desired that the line flows

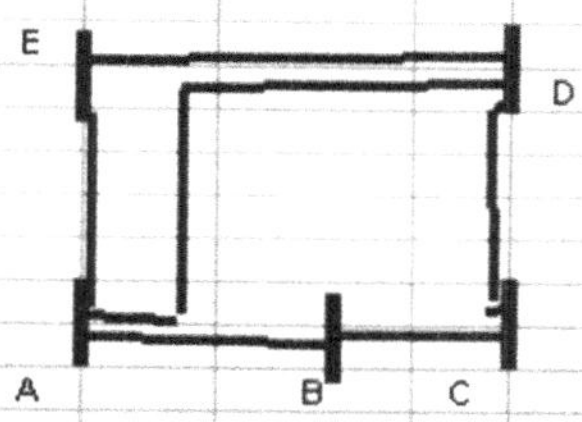

	A	B	C	D	E	F	G	H	I	J	K	L
1		Offers			Offer							
2	Location	Qty.	Price		Choice	Q*$						
3	A	10	5		10	50						
4	A	20	15		20	300						
5	B	15	6		15	90						
6	B	40	7		40	280						
7	C	30	6		30	180						
8	C	40	15		40	600						
9	C	20	25		0	0						
10	D	10	10		10	100						
11	D	20	15		20	300						
12	D	60	30		0	0			P Injection			
13	D	75	18		0	0			A	-40		
14				sum	185	1900			B	5		
15		Demand bids							C	70		
16	Location	Q	$			Index			D	-10		
17	E	20	35			4275			E	-25		
18	E	15	15		Demand							
19	A	25	40		Choice	Q * $						
20	A	45	30		20	700						
21	B	50	25		5	75						
22	C	60	5		25	1000						
23	D	40	45		45	1350						
24					50	1250						
25					0	0						
26					40	1800						
27				sum	185	6175				Line flows		
28			SF Matrix						Branch	SF * P inj		
29	0	-0.6694	-0.54279	-0.1946	-0.034		-100	>=	a-b	-38.53535	<=	100
30	0	-0.18	-0.24921	-0.4395	-0.078		-100	>=	a-d	-12.00477	<=	100
31	0	-0.1514	-0.20962	-0.3697	-0.887		-100	>=	a-e	10.443395	<=	100
32	0	0.32981	-0.54341	-0.1948	-0.034		-100	>=	b-c	-33.5798	<=	30
33	0	0.32981	0.456585	-0.1948	-0.034		-100	>=	c-d	36.420202	<=	100
34	0	0.15166	0.209954	0.37026	-0.111		-100	>=	d-e	14.539895	<=	100

Figure 4.7. Generation dispatch based on offers and demand bids with transmission network.

(positive or negative depending on the direction of flows) remain within those limits entered in cells G29:G34 and L29:L34. Consequently, there are two additional sets of constraints in the Solver to accommodate this. These constraints are that the flows in cells J29:J34 are $\geq$ limits entered in cells G29:G34 and that they are $\leq$ limits in cells L29:L34. In Figure 4.7, $\mathbf{p}_{inj}$ is given by an appropriate summation of supply and demand at each node shown in cells J13:J17. The appropriate formulae for summation can be seen in the spreadsheet.

Mathematical Statement: Maximize

$$\sum_{i=1}^{11} Q_i P_i - \sum_{j=1}^{7} Q_j P_j$$

subject to

$$Q_i, Q_j \geq 0,$$

$$\sum_{i=1}^{11} Q_i = \sum_{j=1}^{7} Q_j,$$

$$Q_i \leq Q_i \quad \text{(offer)}, \qquad i = 1, \ldots, 11$$

$$Q_j \leq Q_j \quad \text{(demand)}, \qquad j = 1, \ldots, 7$$

$$-\text{Lim}_{ij} \geq F_{ij} \leq \text{Lim}_{ij},$$

where Q_i are the quantities procured from offers i, Q_j are the quantities supplied to demands, P_i is the offer price from offers i, and P_j are the prices bid by demands. In the above, F_{ij} indicate power flows in appropriate branches connecting nodes i and j, depending on the network topology, and Lim_{ij} are the maximum flows allowed in these branches in either direction. As mentioned earlier, the product of the shift factor matrix and the vector of power injection at nodes obtain the vector of branches flows. This can be written in compact form as

$$\mathbf{F} = [\mathbf{SF}]\mathbf{p}_{inj}.$$

Solution: See Figure 4.7 and the spreadsheet "Disp-demand-Trans.Xls" that show the solution. In this example, since the line flow limits are set at a very large value, none of the flows are binding constraints. Consequently, the objective function (4275) is the same as the earlier example of Figure 4.4. And, as the associated sensitivity sheet will confirm, the shadow price of the last block of energy (15) is identical to the previous example.[6] Later examples of optimal power flow examines

[6] Although the sensitivity sheet for this example is not shown in the text here, the reader can examine the spreadsheet that is offered as part of the files located on Wiley's ftp site at ftp://ftp.wiley.com/public/sci_tech_med/electric_power/.

transmission constraints in greater detail. For now, the reader is encouraged to reduce the flow limits of some or all lines from their present value and examine what happens to the objective function and the shadow prices. In order to change the flow limits, the entries in cells G29:G34, and L29:L34 must be changed. Such changes will reveal that the shadow prices in all nodes need not be the same.

4.6 TRADEOFF BETWEEN COST OF EMISSIONS AND COST OF DISPATCH

Electric power generation using fossil fuels causes emissions. A Power Generator has access to different fuels that cause varying degrees of pollution. Presumably, the cost of such fuels differs depending on their heat content, potential for pollution, and the source of supply. Because of State and Federal regulations, a Generator is not allowed to exceed a certain quanta of pollution in a given period of time—for example, a year. Consequently, the optimum procurement and burn of different fuels to minimize cost and yet remain within the ceiling of pollution levels becomes an optimization problem.

Minimizing cost of production and minimizing pollution can be disparate objectives in the electricity industry. The following problem is modeled after an exercise in Rao (1996) with some modifications illustrates this point.

Problem: A Generator (a generating company) operates two power-generating facilities A and B from which it is obligated to inject 200 MW into the grid. The power plants burn coal, and there are three types of coals C_1, C_2, and C_3 available to the Generator. Heat contents of three coals are different. Consequently, the table below shows the amounts of various grades of coal required to generate 1MWh at each power plant. The pollution caused by various grades of coal and the costs of coal are given in the following table:

	Tons of Coal Required to Generate 1 MWh		Pollution per MWh Generated at Power Plant		Cost of Coal at Power Plant	
Coal Type	A	B	A	B	A	B
C_1	2.5	1.5	1.0	1.5	20	18
C_2	1.0	2.0	1.5	2.0	25	28
C_3	3.0	2.5	2.0	2.5	18	12

The power plant characteristics are such that the minimum power to be generated in plants A and B is 30 MWh and 80 MWh, respectively. Formulate the problem to compute the amount of different grades of coal to be used at each power plant to minimize (a) the total cost of production and (b) the total pollution level.

Mathematical Statement of Problem: Select quantities at power plants A and B

$$Q_i^A,\ Q_i^B, \qquad i = 1, 3,$$

to minimize

$$\pi = \sum_{i}^{i=1,3} (Q_i^A D_i^A + Q_i^B D_i^B),$$

where D_i^A and D_i^B are costs of coal of type i at power plants A and B, subject to power plant A generation:

$$\frac{Q_1^A}{2.5} + \frac{Q_2^A}{1} + \frac{Q_3^A}{3} \geq 30,$$

power plant B generation

$$\frac{Q_1^B}{1.5} + \frac{Q_2^B}{2} + \frac{Q_3^B}{2.5} \geq 80,$$

$$\frac{Q_1^A}{2.5} + \frac{Q_2^A}{1} + \frac{Q_3^A}{3} + \frac{Q_1^B}{1.5} + \frac{Q_2^B}{2} + \frac{Q_3^B}{2.5} = 200,$$

and the nonnegativity restrictions

$$Q_i^A, Q_i^B \geq 0, \qquad i = 1, 3.$$

Solution—Minimum Production Cost: Open the program "Powerplant-Emissions.Xls" from the Wiley's ftp site. Figure 4.8 shows the corresponding spreadsheet.

Power-plant-Emissions.Xls

The choice is the quantities Q_i of coal of type $i = 1, 3$ to be selected for burns in A and B to minimize the cost of production. The cost of production is given by $\sum_{i=1,3}(Q_i^A D_i^A + Q_i^B D_i^B)$, where D_i^A and D_i^B are the costs of coal of type i at plants A and B. This formula is embedded into cell D13, which is the objective

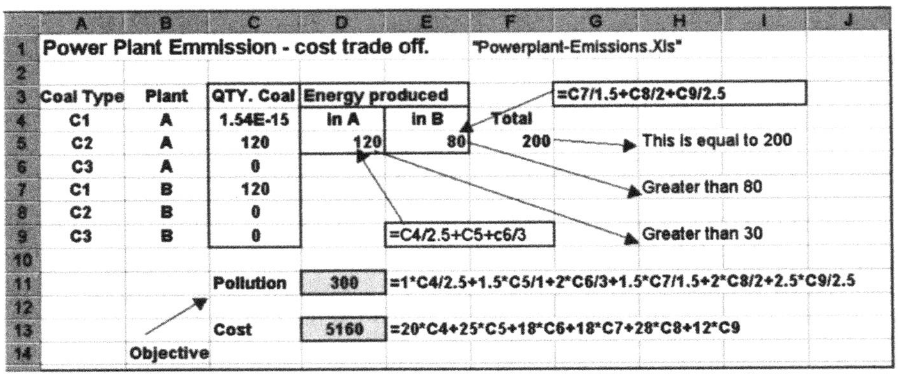

Figure 4.8. Solution indicating tradeoff between emissions and cost.

function. The energy produced by the quanta of coal procured is given by the quantity of coal burned divided by the quantity required to produce 1 MWh for that type of coal. This formula for energy produced in plants A and B is embedded into cells D5 and E5. Finally, the total energy produced (cell F5) has to be equal to 200, and the energy output from plants A and B should be greater than or equal to 30 MWh and 80 MWh, respectively. Worth noting is that in cell D11 the formula for total emissions is also embedded.

The minimum cost dispatch solution obtained by Solver is shown in Figure 4.8 as 5160. The resulting pollution is 300.

Solution—Minimize Pollution: The problem formulation is identical to the above except that the objective function to be minimized is that of pollution given by

$$\pi = 1.0\frac{Q_1^A}{2.5} + 1.5\frac{Q_2^A}{1.0} + 2.0\frac{Q_3^A}{3.0} + 1.5\frac{Q_1^B}{1.5} + 2.0\frac{Q_2^B}{2.0} + 2.5\frac{Q_3^B}{2.5}.$$

The resulting pollution is 240, corresponding to an increased production cost of 8160.[7] Therefore, in order to reduce pollution by 60 (= 300 − 240), the additional cost is 8160 − 5160 = 3000. Consequently, the shadow price or the proxy cost of pollution is $\frac{3000}{60} = \$500$ per unit of pollutant.

The reader is encouraged to examine and interpret the sensitivity report.

4.7 OPTIMAL POWER FLOW (OPF) BASED ON DC LOAD FLOW

The following set of problems describes optimal dispatch in a power network subject to several constraints. The problems address issues such as the most economic energy exchanges between and among subregions constrained by transmission system capabilities.

In the electric power industry, power flow studies are generally conducted by specifying injections of generation and demand at all nodes, except in one called the slack bus, in order to accommodate losses.[8] In the OPF problem, however, the cost associated with generation at each node is specified and the optimization process is asked to pick necessary injections to meet the demand at minimum cost. In that sense, the optimization program dispatches the generation just as a system operator would. In another sense, the problem formulation is similar to that of an auction or a dispatch based on bids as shown in an earlier example.

Problem Formulation: The example in this section is somewhat similar to that of Section 4.5. We consider first a loss-less transmission system between nodes with a topology indicated in Figure 4.9. However, to start with, we assume that the transmission system is of very large capacity, making it unnecessary to calculate

[7] In order to obtain this result, the reader should rerun the Solver optimization routine by choosing cell D11 to be minimized.

[8] In a dc load flow simulation, the system is modeled to have no losses.

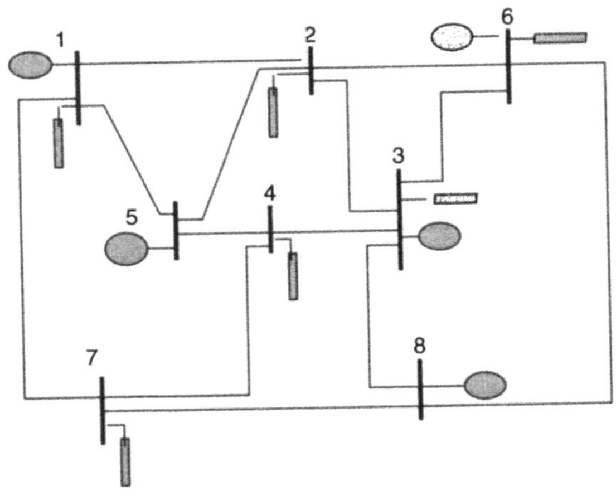

Figure 4.9. Transmission network with eight nodes.

branch flows. The intent of this example is to discuss the sensitivity sheet in greater detail in order to explain the concept of locational marginal prices (LMP). Subsequent examples will simulate realistic flow limits in the transmission system.

The first program of this nature on the ftp site supplied by Wiley is "OPF DC flow Xls." Because of unlimited branch flow capability, no explicit calculations are made to compute them. However, we recognize the location of injections in the network in order that we can compute LMP.

OPF DC flow.Xls

Figure 4.10 shows the spreadsheet of optimization incorporating the Solver routine. The costs of generation at various locations are shown in cells B9:I9. An objective in cell K8 (there can be others) is to minimize the total cost of generation, also called the production cost. In this example, there is no cost (bids) associated with demands (loads). Furthermore, the upper limit for generation at each generator is specified in cells B4:I4.

In order to fulfill the objective of minimizing production cost, quantities of generation selected from each generator become choice variables. The spreadsheet "OPF-DC flow.Xls." indicates the net injection at each node of the network (generation minus load at that node) in cells B16:I16. The only constraints are that the chosen generation cannot exceed the maximum bid capacity and the nonnegativity restrictions.

Results: Results of computation in Figure 4.10 are quite different from the solution of load flow in B.3.1. In the latter, power injections at nodes were explicitly specified. In the present solution, injections are picked to minimize cost. The resulting injections of power at each node are different in these two simulations. Therefore, though not of concern, branch flows in this example would be quite different from those of Section B.3. The marginal generator is P_5 at a cost of \$40/MWh. There is no such thing as a marginal generator in the load flow of Section B.3, as there is no cost associated with outputs from generators.

	A	B	C	D	E	F	G	H	I	J	K	L
1	**Optimal Power Flow Formulation Based On DC Load Flow**											
2	nodes	1	2	3	4	5	6	7	8			
3												
4	P	4	0	3.5	0	2.8	2	0	3	*Specified upper Limit*		
5	P	0	0	3.5	0	1.3	2	0	3	Selected Generation		
6												
7		1	2	3	4	5	6	7	8			
8	Cost (offers) $/MWh											
9		50	0	22	0	40	35	0	32		**295**	
10											**Perf. Index**	
11			Specified loads at nodes									
12		1	1.5	1.7	0.5	0	3	2.1	0		load sum	
13											9.8	
14	Resulting net nodal Injection into network = generation minus load											
15		1	2	3	4	5	6	7	8		Choice Variables	
16		-1.00	-1.50	1.80	-0.50	1.30	-1.00	-2.10	3.00		P1	0
17											P3	3.5
18	Sensitivity Report.										P5	1.3
19											P6	2
20	Adjustable Cells										P8	3
21				Final	Reduced	Objective	Allowable	Allowable			sum	9.8
22		Cell	Name	Value	Cost	Coefficient	Increase	Decrease				
23		$L$15	P1	0	10	50	1E+30	10		Offer Bids	LMP	
24		$L$16	P3	3.5	-18	22	18	1E+30		50	40	
25		$L$17	P5	1.3	0	40	10	5		22	40	
26		$L$18	P6	2	-5	35	5	1E+30		40	40	
27		$L$19	P8	3	-8	32	8	1E+30		35	40	
28										32	40	
29	Constraints											
30				Final	Shadow	Constraint	Allowable	Allowable				
31		Cell	Name	Value	Price	R.H. Side	Increase	Decrease				
32		$L$20	sum	9.8	40	9.8	1.5	1.3				
33												

Figure 4.10. Dispatch solution to minimize cost with ample transmission, along with sensitivity report.

Sensitivity: Before realistic transmission flow capabilities can be considered in the next section, it is necessary to understand the sensitivity report of this spreadsheet shown inserted into Figure 4.10. On the spreadsheet, click "sensitivity sheet" to see this. Recall that the reduced costs indicate how much the objective changes when the bound on the variable is increased. For example, in the sensitivity report of Figure 4.10, the reduced cost associated with P_3 in cell $L16$ is -18. This means that if the upper bound for generator #3 output is increased from 3.5 to 4.5, a reduction of \$18.00 in the objective will result. Since the cost of dispatch from generator #3 is \$22 and since the marginal generator #5 has a cost of \$40, increasing the output of generator #3 with a corresponding reduction from generator #5 results in a saving of \$18, which is equal to the difference between the two generation costs. Similar observations based on reduced costs can be made for other generators as well.

We now turn to the question about the cost of supplying the next unit of demand at these nodes. The shadow price associated with cell L20, the total generation, is \$40.00. Therefore, when load is increased in any node, the objective increases by \$40. This is shown in the sensitivity report (row 32 of Figure 4.10). Hence, the cost of supplying the next unit of energy, the marginal cost of supplying energy at any node, is \$40. This is also called the locational marginal price of energy (LMP) in a network.[9] However, in this case, this price is the same at all locations (nodes).

The LMP at generator nodes can be obtained as follows. Consider an increased demand of 1 at, for example, node 3 of Figure 4.9. Since the generator at this node is producing its maximum energy of 3.5, the additional load would mean that energy of only 2.5 will be available for the rest of the network. We know that the reduced cost, the change in the objective resulting from a reduction of one unit of energy into the network from generator #3, increases the production cost by \$18.00 (see cell E24 of Figure 4.10). The energy to meet this additional demand at node 3 comes from generator #3 at an offer cost of \$22. Therefore, the increase in the production cost for the whole network due to this additional demand is \$22 plus \$18 = \$40. This is the LMP at this node that represents the cost of supplying the next unit of demand. Consequently, we confirm that LMP = bid cost of generation minus the shadow price. This is shown in cells K23:K27 of Figure 4.10, all of which are all equal to \$40.00, the cost of the marginal generator. An obvious but important observation is that in the absence of any transmission limitation, the marginal cost of supplying energy is the same at all nodes and is equal to the cost of the marginal generator.

4.7.1 Effect of Transmission Limitations on Solution

This study is the same as that of the previous problem but with a more realistic transmission system model. The network included is shown in

[9]This concept of marginal cost of supplying the next unit of energy was originally due to Caramanis, Bohn and Schweppe (Caramanis, Bohn and Schweppe, 1982), who called it spot prices. It is the cost of supplying the next unit of energy at each node. The LMP could be different at different nodes because of transmission constraints. If there are no losses as in this DC load flow case, and if there is no transmission restriction, the LMP is the same at all nodes and is equal to the cost of dispatching the marginal generator.

Figure 4.9.[10] Additionally, generation injections P_2, P_4, and P_7 have also been included as choice variables even though there are no generators at these nodes. Constraints requiring generation injections at these nodes to be zero account for the fact that there are no generators at these nodes. The need for this artifice is to permit the sensitivity report to calculate the shadow prices associated with these constraints enabling the display of LMP. However, this artifice is not necessary for the solution of the problem.

A copy of the spreadsheet "OPF-DC flow-Line limits.Xls" is shown in Figure 4.11. The network topology is inserted into the figure for ready reference. As in Appendix B.3.1, branch flows in Figure 4.11 have been computed from the shift factor matrix. The shift matrix $\mathbf{S}_F$ taken from Appendix B.3.1 is copied into cells B20:I32. The vector of net injection into the network (generation minus load) is represented in cells J20:J27, which is a transposed replica of row 16. The product of $\mathbf{S}_F$ and $\mathbf{P}_{inj}$ gives branch flows in cells M20:M32 of Figure 4.11.

It is required to keep the flows in branches below some limit.[11] Since the flow between two nodes can change direction, there may be different limits for the flows in two directions. In this example we assume the flow limit to be the same in both directions (it need not be so). Consequently, the constraint equation for branch flow F_b will have the formulation $-limit \leq F_b \leq limit$. These limits are incorporated into constraints of the Solver. To be sure that it is not constraining, a very high limit of 10 has been chosen for all branches except branch 3–4, for which a limit of 0.5 is chosen.

Sensitivities associated with the study of Figure 4.11 are shown in Figure 4.12. Note that the line flow constraint associated with branch 3–4 is binding indicated by a nonzero shadow price of −59.154. The production cost has increased from a previous value of $295.00 to $305.19. This is because generators had to be redispatched, backing down cheaper generators and dispatching more expensive generation to accommodate transmission limitations (congestion). In this instance, Generators #3 and #6 ($22 and $35) are backed off (not at full output) while generator #5 of $40 is at full output.

OPF-DC-flow linelim-its.Xls

The cell numbers in the sensitivity sheet refer to the main spreadsheet (Figure 4.11). LMP at nodes with generators in Figure 4.12 is shown in cells at the right-hand top corner by subtracting reduced costs from bid values at that node. Because of "congestion in transmission," there is a "spread" in the LMP at nodes. For instance, the LMP at node 3 is $22 while the LMP at node 5 is $48.545. The marginal cost of supplying energy at nodes 2, 4, and 7 that do not have generators (cells N14:N16 of Figure 4.11) is indicated in Figure 4.12 as $40.498. Unlike nodes that have generators connected to them, the calculation of LMP at nodes only with

[10] As shown in Section B.3.1, only the imaginary component of the branch admittances have been included, thus making the network loss-less.

[11] Such flow limitations may arise from "thermal" considerations (over heating of transmission conductors), or from considerations of inability to maintain voltage levels, system stability, and a myriad of other reasons. In all cases, such limitations are finally translated to flow limits in appropriate branches, or across interfaces between subregions. This flow limit, except under dire circumstances, is not exceeded while operating the system.

Optimal Power Flow Formulation Based On DC Load Flow - Line Flow Limits

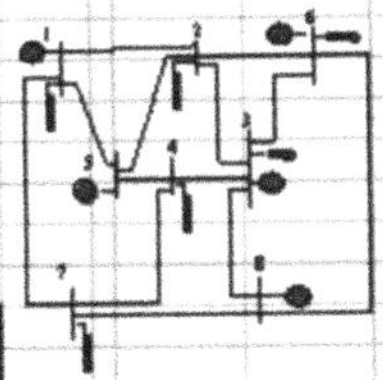

nodes	1	2	3	4	5	6	7	8	
P	4	0	3.5	0	2.8	2	0	3	*Specified upper Limit*
P	0	0	3.2928	0	2.8	0.70720393	0	3	Selected Generation

Nodes	1	2	3	4	5	6	7	8
Cost (offers) $/MWh	50	0	22	0	40	35	0	32
Specified loads at nodes	1	1.5	1.7	0.5	0	3	2.1	0

305.1937 Perf. Index

9.8 load sum

Resulting net nodal injection into network

P1	P2	P3	P4	P5	P6	P7	P8
-1.00	-1.50	1.59	-0.50	2.80	-2.29	-2.10	3.00

Choice Variables

P1	0
P3	3.2928
P5	2.8
P6	0.7072
P8	3
p2	0
p4	0
p7	0
sum	9.8

Shift Factor Matrix

									P inj	limit lower	Branch flow		Limit upper
1-2	0.3285	-0.18314	-0.04301	0.0060037	0.032486205	-0.0567514	0.05369722	0	-1.00	-10	-0.566387	P12	10
1-5	0.27474	-0.02127	-0.0383	-0.0659433	-0.251629993	-0.0160432	0.02098894	0	1.5	-10	-1.046534	P15	10
1-7	0.39677	0.20441	0.08131	0.0599396	0.219143788	0.07279452	-0.0746862	0	1.59	-10	0.612921	P17	10
2-3	0.14561	0.25071	-0.13805	-0.0272618	0.132840079	0.02180734	0.01371556	0	-0.50	-10	0.317351	P23	10
2-5	-0.0672	0.20233	0.00589	-0.0899338	-0.355145246	0.05088526	-0.0408853	0	2.80	-10	-0.600184	P25	10
2-6	0.25008	0.36383	0.08915	0.1231992	0.254791372	-0.129444	0.080867	0	-2.29	-10	1.216447	P26	10
3-4	-0.08073	0.01729	0.31271	-0.3829502	-0.136023082	0.09294772	-0.1417842	0	-2.10	-1	0.5	P34	0.5
3-6	-0.08791	-0.07348	-0.29728	-0.1818989	-0.111280491	0.18154823	-0.075143	0	3.00	-10	-0.974897	P36	10
3-8	-0.13843	-0.15994	-0.25197	-0.1737895	-0.157582671	-0.1104078	-0.0803567	0		-10	-0.43525	P38	10
4-5	0.20754	0.18106	-0.03241	-0.1558771	0.393224761	0.0348421	-0.0198964	0		-10	1.153282	P45	10
4-7	-0.1268	-0.19834	-0.2803	-0.4611727	-0.257201679	-0.1277898	0.16168059	0		-10	-1.153282	P47	10
6-8	-0.338	-0.43731	-0.38643	-0.3050982	-0.366071863	-0.6890078	-0.15601	0		-10	0.101453	P68	10
7-8	-0.52357	-0.40275	-0.36161	-0.5211124	-0.476345467	-0.2005843	-0.7636333	0		-10	0.333797	P78	10

Figure 4.11. Dispatch solution to minimize cost with transmission constraints.

Sensitivity Report.

Adjustable Cells

Cell	Name	Final Value	Reduced Cost	Objective Coefficient	Allowable Increase	Allowable Decrease
$N$9	P1	0	4.7259227	50	1E+30	4.72592271
$N$10	P3	3.2928	0	22	8.20104582	5.97980661
$N$11	P5	2.8	-8.5446178	40	8.54461777	1E+30
$N$12	P6	0.7072	0	35	2.63971777	4.18465363
$N$13	P8	3	-8.4982616	32	8.4982616	1E+30
$N$14	p2	0	-40.498262	0	1E+30	40.4982616
$N$15	p4	0	-40.498262	0	1E+30	40.4982616
$N$16	p7	0	-40.498262	0	1E+30	40.4982616

Offer Bid	LMP
50	45.2740773
22	22
40	48.5446178
35	35
32	40.4982616
0	40.4982616
0	40.4982616
0	40.4982616

Constraints

Cell	Name	Final Value	Shadow Price	Constraint R.H. Side	Allowable Increase	Allowable Decrease
$N$17	sum	9.8	40.498262	9.8	0.90853666	0.4899096
$M$20	1-2 Branch	-0.56639	0	10	1E+30	10.5663869
$M$21	1-5 Branch	-1.04653	0	10	1E+30	11.0465343
$M$22	1-7 Branch	0.61292	0	10	1E+30	9.38707882
$M$23	2-3 Branch	0.31735	0	10	1E+30	9.68264932
$M$24	2-5 Branch	-0.60018	0	10	1E+30	10.6001842
$M$25	2-6 Branch	1.21645	0	10	1E+30	8.78355343
$M$26	3-4 Branch	0.5	-59.154344	0.5	0.04553598	0.28411014
$M$27	3-6 Branch	-0.9749	0	10	1E+30	10.9748967
$M$28	3-8 Branch	-0.43525	0	10	1E+30	10.43525
$M$29	4-5 Branch	1.15328	0	10	1E+30	8.84671843
$M$30	4-7 Branch	-1.15328	0	10	1E+30	11.1532816
$M$31	6-8 Branch	0.10145	0	10	1E+30	9.89854722
$M$32	7-8 Branch	0.3338	0	10	1E+30	9.66620275
$M$20	1-2 Branch	-0.56639	0	-10	9.43361308	1E+30
$M$21	1-5 Branch	-1.04653	0	-10	8.95346574	1E+30
$M$22	1-7 Branch	0.61292	0	-10	10.6129212	1E+30
$M$23	2-3 Branch	0.31735	0	-10	10.3173507	1E+30
$M$24	2-5 Branch	-0.60018	0	-10	9.39981583	1E+30
$M$25	2-6 Branch	1.21645	0	-10	11.2164466	1E+30
$M$26	3-4 Branch	0.5	0	-0.5	1	1E+30
$M$27	3-6 Branch	-0.9749	0	-10	9.02510328	1E+30
$M$28	3-8 Branch	-0.43525	0	-10	9.56474998	1E+30
$M$29	4-5 Branch	1.15328	0	-10	11.1532816	1E+30
$M$30	4-7 Branch	-1.15328	0	-10	8.84671843	1E+30
$M$31	6-8 Branch	0.10145	0	-10	10.1014528	1E+30
$M$32	7-8 Branch	0.3338	0	-10	10.3337972	1E+30

Figure 4.12. Sensitivity analysis of transmission constraints.

loads is straightforward; it is the reduced cost associated with choice variables in cells N14:N16. In this instance, all reduced costs are equal to the shadow price related to the constraint of total demand in cell N17 of Figure 4.11 (see the first line under constraints in Figure 4.12). A word of caution is that in the presence of ohmic losses in the transmission system, the LMP at all nodes will not be equal. Therefore, the difference between shadow prices of nodes arises because of transmission constraints and ohmic losses, if any. This statement is the same as that of saying that, in a loss-less system, with no transmission flow constraints, the LMP at all nodes are the same.

The sensitivity report of Figure 4.12 is related to the optimization in Figure 4.11. The shadow price for cell M26 of Figure 4.11 is \$−59.15. Therefore, if the constraint of 0.5 for the flow on branch 3–4 is relaxed (increased), the objective function will decrease by \$59.15. This is called the marginal cost associated with the transmission constraint.

Additional Transmission Constraints — Effect on Solution. In the example above, only one branch is shown to be limiting and, therefore, exhibits a nonzero shadow price. The reader is encouraged to input lower flow limits for several other branches simultaneously and study the results. It is important to make several such studies in order to fully comprehend and retain ideas about LMP, and the effects of congestion.

In real power systems, it is often the case that a certain interface is a limiting constraint rather that a single branch or a line. An interface is the sum of flows on some lines connecting one subarea to another. Such constraints are easy to accommodate into the Solver program. For example, if the sum of flows on branches 3–6, and 3–2 should not exceed a certain value, the constraint can be expressed as: $F_{3-6} + F_{3-2} \leq$ the desired limit. Correspondingly, an additional entry into the Solver routine would be the constraint that the value of cell M27 minus[12] M23 in Figure 4.11 is less than or equal to some chosen limit.

In exercise problem 2 of Section 4.13, the reader is asked to simulate a situation where the flow in a tie line is related to the flow in another according to some relation.

4.8 INTER-REGIONAL DISPATCH

Formulation and Discussion of Problem: A matter of considerable interest in the electric power industry is that of adequacy of transmission. Within the territory of a system operator (such as California, New York, or the PJM interconnection), there are regions or subregions of generation connected to other regions by transmission network. Certain regions may have cheaper generation, which cannot supply the need in another region because of inadequate transfer capability from that region to the other region. That is not to say that transmission capability between such subregions should be increased automatically, because such decisions are based on the economics of cost and benefit. For example, only for a few hours of the year, the demand and supply situation may represent "bottling up" of cheaper generation in some subregions. Such situations may not warrant the improvement of transmission.

In the deregulated electric industry, the issue of transmission adequacy between subregions takes on an added dimension because of bidding strategy and the exercise of market power by Generators. Because of the bidding volatilities for energy as affected by transmission limitation, the system operator may not be able to dispatch all of cheaper generation, thus creating a difference in the LMP between subregions. One would like a model to study and analyze such situations.

Inter-region.Xls

The spreadsheet of "Inter-region.Xls" simulates five subregions A, B, C, D, and E connected by a transmission system with the indicated shift factor matrix. Equation (B.24) in Section B.3 of Appendix B shows the process of obtaining the

[12]Although the interface flow is the sum of flows in lines, a negative sign is used here because the convention is positive for flow in the direction 2 to 3.

shift factor matrix. A printout of the spreadsheet "Inter-region.Xls" is shown in Figure 4.13.

The network between regions is identical to that of the earlier problem in Figure 4.7 of Section 4.5. In this example, however, it is assumed that the subregions A to E can be further represented by a detailed transmission network without any constraining flow limitations within the subregions themselves. Within each subregion, there can be several generators who submit offer bids for the supply of energy. In our example, we have considered three generators in each subregion whose offers are shown in cells F5:F19. Each subregion also has loads whose demands are indicated in cells C7:C19.

The generators have a claimed capability shown as the upper limit in column E. They also have a lower limit of output, either because of the characteristics of the network such as voltage limits, or because of the generator characteristics. These limits are in column D. Given these data, the objective of the system operator is to minimize the cost of dispatch, which is the sum of products of outputs and offer bids—this is in cell H20. The decision variables are the generator outputs and, correspondingly, how much to import or export from one subregion to another, subject to the limitations of the transmission system. Consequently, choice variables are the desired outputs of generators (cells G5:G19) and the desired imports to and exports from subregions in cells C24:G28.[13] By taking the appropriate row sums and column sums, the export from and import into subregions are computed in row 30, and column I. From these sums, the net injection to the subregional network is calculated in cells M24:M28. Since the system is loss-less, the generation in a subregion, plus imports, minus exports must be equal to the demand of the subregion. This formula is put into cells J7, J10, J13, J16, and J19 and is presented as an equality constraint to the Solver.

The only remaining constraints, in addition to the nonnegativity conditions, are those relating to transfer capabilities between subregions. These are indicated in rows 32 to 37 in columns I, K, and L for all the six branches of the network. The flow limits need not necessarily be the same for flows in either direction of a branch. In addition, two interface limits have also been included. Interface limits are those that specify a limit the sum of flows between certain regions or subregions. In this example, the limit in cell N33 requires import into subregion E not to exceed 100. The constraint in cell N36 requires the export from subregion C not to exceed 160.

Mathematical Formulation: Given the costs of generations C_i^r and quantities offered Q_{iO}^r from the ith generator in regions $r = 1, \ldots, R$ from $i = 1, \ldots, n_r$ generators in each region r, do the following:

Choose Q_{ig}^r, quantities to be generated from generators $i = 1, \ldots, i_r$ in regions $r = 1, \ldots, R$, and let I^r, E^r be the quantities to be imported into and exported from regions r, $r = 1, \ldots, R$, in order to minimize

$$\sum_{r=1}^{R} \sum_{i=1}^{n_r} C_i^r Q_{ig}^r,$$

[13]Evidently, diagonal entries in cells C24, D25, and so on, are zero.

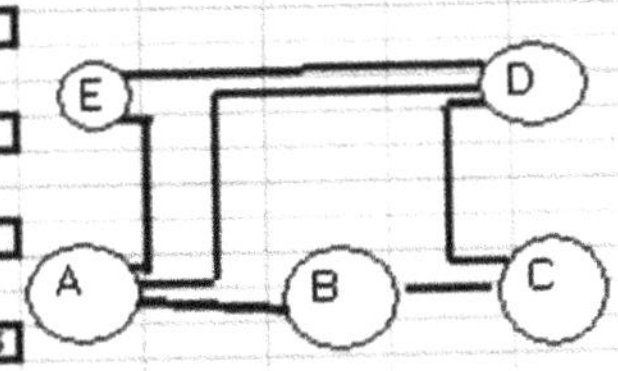

Energy Interchange Between Sub-Regions

Region	Demand	Generators Lower MW-limit	Generators Upper MW-limit	cost/offer	output	prod. cost
A		20	50	13.5	20	270
A		0	150	2	150	300
A	350	0	100	1.5	100	150
B		0	10	5	10	50
B		0	60	3	60	180
B	201	20	85	8.1	85	688.5
C		0	30	9	30	270
C		0	300	10	280	2800
C	300	0	150	7	150	1050
D		0	130	16	0	0
D		0	20	11	15.89472	174.8419
D	150	0	160	14	0	0
E		0	40	8	30.10528	240.84225
E		30	120	6	120	720
E	50	0	20	8.5	0	0
	1051			Totals	1051	6894.18
						INDEX

Constraint gen+imp-exp-load

Area gen.			Net imp/exp
A 270	5.67E-11	Import(-), Export(+)	-80
B 155	0	Import(-), Export(+)	-46
C 460	0	Import(-), Export(+)	160
D 15.894719	7.95E-11	Import(-), Export(+)	-134.105
E 150.10528	0	Import(-), Export(+)	100.105
1051			

flow	to a	to b	to c	to d	to e
From a	0	-0.737	-51.868	10.8687355	79.24505
From b	15.846	0	20.7555	20.3360387	-12.8536
From c	13.445	73.757	0	41.642656	10.76441
From d	35.963	-10.88	-8.3397	0	-17.4427
From e	52.249	28.6	19.1208	59.8426038	0
Total Import to: a,b,c,d or e	117.51	30.738	-20.391	133.3501	59.70719

Total Exp from	
37.509143	a
44.737711	b
139.6089	c
-0.755181	d
159.81247	e

NET Injection	
A	-80
B	-46
C	160
D	-134.105
E	100.105

SHIFT FACTOR

-0.669	-0.543	-0.195	-0.034
-0.18	-0.249	-0.433	-0.078
-0.151	-0.21	-0.37	-0.887
0.3298	-0.543	-0.195	-0.034
0.3298	0.4566	-0.195	-0.034
0.1517	0.21	0.3703	-0.111

Constraints

>=	Branch Flow		<=
-50	A_B	-33.40848	40
-40	A-D	19.555381	20
-100	A-E	-65.80093	70
-30	B-C	-79.44701	100
-100	C-D	80.552386	100
-50	D-E	-34.19307	100

A-E+D-E	>= -100	-100	Export From E
(C-D)-(B-	<= 160	160	Export from C

Figure 4.13. Inter-regional dispatch.

subject to

$$I_r - E_r + \sum_{i=1}^{n_r} Q_{ig}^r - D_r = 0, \qquad r = 1, R,$$

where D_r is the demand of region r and I_r, E_r indicate the import and export into regions r,

$$Q_{ig}^r(\text{min}) \leq Q_{ig}^r \leq Q_{ig}^r(\text{max}), \qquad i = 1, \ldots, n_r, \qquad r = 1, \ldots, R,$$

in which (max) and (min) indicate the maximum and minimum output capabilities of generators, and the regional transfer limits

$$I_{rj} \leq I\text{max}_{rj}, \qquad r = 1, \ldots, R, \quad \text{and} \quad j = 1, \ldots, R(j \neq r),$$

$$E_{rj} \leq E\text{max}_{rj}, r = 1, \ldots, R \quad \text{and} \quad j = 1, \ldots, R(j \neq r),$$

where $I\text{max}_{rj}$ and $E\text{max}_{rj}$ are the maximum permitted imports to and exports from regions $r = 1, R$, and the limits

$$Q_{ig}^r \leq Q_{iO}^r, \quad r = 1, \ldots, R, \qquad i = 1, \ldots, n_r,$$

in which the subscript O represents offered quantity.

We also have

$$I_r = \sum_{j=1}^{R,\ j \neq r} I_{rj}, \qquad r = 1, \ldots, R,$$

$$E_r = \sum_{j=1}^{R,\ j \neq r} E_{rj}, \qquad r = 1, \ldots, R,$$

Solution: The solution obtained by the Solver in Figure 4.13 shows the minimum cost of dispatch to be \$6894.18. Only some constraints are binding. For example, the upper limit of the third generator in subregion B and the second generator in subregion E are binding; also, the lower limit of the first generator in subregion A is binding.

The reader will gain further knowledge from an exercise of changing the limits for generation and transmission flows, and by examining the associated sensitivity reports. In the following, the sensitivity report associated with Figure 4.13 illustrates the type of information that can be obtained therefrom. In fact, such information is used to price energy in subregions. Results of sensitivity are essential for the settlement of markets of the deregulated electrical system.

4.8.1 Sensitivity Report

Figure 4.14 shows the sensitivity report. This report refers to cells in the study of Figure 4.13. The reader should examine and ponder the reduced costs associated with choice variables and shadow prices of constraints.

The most expensive generator feeding energy has a cost of $13.50 in region A, despite some cheaper generators not being at their full outputs in other regions. This is because the lower limit of output from that generator is a binding constraint.

In terms of shadow prices, the constraint regarding export from subregions C and E are binding. Consequently, cheaper generators are backed off in C and E. Therefore, the cost of supplying an additional unit of energy in subregion E is $8 (since the output of the $8.00 machine is only 30.105, well below its maximum capability). This is shown in cell J19. Similarly, the cost of supplying the next increment of energy in subregion C is $10. This is shown in cell J13. Continuing the same logic, it can be confirmed that the shadow prices associated with cells J7 to J19 are the costs of supplying the next unit of energy in the subregions. Hence, they represent the LMP of the subregions. The shadow price associated with cells N33 and N36 indicate the penalty associated with the transfer limit constraint. For example, the shadow price of $3.002 associated with cell N33 indicates that if the constraint value for flow across this interface is increased by one (to −101), the objective will decrease by $3.0024.

The penalties associated with shadow prices have to be examined for several hours of system dispatch in order to ascertain the benefit that would accrue with any transfer capacity upgrades. The cost of such upgrades, of course, has to be juxtaposed with the benefits.

4.9 EXAMPLES OF TRANSPORTATION PROBLEMS

The following two examples typify problems under the category of transportation problems.

4.9.1 Warehouse Distribution Problem

Given three warehouses 1, 2, and 3, and three destinations 4, 5, and 6, for the given costs of transportation between them, find the optimal transportation policy from warehouse to destination. The quantities available at warehouses, as well as quantities needed at destinations, are given as indicated in the spreadsheet "Transportation.Xls."

Transportation.Xls

A printout of the spreadsheet and its solution is shown in Figure 4.15. The quantities needed in destinations are shown in row 17 as 10, 20, and 15. The quantity available at the three warehouses is indicated in column I as 5, 17, and 23. The costs of transportation from each warehouse to destinations are indicated in rows 5, 9, and 13. Choice variables are the quantities transported from the warehouses to destinations shown in rows 4, 8, and 12. The objective of transportation cost is in cell F16.

Adjustable Cells

Cell	Name	Final Value	Reduced Cost	Objective Coefficient	Allowable Increase	Allowable Decrease
$G$5	A output	20	2.5017759	13.5	1E+30	2.501776
$G$6	A output	150	-8.998224	2	8.998224	1E+30
$G$7	A output	100	-9.438224	1.5	9.438224	1E+30
$G$8	B output	10	-5.998952	5	5.998952	1E+30
$G$9	B output	60	-7.998952	3	7.998952	1E+30
$G$1(	B output	85	-2.898952	8.1	2.898952	1E+30
$G$11	C output	30	-1	9	1	1E+30
$G$1	C output	280	0	10	0.999231	1
$G$1	C output	150	-3	7	3	1E+30
$G$1	D output	0	5	16	1E+30	5
$G$1	D output	15.894719	0	11	0.565958	0.999487
$G$1	D output	0	3	14	1E+30	3
$G$1	E output	30.105281	0	8	0.5	0.392545
$G$1	E output	120	-2	6	2	1E+30
$G$1	E output	0	0.5	8.5	1E+30	0.5
$C$2	From a a	0	0	0	1E+30	0
$D$2	From a b	-0.736973	0	0	0	0
$E$2	From a c	-51.86773	0	0	0	0
$F$2	From a d	10.868796	0	0	0	0
$G$2	From a e	79.245048	0	0	0	0
$C$2	From b a	15.845724	0	0	0	1E+30
$D$2	From b b	0	0	0	1E+30	0
$E$2	From b c	20.755508	0	0	0	1E+30
$F$2	From b d	20.996039	0	0	0	1E+30
$G$2	From b e	-12.85956	0	0	0	1E+30
$C$2	From c a	13.444892	0	0	0	1E+30
$D$2	From c b	73.756944	0	0	0	1E+30
$E$2	From c c	0	0	0	1E+30	0
$F$2	From c d	41.642656	0	0	0	1E+30
$G$2	From c e	10.764406	0	0	0	1E+30
$C$2	From d a	35.969387	0	0	0	1E+30
$D$2	From d b	-10.88218	0	0	0	1E+30
$E$2	From d c	-8.339683	0	0	0	1E+30
$F$2	From d d	0	0	0	1E+30	0
$G$2	From d e	-17.44271	0	0	0	1E+30
$C$2	From e a	52.24914	0	0	0	1E+30
$D$2	From e b	28.599917	0	0	0	1E+30
$E$2	From e c	19.1208	0	0	0	1E+30
$F$2	From e d	59.84261	0	0	0	1E+30
$G$2	From e e	0	0	0	1E+30	0

Constraints

Cell	Name	Final Value	Shadow Price	Constraint R.H. Side	Allowable Increase	Allowable Decrease
$J$7	A gen+ir	6E-11	10.9982241	0	4.1077	1.0108
$J$13	C gen+ir	0	10	0	20	280
$J$16	D gen+ir	8E-11	11	0	4.1053	15.895
$J$19	E gen+ir	0	8	0	9.8947	30.105
$J$10	B gen+ir	0	10.9989515	0	4.1067	1.712
$K$32	A_B Imp	-33.41	0	40	1E+30	73.408
$K$33	A-D Imp	19.556	0	20	1E+30	0.444
$K$34	A-E Imp	-65.8	0	70	1E+30	135.8
$K$35	B-C Imp	-79.45	0	100	1E+30	179.45
$K$36	C-D Imp	80.553	0	100	1E+30	19.447
$K$37	D-E Imp	-34.2	0	100	1E+30	134.2
$K$32	A_B Imp	-33.41	0	-50	16.592	1E+30
$K$33	A-D Imp	19.556	0	-40	59.556	1E+30
$K$34	A-E Imp	-65.8	0	-100	34.199	1E+30
$K$35	B-C Imp	-79.45	0	-90	10.553	1E+30
$K$36	C-D Imp	80.553	0	-100	180.55	1E+30
$K$37	D-E Imp	-34.2	0	-50	15.801	1E+30
$N$33	A-E+D-E	-100	3.00248564	-100	4.1019	1.2266
$N$36	(C-D)-(E	160	-0.9992311	160	2.3346	4.1063

Figure 4.14. Sensitivity analysis of inter-regional dispatch.

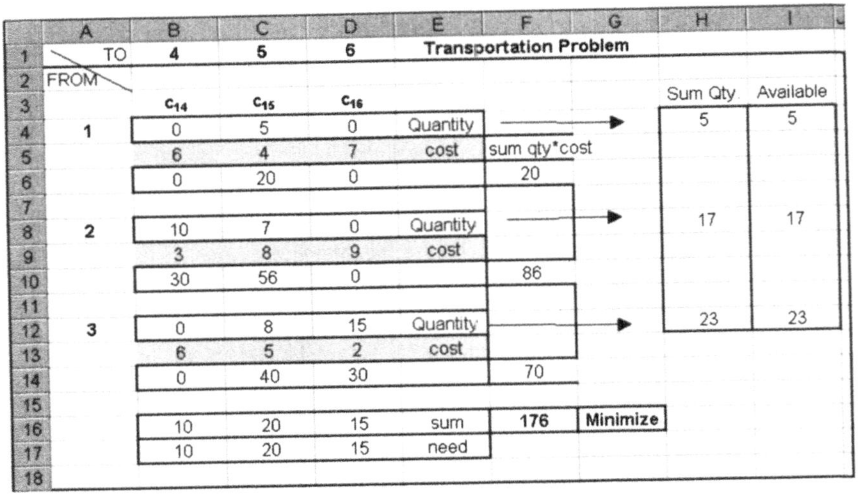

	A	B	C	D	E	F	G	H	I
1	TO	4	5	6	Transportation Problem				
2	FROM								
3		c_{14}	c_{15}	c_{16}				Sum Qty	Available
4	1	0	5	0	Quantity			5	5
5		6	4	7	cost	sum qty*cost			
6		0	20	0		20			
7									
8	2	10	7	0	Quantity			17	17
9		3	8	9	cost				
10		30	56	0		86			
11									
12	3	0	8	15	Quantity			23	23
13		6	5	2	cost				
14		0	40	30		70			
15									
16		10	20	15	sum	176	Minimize		
17		10	20	15	need				
18									

Figure 4.15. Solution to transportation problem.

Mathematical Formulation of the Problem: Choose c_{ij}, $i = 1, \ldots, n$; $j = 1, \ldots, m$, the quantities to be moved from n warehouses to m destinations in order to minimize

$$\sum_{i=1}^{n} \sum_{j=1}^{m} t_{ij} c_{ij},$$

where t_{ij} is the cost of transportation from i to j, and the quantities moved from n warehouses to j destinations should be such that

$$\sum_{i=1}^{n} c_{ij} = Q_j, \qquad j = 1, \ldots, m,$$

where Q_j is the demand at destination j, and

$$\sum_{j=1}^{m} c_{ij} = Q_i, \qquad j = 1, \ldots, m,$$

where Q_i is the quantity available at warehouse i, and, of course,

$$\sum_{i=1}^{n} Q_i = \sum_{j=1}^{m} Q_j,$$

and

$$c_{ij} \geq 0.$$

Solution: The final solution using the Solver is shown in the figure under discussion.

4.9.2 Integrating Natural Gas Transportation and Electricity Generation

We now extend the above problem to include a case of fuel supply (natural gas) and electricity generation. Even this extension is somewhat simplistic. In "real world" situations, optimal fuel procurement by electricity generating companies is a complex task. One enters into longer-term contracts with "take or pay" clause, and one also deals in Options, Futures, and Spot markets[14] for fuel procurement. Furthermore, a diversification in sources of supply as well as in types of fuel to avoid risks of shortage is common practice. Understandably, it is not possible to simulate all the complications of fuel procurement in an example of a textbook. Nevertheless, our intent here is to demonstrate how different types of optimization problems can be integrated to simulate practical situations.

Toward that end, in this example consider three sources of natural gas connected by a pipeline transportation network to three generating units owned by a Generator. The spreadsheet "Gas-Electric-Transportation.Xls" is shown in Figure 4.16. Let the heat content of the gas from the three wellheads be the same. However, the heat rates of generators are different; G1 generates 1 MWh per MBTU, G2 produces 0.95 MWh per MBTU, and G3 produces 1.05 MWh per MBTU. The data shown in row 3 incorporate these heat rates.

Gas-electric-transportation.Xls

Discussion of Problem: The cost of transporting gas from the three wellheads to the three generators are in rows 7, 15, and 23 in units of $/MBTU. The quantity that can be delivered per hour (MBTU) is dependent on pipeline capacity as indicated in rows 6, 14 and 22. The total amount of gas (in MBTU) that can be delivered from each wellhead (per hour) is shown in column I. This represents pipeline capacity.

In terms of pollution arising from electricity generation, each generating unit has different types of burners. Further there are differences in the composition of natural gas from the three wellheads. Consequently, different rates of pollution (thousand lbs per MBTU) are indicated for the generators in rows 10, 19, and 27, respectively.

In this simple problem, we assume that there is no demand charge for gas delivery, and that the cost of gas per MBTU, apart from the transportation charges, is the same from three wellheads. The problem now is to determine the optimum deliveries in

[14]Take or pay clauses are written into the fuel delivery contract, obligating the buyer to buy a certain minimum quantity. Failure to buy the minimum quantity would still entail a certain payment or penalty. Options and futures are more complicated. The buyer in a secondary market, or in a commodity exchange, buys a paper contract assuring the delivery of a commodity at some future date at a certain fixed price. The commodity is seldom delivered, but the papers are traded and used as hedging instruments to insulate the holder from price fluctuations.
The field of hedging is vast, and organizations employ professionals to hedge risks. An excellent source of reference on these matters is Hull (1999).

	A	B	C	D	E	F	G	H	I	J
1	TO	G1	G2	G3	**Optimum Gas Supply**					
2	FROM									
3		1	0.95	1.05	Heat rate for Energy					
4									available	
5	WH1	0	0	150	Quantity delivered		Sum	150	250	
6		150	150	150	Deliveribility					
7		6	4	7	cost					
8		0	0	1050		1050	sum qty*cost			
9		0	0	142.8571	Energy produced		-------Total	142.8571		
10		0.6	0.6	0.45	Pollution rate					
11		0	0	67.5	Pollution ------------		Total	67.5		
12										
13	WH2	0	200	0	Quantitydelivered		Sum	200	200	
14		150	250	250	Deliveribility					
15		3	8	9	cost					
16		0	1600	0		1600	sum qty*cost			
17		0	210.5263	0	Energy produced		-------Total	210.5263		
18		0.65	0.6	0.55	Pollution rate					
19		0	120	0	Pollution ------------		Total	120		
20										
21	WH3	0	91.78571	0	Quantitydelivered		Sum	91.78571	220	
22		150	100	300	Deliveribility					
23		6	5	2	cost					
24		0	458.9286	0		458.9286	sum qty*cost			
25		0	96.61654	0	Energy produced		-------Total	96.61654		
26		0.65	0.6	0.65	Pollution rate					
27		0	55.07143	0	Pollution ------------		Total	55.07143		
28										
29		0	291.7857	150		Total Gas delivery		441.7857		
30										
31						Total Energy		450		
32										
33						Total Pollution		242.5714		
34										
35					Cost of transportation			3108.929		

Figure 4.16. Optimization of natural gas delivery through transportation network to electric generators.

order to a) minimize the transportation charge, b) minimize the pollution given that the owner of generating units is obligated to deliver 450 MW into the grid.

Mathematical Statement of Problem: Choose gas flows f_{1i}, f_{2i}, f_{3i} for $i = 1, 2, 3$ from wellheads 1, 2, and 3 in order to minimize

$$\sum_{j=1}^{3}\sum_{i=1}^{3} f_{ji}t_{ji}, \tag{4.1}$$

subject to

$$\sum_{j=1}^{3} \frac{f_{j1}}{1} + \frac{f_{j2}}{0.95} + \frac{f_{j3}}{1.05} = 450,$$

$$\sum_{i=1}^{3} f_{1i} \leq 250,$$

$$\sum_{i=1}^{3} f_{2i} \leq 200,$$

$$\sum_{i=1}^{3} f_{3i} \leq 220,$$

and

$$f_{ji} \geq 0, \qquad j = 1, 2, 3; \qquad i = 1, 2, 3,$$

where f_{ji} is the gas flow from wellhead j to generator i, and t_{ji} is the transportation cost associated with this flow.

For the case of minimizing pollution, (4.1) has to be replaced with a corresponding objective taking into account the pollution rates at generators for different fuels as depicted in Figure 4.16.

Solution: The solution to the problem of minimizing transportation cost using the Excel Solver is shown in Figure 4.16 to be \$1233.81. The resultant pollution with this strategy of gas purchase is 292.07×10^3 lb. Similarly, the reader will find that the optimum solution to minimize pollution (choose cell H33 to be minimized) results in a total pollution of 242.5714×10^3 lb. Correspondingly, the cost of gas transportation increases. The cost of gas transportation is a degenerate solution as explained in Section 3.4. This cost can vary from a cost of \$2984.28 to \$3018.92.

The reader is expected to examine sensitivity sheets resulting from these programs and ponder the reasons for the resulting shadow prices and reduced costs.

4.10 SOLUTION OF PRIMAL–DUAL PROBLEMS

In the example problems of Section 3.10.1, for simplicity, only less than or equal type of inequalities were considered in the primal problem. However, Table 3.1 outlined rules to formulate the dual problem for all types of inequalities and equality constraints. In the following examples, we formulate and solve the dual when the primal problem includes greater than and less than type of inequalities, as well as equality constraints. We study sensitivity reports of these two solutions and make observations that correspond to the duality theorems of Section 3.10.2.

Consider the following primal problem:
Choose x_1, x_2, x_3 to minimize

$$\pi = 82x_1 + 100x_2 + 200x_3, \tag{4.2}$$

subject to

$$2x_1 + 4x_2 \qquad \leq 120, \tag{4.3}$$

$$3x_1 + 4x_2 + 6x_3 \leq 13, \tag{4.4}$$

$$2x_1 \qquad - 4x_3 \geq 200, \tag{4.5}$$

$$10x_2 + 5x_3 = 11, \tag{4.6}$$

and

$$x_1, x_2 \geq 0$$

and

$$x_3, \qquad \textit{unspecified}.$$

The corresponding dual of this problem in view of the relations expressed in Table 3.1 is obtained as follows.

Since the primal has three choice variables, there will be three constraint equations in the dual. There are four constraint equations for the objective of (4.2). Therefore, there will be four choice variables $y_1 \cdots y_4$ in the dual. Using the right-hand terms of (4.3) to (4.6), we obtain the objective function to be maximized as

$$\pi = 120y_1 + 13y_2 + 200y_3 + 11y_4. \tag{4.7}$$

The relationships shown in Table 3.1 stipulate constraints for the dual problem as follows. In the primal, since $x_1, x_2 \geq 0$, two "less than or equal to" constraints equations result. The coefficients of choice variables **y** of the dual are the transpose of the first two columns of (4.3) to (4.6) giving

$$2y_1 + 3y_2 + 2y_3 \leq 82, \tag{4.8}$$

and

$$4y_1 + 4y_2 + 10y_4 \leq 100. \tag{4.9}$$

In above constraint equations, the right-hand-side values are the first two coefficients of the primal objective.

Since x_3 is unspecified in the primal, an equality constraint is obtained using the third column of equations (4.3) to (4.6) as

$$6y_2 - 4y_3 + 4y_4 = 200. \tag{4.10}$$

Since the first two constraints in the primal are "less than or equal to" and since the third constraint is "greater than or equal to," we get

$$y_1, y_2, \leq 0; \qquad y_3 \geq 0. \tag{4.11}$$

Since the fourth constraint in the primal is an equality, y_4 is unspecified. **Solved Primal–Dual.Xls**

The reader should examine Table 3.1 carefully and correlate the relations defined therein to the dual equations developed above.

Figures 4.17 and 4.18 indicate solutions for the primal and the dual in the Solved "Primal–Dual.Xls" spreadsheet. Sheet 1 contains the Primal problem and Sheet 2

	A	B	C	D	E
1	**Primal Problem**				
2	x1	x2	x3		
3	0	30	-57.8	Objective	
4				-8560	
5	Constraints				
6	2 x1 + 4 x2		120	<=	120
7	3 x1 + 4 x2 + 6 x3		-226.8	<=	13
8	2 x1 - 4 x3		231.2	>=	200
9	10 x 2 + 5 x3		11	=	11

Figure 4.17. Solution to the primal problem.

	A	B	C	D	E
1	**Dual Problem Formulation**				
2	y1	y2	y3	y4	objective
3	-75	0	0	40	-8560
4	Constraints				
5	2y1+3y2+2y3		-150	<=	82
6	4y1+4y2+10Y4		100	<=	100
7	6y2-4Y3+5Y4		200	=	200
8					
9	y1	<=	0		
10	y2	<=	0		
11	y3	>=	0		
12	y4		un specified		

Figure 4.18. Solution to the dual problem.

Sensitivity Analysis of Primal Problem

Adjustable Cells

Cell	Name	Final Value	Reduced Cost	Objective Coefficient	Allowable Increase	Allowable Decrease
$A$2	x1	0	232	82	1E+30	232
$B$2	x2	30	0	100	300	1E+30
$C$2	x3	-57.8	0	200	1E+30	150

Constraints

Cell	Name	Final Value	Shadow Price	Constraint R.H. Side	Allowable Increase	Allowable Decrease
$C$5	2x1+4x2 x3	120	-75	120	1E+30	15.6
$C$6	3x1+4x2+6x3 x3	-226.8	0	13	1E+30	239.8
$C$7	2x1-4x3 x3	231.2	0	200	31.2	1E+30
$C$8	10x2+5x3 x3	11	40	11	39	1E+30

Figure 4.19. Sensitivity of the primal problem.

Sensitivity of Dual Formulation

Adjustable Cells

Cell	Name	Final Value	Reduced Cost	Objective Coefficient	Allowable Increase	Allowable Decrease
$A$2	y1	-75	0	120	1E+30	15.6
$B$2	y2	0	239.8	13	1E+30	239.8
$C$2	y3	0	-31.2	200	31.2	1E+30
$D$2	y4	40	0	11	39	1E+30

Constraints

Cell	Name	Final Value	Shadow Price	Constraint R.H. Side	Allowable Increase	Allowable Decrease
$C$4	2y1+3y2+2y3 y3	-150	0	82	1E+30	232
$C$5	4y1+4y2+10Y4 y3	100	30	100	300	1E+30
$C$6	6y2-4Y3+5Y4 y3	200	-57.8	200	1E+30	150

Figure 4.20. Sensitivity of the dual problem.

contains its dual formulation. The optimized objective in both solutions is the same, namely -8560.

It is now instructive to examine the sensitivity reports of solutions, which are reproduced in Figures 4.19, and 4.20.

For the primary problem, optimal choice variables are 0, 30, and −57.8. The shadow prices corresponding to the four constraints are −75, 0, 0, and 40. In the dual problem, values of shadow prices and choice variables are reversed: Final values of choice variables are −75, 0, 0, and 40, and shadow prices corresponding to three constraints are 0, 30, and −57.8. The reader is asked to compare the reduced costs and the allowable increase and decrease for variables for the two solutions. What other conclusions can be drawn? Further by relating the solutions to the duality theorems in Section 3.10.2, the reader will be in a position to discern relations between the dual and the primal problems and will be able to draw some interesting conclusions.

4.11 RESERVOIR MANAGEMENT

Linear programs are ideally suited to optimize problems associated with water flow management, maximization of electricity generation in hydroelectric systems, and a variety of other issues related to water resources management. Some problems, particularly those involving the time of travel for water from one dam to another, can be very complicated. The following modified example modeled after an exercise in Rao (1996) is an illustration of a simple problem in water management.

Problem Statement: It is proposed to build a reservoir to manage water in an irrigation district. The object is to compute the optimum storage capacity, s, of this reservoir, given the following data.

Expected inflow to reservoir: dry season (summer) 1.5×10^6 acre-feet, wet season 4.9×10^6 acre-feet. Water released for irrigation from reservoir during dry and wet seasons are x_{outdry} and x_{outwet}, respectively. Further, it is required that the ratio of wet and dry season outflows is 30:70. Annualized cost of building and maintaining reservoir is equal to $20s$. The outflow of reservoir, before reaching the irrigation district, is augmented by another stream. This stream adds 0.7×10^6 acre-feet in wet season and 0.1×10^6 acre-feet in dry season. The cost of operating the outflow is 18 times the flow into the irrigation district in the wet season, and 12 times the flow in the dry season.

Compute the optimum reservoir storage capability and the discharges from it during the wet and dry seasons. **Reservoir problem .Xls**

Solution: Choice variables of the problem are s, x_{outwet}, and x_{outdry}.

The assumption is that there is no annual carryover of water. This means that the flow during a year is all utilized and not stored for subsequent years. Therefore, for storage capacity s, we have

$$s \geq 4.9 - x_{\text{outwet}},$$

$$s \geq 1.5 - x_{\text{outdry}}.$$

	A	B	C	D	E	F
1	Reservoir Optimization					
2	Choice Variables		Objective			
3	s	2.57998	166.61964			
4	X out wet	1.92002	=25 s+18(xout wet+0.7) + 12(xout dry+0.1)			
5	X out dry	4.47998		Constraints		RHS
6			s+X out wet>=	4.5		4.5
7			s+X out dry>=	7.0599616	>	1.5
8			X out dry - 2.33 Xout wet=	-3.38E-14	=	0
9			X out dry + X out wet=	6.4	=	6.4
10						

Figure 4.21. Solution to reservoir optimization problem.

Finally, from the ratio of outflows during wet and dry seasons, we obtain the equality constraint as

$$x_{\text{outwet}} - 2.33x_{\text{outdry}} = 0.$$

The net flow into the irrigation district during dry and wet seasons are $x_{\text{outwet}} + 0.7$, and $x_{\text{outdry}} + 0.1$ respectively.

The objective is to minimize the cost given by

$$\pi = (x_{\text{outwet}} + 0.7)12 + (x_{\text{outdry}} + 0.1)18 + 20s,$$

subject to constraints outlined above.

Results of computation are shown in Figure 4.21. As can be seen therein, the optimum storage capacity of the reservoir is 2.57910^6 acre-feet, with wet and dry season outflows of 1.92, and 4.479 million acre-feet.

4.12 SUMMARY

The above examples and their setup on Excel sheets have exposed the reader to ways of conceiving and structuring physical problems as optimization problems. Clearly, one can set up one's spreadsheets in a manner different from those in these spreadsheets. Nonetheless, the mathematical formulation of problems arising from a physical phenomena is what the reader is expected to see and formulate in order to obtain a solution.

As in all disciplines, "practice makes perfect." It is with further practice on variants of the above problems, along with the analysis of sensitivity sheets, that one becomes better in conceiving and solving problems.

4.13 EXERCISE PROBLEMS

1. The following problem illustrates an optimal mixing problem discussed in Section 4.1.

   A generating station with several generators can obtain coal from three sources with the following attributes:

Coal Type	% Sulfur	% Hg	MWh/ton	Cost, $/ton
A	2.2	0.01	2	42
B	1.6	0.02	1.9	45
C	2.2	0.008	2.1	40

   The table indicates the percentage of pollutants, amount of energy (heat content) per ton and the cost of coal.

   (a) Find the optimal blending (procurement from) of supply from three sources in order to ensure that:

      i. The percent of Hg does not exceed 0.015% per ton of the blend.
      ii. Sulfur content does not exceed 1.8% per ton of the blend.
      iii. Sulfur content is equal to 1.8% per ton of the blend.

   (b) Find the optimum mixing strategy if:

      i. The Hg content cannot exceed 0.007% in the coal burned to produce 1 MWh.
      ii. The sulfur content cannot exceed 0.09% in the coal burned to produce 1 MWh.

2. Consider the problem of inter-regional flows in Section 4.8. The system operator is generally guided by nomograms for flows in certain branches and interfaces. Let Figure 4.22 represent nomograms that require the flows to be within the bounded region indicated. Introduce these constraints to the Excel Solver spreadsheet and obtain a solution to the problem. Compare this solution with that obtained in Section 4.8. Examine the sensitivity sheet.

3. For the problem of Section 4.9.2, formulate its dual. Examine the sensitivity sheet of the primal problem in Section 4.9.2 and compare it to the solution of the dual problem. What is the meaning of the dual variables in this problem?

4. What is the increased cost of transporting the goods in the problem of Section 4.9.1 if:

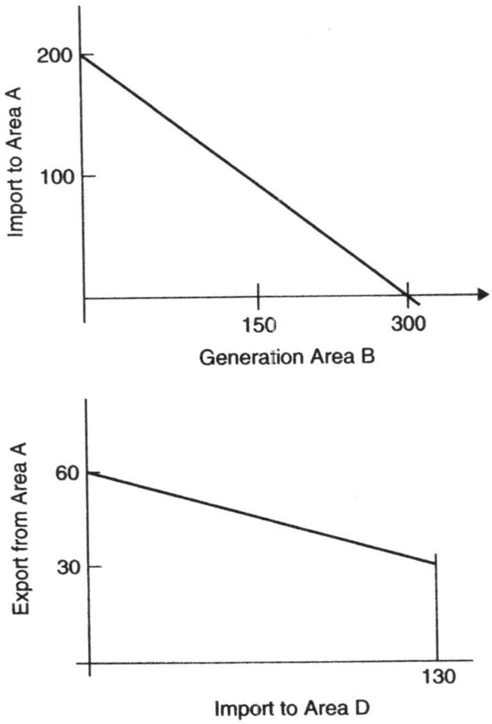

Figure 4.22. Inter-regional flow constraints.

(a) The path from warehouse 1 to destination 6 is unavailable.

(b) Warehouse 3 became inoperative.

5. Consider the problem of procuring gas to generate electricity while minimizing transportation cost in Section 4.9.2. If the regulations required the pollution not to exceed 250, what are the new procured quantities? What is the incremental cost of transportation? Consider the gas transporting entity that collects tariff. How can it maximize its profits by varying the tariff given the above situation? (This assumes that the entity is aware of all the above information.)

PART III

NONLINEAR OPTIMIZATION

CHAPTER 5

MATHEMATICAL BACKGROUND TO NONLINEAR PROGRAMS

5.1 FORM OF NONLINEAR OPTIMIZATION PROBLEM

Nonlinear optimization problems have a characterization similar to that of linear optimization problems previously discussed. They have a goal, or an objective function, to be minimized or maximized subject to some constraints. However, either the objective function or constraints, or both, can be nonlinear relationships. If there are no constraints, the problem becomes one of optimizing a nonlinear function. Such an exercise is called unconstrained optimization. If constraints are present, the procedure is called constrained optimization. The solution of nonlinear constrained problem is generally referred to as nonlinear programming.

The general nature of a nonlinear program is as follows.

Select m choice or design variables $x_1, x_2, \dots, x_m$ in order to minimize

$$\pi = f(x_1, x_2, \dots, x_m),$$

subject to n inequalities

$$g_1(x_1, x_2, \dots, x_m) \le r_1$$

$$g_2(x_1, x_2, \dots, x_m) \le r_2$$

$$\vdots$$

$$g_n(x_1, x_2, \dots, x_m) \le r_n$$

Optimization Principles, by Narayan S. Rau
ISBN 0-471-45130-4

and p equalities

$$h_1(x_1, x_2, \ldots, x_m) = k_1$$
$$h_2(x_1, x_2, \ldots, x_m) = k_2$$
$$\vdots$$
$$h_p(x_1, x_2, \ldots, x_m) = k_p$$

and

$$x_1, x_2, \ldots, x_m \geq 0.$$

This can be written in compact form as follows:
Minimize

$$f(\mathbf{x})$$

subject to

$$g(\mathbf{x}) \leq \mathbf{r},$$
$$h(\mathbf{x}) = \mathbf{k},$$
$$x_i \geq 0, \quad i = 1, \ldots, m.$$

The convention used in this book is that m represents the number of choice variables, n inequalities are represented as functions $g_1, \ldots, g_n$, and p equalities are represented as functions $h_1, \ldots, h_p$.

Despite the common characteristic of problem formulation, the nonlinearities introduce a complexity making a straightforward modification of any method of linear optimization unsuitable to solve nonlinear problems. Consequently, the solution algorithms are complex, and most algorithms for the solution of nonlinear programs are based on the Taylor's expansion of functions.

Our purpose in this chapter is first to discuss some mathematical preliminaries such as the gradient, Hessian, and the Taylor's expansion of functions. Subsequent chapters discuss solution methods for constrained and unconstrained optimization problems.

5.2 PRELIMINARIES

5.2.1 The Gradient

Let $f(x, y, z)$ be a differentiable function defined at every point in a certain region of space. The gradient of a function f, written as ∇f, for the three-dimensional case is given by the vector

$$\nabla f = \left[\frac{\partial f}{\partial x}, \frac{\partial f}{\partial y}, \frac{\partial f}{\partial z}\right]^T. \tag{5.1}$$

Our convention is to represent the gradient of a function as $\nabla f(\mathbf{x})$. The gradient is also represented as $f'(\mathbf{x})$.

The application of the above to multivariate function is obvious. For example, if an objective function is given by

$$\pi = x_1^3 - 2x_2,$$

the gradient of the objective function is given by

$$\nabla \pi = \begin{pmatrix} \frac{\partial \pi}{\partial x_1} \\ \frac{\partial \pi}{\partial x_2} \end{pmatrix} = \begin{pmatrix} 3x_1^2 \\ -2 \end{pmatrix},$$

and the value of the gradient vector at $x_1 = 1, \ x_2 = 0$ is

$$(\nabla \pi)_{(1,0)} = \begin{pmatrix} 3 \\ -2 \end{pmatrix}.$$

In two-dimensional case, the component of any vector $\mathbf{x}$ along the direction of another vector at angle θ to it is $\mathbf{x}\cos\theta$. For multidimensions, the component of a vector $\mathbf{x}$ along $\mathbf{y}$ can be expressed as $\mathbf{x}^T\mathbf{y}/\|\mathbf{y}\|$. For example, if $\mathbf{x} = (3, 3)^T$ and $\mathbf{y} = (1, -0.5)^T$, component of $\mathbf{x}$ in the direction of $\mathbf{y}$ is $(3, 3)\begin{pmatrix} 1 \\ -0.5 \end{pmatrix} \Big/ \|\mathbf{y}\|$. Since $\|\mathbf{y}\| = (1 + 0.025)^{\frac{1}{2}} = 1.118$, we obtain a value of 1.3416 for this component.

The ith element of the gradient vector of a function f, $\partial f/\partial x_i$, is the first derivative of the function π along the direction of the x_i axis.

From an extension of above concepts, we note that the component of the gradient vector ∇f in the direction of a unit vector $\mathbf{a}$ is given by the dot product $\nabla f \cdot \mathbf{a}$. Consequently, for any unnormalized vector $\mathbf{p}$, the component of the gradient along its direction is given by $\nabla f \cdot \mathbf{p}/\|\mathbf{p}\|^2$. This is called the variation of the function f in the direction of $\mathbf{p}$, or the *directional derivative*[1] of f.

[1]The gradient vector ∇f is a vector perpendicular to the surface $f(x, y, z) = c$. The proof is trivial: Since the vector $[x, y, z]^T$ represents the radius vector at the point (x, y, z) on the surface of f, the vector $[dx, dy, dz]^T$ represents the change in the radius of curvature, $d\mathbf{r}$. By definition, $d\mathbf{r}$ is a vector normal to f. Direction normal to f represents the direction of a contour line. Therefore, $\mathbf{dr}$ is along a contour line. Since the value of a function is constant along a contour line, the change in the value of the function f (total variation) along this vector normal to f is zero. The directional derivative along

It is intuitively obvious from the above that:

1. The gradient vector represents the direction of steepest increase in function value (*called steepest ascent*).
2. The gradient vector at any point $\mathbf{x}$ indicates the maximum rate of change of a function f at that point.

For a formal proof of these two attributes of the gradient vector, see references (Rao, 1996) and (Pierre, 1969). We make use of this property in a subsequent section to develop so called *descent methods* for the solution of minimization problems.

Examples: Figure 5.1 shows the contours of the function $\pi = x_1^3 - 2x_2$ along with the gradient vector computed earlier (shown to half-scale) which is given by $x_1 = 3,\ x_2 = -2$ at (1,0). The component of this vector along the x_1 axis is 3 which is equal to the value of the first element of gradient vector, $\partial\pi/\partial x_1$, at (1,0). Since a unit vector in the direction of x_1 is $(1, 0)$, the directional derivative along x_1 can also be obtained as the product of vectors $(3, -2) \cdot \begin{pmatrix} 1 \\ 0 \end{pmatrix} = 3$. Consequently, the value of π increases if we move in the direction of the gradient, or reduces if we move in an opposite direction.

As another example, consider the contours of simple concentric circles given by the function $\pi = x^2 + y^2$ which has $\nabla\pi = [2x, 2y]^T$. The value of this gradient vector at a point $A(3, 0)$ is $\begin{pmatrix} 6 \\ 0 \end{pmatrix}$, which represents a vector of Euclidean length 6 pointing to the right along the x axis. The value of π at $A(3, 0) = 9$, and its value a small distance away at $A'(3.01, 0) = 9.0601$. The increase in the value of the function is $d\pi = 9.0601 - 9 = 0.0601$ in a distance of $\Delta x = 0.01$. Hence the rate of increase is given by $d\pi/\Delta x = 0.0601/0.01 = 6.01$ which corresponds closely to the value of the gradient.[2]

5.2.2 Differential of a Function

First Differential. The above has indicated that the dot product of the gradient vector and another vector gives the directional derivative of the function. An obvious corollary of this is that the product of the gradient vector and the perturbation of variables $(\Delta x_1, \Delta x_2, \ldots, \Delta x_n)$ in an n-dimensional space gives the

$\mathbf{dr}$ is expressed as the dot product of vectors

$$df = \left(\frac{df}{dx}, \frac{df}{dy}, \frac{df}{dz}\right)\begin{pmatrix} dx \\ dy \\ dz \end{pmatrix} = 0.$$

The first vector in the above expression being the transpose of the gradient vector proves that it is orthogonal to $d\mathbf{r}$, the change in the radius vector, and therefore to the surface of f.

[2]The small difference is due to the nonlinearity of the function π. If Δx is made infinitesimally small, in the limit, the value for the gradient is exactly equal to 6.

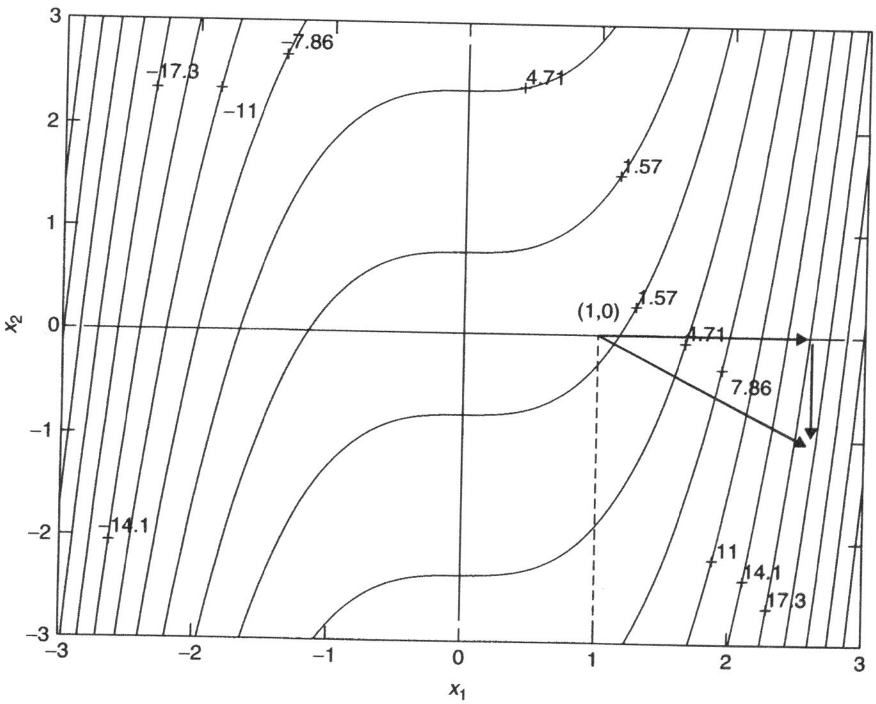

Figure 5.1. Contour of function $\pi = x_1^3 - 2x_2$.

first differential of the function π—that is, by how much the value of the function changes for changes in $(\Delta x_1, \Delta x_2, \ldots, \Delta x_n)$.

Then, for a function $f(x_1, x_2, \ldots, x_n)$ and for a change in the value of $\Delta \mathbf{x} = (\mathbf{x} - \mathbf{x}_n)$, one has

$$[\nabla \mathbf{f}]^T \cdot [\mathbf{x} - \mathbf{x_n}] = \left[\frac{\partial f}{\partial x_1}, \frac{\partial f}{\partial x_2} \cdots \frac{\partial f}{\partial x_n}\right] \begin{bmatrix} x - x_1 \\ x - x_2 \\ \vdots \\ x - x_n \end{bmatrix}$$

$$= \frac{\partial f}{\partial x_1}(x - x_1) + \frac{\partial f}{\partial x_2}(x - x_2) + \cdots + \frac{\partial f}{\partial x_n}(x - x_n).$$

If $x - x_n = h_n$ for a vector $\mathbf{h}$ of perturbations $[h_1, h_2, \ldots, h_n]$, we get

$$\nabla \mathbf{f}^{\mathbf{T}} \cdot \mathbf{h} = \frac{\partial f}{\partial x_1} h_1 + \frac{\partial f}{\partial x_2} h_2 + \cdots + \frac{\partial f}{\partial x_n} h_n. \tag{5.2}$$

For two dimensions, this reduces to

$$\nabla \mathbf{f}^{\mathrm{T}} \cdot \mathbf{h} = \frac{\partial f}{\partial x_1} h_1 + \frac{\partial f}{\partial x_2} h_2. \tag{5.3}$$

Note that above equations for the first differential represent a scalar.

pth Differential of f. If a function $f(x_1, x_2, \ldots, x_n)$ is continuous at $\mathbf{x}^*$ and if all partial differentials up to order p exist, the pth differential is given by the polynomial of p summations as

$$d^p f(\mathbf{x}^*) = \overbrace{\sum_{i=1}^{n} \sum_{j=1}^{n} \cdots \sum_{q=1}^{n}}^{p \text{ summations}} h_i h_j \cdots h_q \frac{\partial^p f(\mathbf{x}^*)}{\partial x_i \partial x_j \cdots \partial x_q}. \tag{5.4}$$

In the above equation, note that $p \geq 1$ and that there are p summations with one h_p associated with each summation.

Illustrations. If $p = 1, n = 1$, at any point $\mathbf{x}^*$ we get?

$$df(\mathbf{x}^*) = h_1 \frac{\partial f(\mathbf{x}^*)}{\partial x_1}. \tag{5.5}$$

This is the familiar expression for the change in value of a function expressed as the slope times the displacement h.

If $p = 1, n = 2$, we get

$$df(\mathbf{x}^*) = df(x_1^*, x_2^*) = \sum_{i=1}^{2} h_i \frac{\partial f(\mathbf{x}^*)}{\partial x_i}, \tag{5.6}$$

$$= \frac{\partial f(\mathbf{x}^*)}{\partial x_1} h_1 + \frac{\partial f(\mathbf{x}^*)}{\partial x_2} h_2. \tag{5.7}$$

Equation (5.7) is identical to (5.3).

If $p = 2, n = 2$, we get

$$d^2 f(\mathbf{x}^*) = d^2 f(x_1^*, x_2^*) = \sum_{i=1}^{2} \sum_{j=1}^{2} h_i h_j \frac{\partial^2 f(\mathbf{x}^*)}{\partial x_i \partial x_j} \tag{5.8}$$

$$= h_1^2 \frac{\partial^2 f(\mathbf{x}^*)}{\partial x_1^2} + h_2^2 \frac{\partial^2 f(\mathbf{x}^*)}{\partial x_2^2} + 2 h_1 h_2 \frac{\partial^2 f(\mathbf{x}^*)}{\partial x_1 \partial x_2}. \tag{5.9}$$

If $p = 2, n = 3$, we obtain

$$d^2 f(\mathbf{x}^*) = d^2 f(x_1^*, x_2^*, x_3^*) = \sum_{i=1}^{3}\sum_{j=1}^{3} h_i h_j \frac{\partial^2 f(\mathbf{x}^*)}{\partial x_i \partial x_j} \tag{5.10}$$

$$= h_1^2 \frac{\partial^2 f(\mathbf{x}^*)}{\partial x_1^2} + h_2^2 \frac{\partial^2 f(\mathbf{x}^*)}{\partial x_2^2} + h_3^2 \frac{\partial^2 f(\mathbf{x}^*)}{\partial x_3^2} + 2h_1 h_2 \frac{\partial^2 f(\mathbf{x}^*)}{\partial x_1 \partial x_2}$$
$$+ 2h_2 h_3 \frac{\partial^2 f(\mathbf{x}^*)}{\partial x_2 \partial x_3} + 2h_1 h_3 \frac{\partial^2 f(\mathbf{x}^*)}{\partial x_1 \partial x_3}. \tag{5.11}$$

Example: Calculate the second differential of the function $x_2 x_3^2 + x_1 e^{x_2}$ at a point $\mathbf{x}^* = [1, 0, -1]^T$.

Solution: The second differential is given by (5.11). We substitute the partial derivatives of $f = x_2 x_3^2 + x_1 e^{x_2}$ into this equation to obtain

$$\left\|\left[h_1^2(0) + h_2^2(x_1 e^{x_2}) + h_3^2(2x_2) + 2h_1 h_2(e^{x_2}) + 2h_2 h_3(2x_3) + 2h_1 h_3(0)\right]\right\|_{[1,0,-1]^T}. \tag{5.12}$$

A substitution of $x_1 = 1, x_2 = 0, x_3 = -1$ into the above gives the second differential as

$$d^2 f(\mathbf{x}) = h_2^2 + 2h_3^2 + 2h_1 h_2 - 4h_2 h_3. \tag{5.13}$$

An observation is that given the value of a function (and its derivatives) at a point ($[1, 0, -1]^T$ in this case), its second variation can be computed in terms of perturbations or movements from that point. Subsequently, we will show that this theme is embedded in the Taylor expansion, which is exploited to search for the extrema of a function.

5.2.3 The Jacobian Matrix

A collection of gradient vectors, generally used for the solution of a set of equations, is called the Jacobian matrix or the Jacobian. Consider a system of n nonlinear equations in n unknowns $x_1, x_2, \ldots, x_n$ as

$$\begin{aligned} f_1(x_1, x_2, \ldots x_n) &= k_1 \\ f_2(x_1, x_2, \ldots x_n) &= k_2 \\ \vdots \quad &\vdots \quad \vdots \\ f_n(x_1, x_2, \ldots x_n) &= k_n. \end{aligned} \tag{5.14}$$

The Jacobian matrix for the system of equations is given by

$$\mathbf{J} = \begin{bmatrix} \dfrac{\partial f_1}{\partial x_1} & \dfrac{\partial f_1}{\partial x_2} & \cdots & \dfrac{\partial f_1}{\partial x_n} \\ \dfrac{\partial f_2}{\partial x_1} & \dfrac{\partial f_2}{\partial x_2} & \cdots & \dfrac{\partial f_2}{\partial x_n} \\ \vdots & \vdots & \vdots & \vdots \\ \dfrac{\partial f_n}{\partial x_1} & \dfrac{\partial f_n}{\partial x_2} & \cdots & \dfrac{\partial f_n}{\partial x_n} \end{bmatrix}. \tag{5.15}$$

In (5.15), each row represents the transpose of the gradient vector of the individual equations in the system of (5.14). This point will be discussed later in Section 5.3.2 to show that the solution of a system of nonlinear equations is a gradient search procedure.

5.2.4 The Hessian Matrix

The Hessian matrix associated with a function $f(x_1, x_2, \ldots, x_n)$ represented by $\mathbf{H}$ is

$$\mathbf{H} = \begin{bmatrix} \dfrac{\partial^2 f}{\partial x_1^2} & \dfrac{\partial^2 f}{\partial x_1 \partial x_2} & \cdots & \dfrac{\partial^2 f}{\partial x_1 \partial x_n} \\ \dfrac{\partial^2 f}{\partial x_2 \partial x_1} & \dfrac{\partial^2 f}{\partial x_2^2} & \cdots & \dfrac{\partial^2 f}{\partial x_2 \partial x_n} \\ \vdots & \vdots & \vdots & \vdots \\ \dfrac{\partial^2 f}{\partial x_n \partial x_1} & \dfrac{\partial^2 f}{\partial x_n \partial x_2} & \cdots & \dfrac{\partial f}{\partial x_n^2} \end{bmatrix}. \tag{5.16}$$

The use of the Hessian matrix in nonlinear optimization will be discussed in later sections.

5.2.5 Directional Second Derivative

Note from the Hessian matrix that the component of the second derivative along the x_1 axis is given by $\partial^2 f/\partial x_1^2$. The component of the second derivative along arbitrary direction $\mathbf{p}$ is given by

$$\frac{\mathbf{p}^T \mathbf{H} \mathbf{p}}{\| \mathbf{p} \|^2}.$$

For example, consider $f(\mathbf{x}) = x_1^2 + 2x_2^2$, which has $\nabla f(\mathbf{x}) = [2x_1, 4x_2]^T$ and $\mathbf{H} = \begin{bmatrix} 2 & 0 \\ 0 & 4 \end{bmatrix}$. The second derivative along the direction $[2, -1]^T$ is given by

$$\frac{(2, -1)\begin{bmatrix} 2 & 0 \\ 0 & 4 \end{bmatrix}\begin{pmatrix} 2 \\ -1 \end{pmatrix}}{\sqrt{5}} = \frac{12}{\sqrt{5}}.$$

The directional second derivative is a scalar.

5.3 TAYLOR SERIES EXPANSION OF FUNCTIONS

The Taylor expansion for a function $f(\mathbf{x})$ about a point $\mathbf{x}^*$ is given in terms of the differentials as

$$f(\mathbf{x}) = f(\mathbf{x}^*) + df(\mathbf{x}^*) + \frac{1}{2!}d^2 f(\mathbf{x}^*) + \frac{1}{3!}d^3 f(\mathbf{x}^*) + \cdots + \frac{1}{n!}d^n f(\mathbf{x}^*) + R_n(\mathbf{x}^*, \mathbf{h}), \tag{5.17}$$

in which the last term called the *remainder* is given by

$$R_n(\mathbf{x}^*, \mathbf{h}) = \frac{1}{(n+1)}d^{n+1} f(\mathbf{x}^* + \theta\mathbf{h}) \tag{5.18}$$

where $0 < \theta < 1$, and $\mathbf{h} = (\mathbf{x} - \mathbf{x}^*)$.

Example: Find the second-order Taylor series approximation for

$$f(x_1, x_2, x_3) = x_1 e^{x_2} + x_2 x_3^2, \tag{5.19}$$

about the point $\mathbf{x}^* = [1, 0, -1]^T$.

Solution: The second-order approximation is given by

$$f(\mathbf{x}) = f|_{[1,0,-1]^T} + df|_{[1,0,-1]^T} + \frac{1}{2!}d^2 f|_{[1,0,-1]^T} + R_2\left(\begin{pmatrix} 1 \\ 0 \\ -1 \end{pmatrix} + \theta h\right). \tag{5.20}$$

The first term is given by

$$f|_{[1,0,-1]^T} = 1e^0 = 1.$$

The second term is

$$df|_{[1,0,-1]^T} = [h_1 \frac{\partial f}{\partial x_1} + h_2 \frac{\partial f}{\partial x_2} + h_3 \frac{\partial f}{\partial x_3}]|_{[1,0,-1]^T}. \tag{5.21}$$

Observe that the right-hand side of the above is the component of the gradient in the direction of the vector $[h_1, h_2, h_3]^T$. It is also called the directional derivative (gradient) of the function along this direction. Substituting partial derivatives of (5.19) into (5.21), we get

$$df = [h_1 e^{x_2} + h_2(x_3^2 + x_1 e^{x_2}) + h_3 2x_2 x_3]|_{[1,0,-1]^T}, \tag{5.22}$$

which upon substitution of 1, 0, and −1 for x_1, x_2, and x_3 gives

$$df = h_1 + 2h_2. \tag{5.23}$$

Next, the second differential of (5.19) has already been computed earlier in (5.13) as

$$d^2 f(\mathbf{x}) = h_2^2 + 2h_3^2 + 2h_1 h_2 - 4h_2 h_3. \tag{5.24}$$

Thus the Taylor's expansion for the function of (5.19) is given by

$$f(\mathbf{x}) = 1 + h_1 + 2h_2 + h_2^2 + 2h_3^2 + 2h_1 h_2 - 4h_2 h_3, \tag{5.25}$$

in which $h_1 = x_1 - 1$, $h_2 = x_2 - 0 = x_2$, and $h_3 = x_3 + 1$.

Finally, we can calculate the remainder since we know the equation for the function. Consider a change in value of +0.1 for the components of $\mathbf{x}^*$. Then we have $\mathbf{x} = [1.1, 0.1, -0.9]^T$. The value of (5.19) at this $\mathbf{x}$ turns out to be 1.29668. If we evaluate (5.25) with $h_1 = 0.1$, $h_2 = 0.1$, and $h_3 = 0.1$, we obtain 1.31. Hence the remainder, or the truncation error, is $1.31 - 1.39668 = -0.01331$.

Clearly, magnitude of the remainder decreases as the number of terms in Taylor's expansion are increased, or if h is decreased.

5.3.1 Compact Form of Taylor's Expansion

Most algorithms in practice use the first three terms of the Taylor's expansion, up to terms including the second difference. Despite such a truncation, one can device very efficient algorithms to solve practical problems.

Ignoring the remainder, the first three terms of the Taylor's expansion in (5.17) are

$$f(\mathbf{x}) = f(\mathbf{x}^*) + df(\mathbf{x}^*) + \frac{1}{2!} d^2 f(\mathbf{x}). \tag{5.26}$$

Substituting the expressions for the first and second differences, we get

$$f(\mathbf{x}) = f(\mathbf{x}^*) + \frac{\partial f(\mathbf{x}^*)}{\partial x_1} h_1 + \frac{\partial f(\mathbf{x}^*)}{\partial x_2} h_2 + \frac{1}{2!} \times \left(h_1^2 \frac{\partial^2 f(\mathbf{x}^*)}{\partial x_1^2} + h_2^2 \frac{\partial^2 f(\mathbf{x}^*)}{\partial x_2^2} + 2h_1 h_2 \frac{\partial^2 f(\mathbf{x}^*)}{\partial x_1 \partial x_2} \right), \tag{5.27}$$

where $\mathbf{h}=(\mathbf{x} - \mathbf{x}^*)$; that is, $[h_1, h_2] = [(x_1 - x_1^*), (x_2 - x_2^*)]$.

The above expansion can be written as

$$f(\mathbf{x}^* + \mathbf{h}) = f(\mathbf{x}^*) + \mathbf{h}\nabla f(\mathbf{x}^*) + \frac{1}{2}\mathbf{h}^T \mathbf{H}(\mathbf{x}^*)\mathbf{h}, \tag{5.28}$$

where

$$\mathbf{H} = \begin{bmatrix} \dfrac{\partial^2 f(\mathbf{x}^*)}{\partial x_1^2} & \dfrac{\partial^2 f(\mathbf{x}^*)}{\partial x_1 \partial x_2} \\ \dfrac{\partial^2 f(\mathbf{x}^*)}{\partial x_1 \partial x_2} & \dfrac{\partial^2 f(\mathbf{x}^*)}{\partial x_2^2} \end{bmatrix}. \tag{5.29}$$

It is expected of the reader to confirm that this indeed is the case by working out details of the algebra, if necessary.

Example: Obtain the first three terms of the Taylor series expansion for

$$f(x_1, x_2, x_3) = x_2^2 x_3 + x_1 x_2 e^{x_3}$$

at a point $\mathbf{x}^* = [1, 2, -1]^T$, using the relation

$$f(\mathbf{x}) = f(\mathbf{x}^*) + \nabla f(\mathbf{x}^*)(\mathbf{x} - \mathbf{x}^*) + \frac{1}{2}(\mathbf{x} - \mathbf{x})^T H(\mathbf{x}^*)(\mathbf{x} - \mathbf{x}).$$

Solution: The gradient vector is

$$\nabla f(\mathbf{x}) = \begin{bmatrix} x_2 e^{x_3} \\ 2x_2 x_3 + x_1 e^{x_3} \\ x_2^2 + x_1 x_2 e^{x_3} \end{bmatrix}.$$

The Hessian matrix (a symmetric matrix) is given by

$$\mathbf{H} = \begin{bmatrix} \dfrac{\partial^2 f(\mathbf{x}^*)}{\partial x_1^2} & \dfrac{\partial^2 f(\mathbf{x}^*)}{\partial x_1 \partial x_2} & \dfrac{\partial^2 f(\mathbf{x}^*)}{\partial x_1 \partial x_3} \\ \dfrac{\partial^2 f(\mathbf{x}^*)}{\partial x_1 \partial x_2} & \dfrac{\partial^2 f(\mathbf{x}^*)}{\partial x_2^2} & \dfrac{\partial^2 f(\mathbf{x}^*)}{\partial x_2 \partial x_3} \\ \dfrac{\partial^2 f(\mathbf{x}^*)}{\partial x_3 \partial x_1} & \dfrac{\partial^2 f(\mathbf{x}^*)}{\partial x_3 \partial x_2} & \dfrac{\partial^2 f(\mathbf{x}^*)}{\partial x_3^2} \end{bmatrix} = \begin{bmatrix} 0 & e^{x_3} & x_2 e^{x_3} \\ e^{x_3} & 2x_3 & 2x_2 + x_1 e^{x_3} \\ x_2 e^{x_3} & 2x_2 + x_1 e^{x_3} & x_1 x_2 e^{x_3} \end{bmatrix}.$$

Assume that we are at a point (1, 2, −1) and that we want an expression to obtain the value of the function at (1.5, 2.5, −0.9). Then, we get[3]

$$(\mathbf{x} - \mathbf{x}^*) = \begin{bmatrix} h_1 \\ h_2 \\ h_3 \end{bmatrix} = \begin{bmatrix} 0.5 \\ 0.5 \\ 0.1 \end{bmatrix},$$

$$\nabla f^T(\mathbf{x}^*)(\mathbf{x}^* - \mathbf{x}_i) = |[x_2 e^{x_3}, \quad 2x_2 x_3 + x_1 e^{x_3}, \quad x_2^2 + x_1 x_2 e^{x_3}]|_{[1,2,-1]^T} \begin{bmatrix} h_1 \\ h_2 \\ h_3 \end{bmatrix},$$

which simplifies to

$$\nabla f^T(\mathbf{x}^*)(\mathbf{x}^* - \mathbf{x}_i) = 0.73575 h_1 - 3.632 h_2 + 4.7357 h_3.$$

Proceeding in a similar manner, we compute

$$\frac{1}{2}(\mathbf{x} - \mathbf{x}^*)^T \mathbf{H} (\mathbf{x} - \mathbf{x}^*)$$

$$= \frac{1}{2}\left\{ (h_1, h_2, h_3) \begin{bmatrix} 0 & e^{x_3} & x_2 e^{x_3} \\ e^{x_3} & 2x_3 & 2x_2 + x_1 e^{x_3} \\ x_2 e^{x_3} & 2x_2 + x_1 e^{x_3} & x_1 x_2 e^{x_3} \end{bmatrix} \begin{bmatrix} h_1 \\ h_2 \\ h_3 \end{bmatrix} \right\}.$$

For $(x_1, x_2, x_3) = (1, 2, -1)$, the above simplifies to

$$\frac{1}{2}(\mathbf{x} - \mathbf{x}^*)^T \mathbf{H} (\mathbf{x} - \mathbf{x}^*)$$

$$= \frac{1}{2}(-2h_2^2 - 0.73575 h_3^2 + 0.73575 h_1 h_2 + 8.73575 h_2 h_3 + 1.4715 h_1 h_3).$$

[3]The reader is expected to check the arithmetic. The best way of doing so is by setting up an Excel spreadsheet or by using MATLAB.

As a result, we obtain the following for $f(\mathbf{x})$ in terms of h:

$$f(\mathbf{x}) = 0.73575h_1 - 3.632h_2 + 4.7357h_3 + 0.5(-2h_2^2 - 0.73575h_3^2 + 0.73575h_1h_2 + 8.73575h_2h_3 + 1.4715h_1h_3).$$

Substituting values of $\mathbf{h} = [0.5, 0.5, 0.1]^T$ in the above expression, we get -4.1379 as the function value at (1.5, 2.5, -0.9). The value of the function at this point computed by using $f(x_1, x_2, x_3) = x_2^2x_3 + x_1x_2e^{x_3}$ is -4.10036.

The reader may find it advantageous to set up an Excel spreadsheet to perform and confirm the above arithmetic.

5.3.2 Solution of Nonlinear Equations Using the Jacobian Matrix

Consider a set of nonlinear equations for n variables $x_1, x_2, \ldots, x_n$ given by

$$f_1(x_1, x_2, \ldots, x_n) = y_1 \tag{5.30}$$

$$f_2(x_1, x_2, \ldots, x_n) = y_2 \tag{5.31}$$

$$\vdots = \vdots$$

$$f_n(x_1, x_2, \ldots, x_n) = y_n. \tag{5.32}$$

For an initial estimated solution $\mathbf{x}^0 = x_1^0, x_2^0, \ldots, x_n^0$, the two-term Taylor expansion to the set of equation for perturbations $\mathbf{x} = (\Delta x_1, \Delta x_2, \ldots, \Delta x_n)$ is

$$f_1(\mathbf{x}^0 + \Delta\mathbf{x}) = f_1(\mathbf{x}^0) + \nabla f_1^T(\mathbf{x}^0)\ \Delta x_1 = y_1 \tag{5.33}$$

$$f_2(\mathbf{x}^0 + \Delta\mathbf{x}) = f_2(\mathbf{x}^0) + \nabla f_2^T(\mathbf{x}^0)\ \Delta x_2 = y_2 \tag{5.34}$$

$$\vdots = \vdots \tag{5.35}$$

$$f_n(\mathbf{x}^0 + \Delta\mathbf{x}) = f_n(\mathbf{x}^0) + \nabla f_n^T(\mathbf{x}^0)\ \Delta x_n = y_n. \tag{5.36}$$

Therefore, we have

$$y_1 - f_1(\mathbf{x}^0) = \nabla f_1^T\ \Delta x_1 \tag{5.37}$$

$$y_2 - f_2(\mathbf{x}^0) = \nabla f_2^T\ \Delta x_2 \tag{5.38}$$

$$\vdots = \vdots \tag{5.39}$$

$$y_n - f_n(\mathbf{x}^0) = \nabla f_n^T\ \Delta x_n, \tag{5.40}$$

which can be written as

$$\mathbf{m} = \mathbf{J}\Delta\mathbf{x}, \tag{5.41}$$

where **m** represents the vector of errors given by the difference between the desired values $y_1, y_2, \ldots, y_n$ on the left-hand side of (5.37)–(5.40) and the computed values at the estimated value for **x**, and **J** is the Jacobian matrix as in (5.15). The above is usually written as

$$\Delta \mathbf{x} = \mathbf{J}^{-1}\mathbf{m}. \tag{5.42}$$

The iterative method to obtain a solution to the set of equations is to correct the values of $\mathbf{x}^{(i)}$ of the ith iteration by adding $\Delta\mathbf{x}$ obtained from (5.42) to get $\mathbf{x}^{(i+1)} = \mathbf{x}^{(i)} + \Delta\mathbf{x}$. Then the Jacobian matrix is recomputed and (5.42) is used to find the new error and the update for **x**. The process is continued until the error is below a specified tolerance.

5.4 EXTREME POINTS OF MULTIVARIATE FUNCTIONS

In Section 3.1 we have discussed the condition of extremum for univariate functions. For multivariate functions, the necessary and sufficient conditions are discussed below.

Necessary Condition. If $f(\mathbf{x})$ has an extreme point (minimum or maximum) at $\mathbf{x} = \mathbf{x}^*$, if the partial derivatives exist at $\mathbf{x} = \mathbf{x}^*$, then

$$f'(\mathbf{x}) = \frac{\partial f}{\partial x_1}(\mathbf{x}^*) = \frac{\partial f}{\partial x_2}(\mathbf{x}^*) = \cdots \frac{\partial f}{\partial x_n}(\mathbf{x}^*) = 0. \tag{5.43}$$

Another way of saying this is that

$$\nabla f(\mathbf{x}) = 0. \tag{5.44}$$

Sufficient Condition. The sufficient condition to ensure that a stationary point $\mathbf{x}^*$ is an extreme point is as follows:

1. The Hessian matrix of $f(\mathbf{x})$ (matrix of second partial derivatives) evaluated at $\mathbf{x}^*$ is positive definite when $\mathbf{x}^*$ is a relative minimum, or
2. The Hessian matrix of $f(\mathbf{x})$ is negative definite when $\mathbf{x}^*$ is a relative minimum.

For a proof of this[4] we refer the reader to Rao (1996), Pierre (1969), Chiang (1984), and Gill et al. (1981).

[4]The general nature of the proof in the references is as follows. If an extreme point is a minimum, it should satisfy relations $f(\mathbf{x}^* + h) - f(\mathbf{x}) > 0$, and $f(\mathbf{x}^* - h) - f(\mathbf{x}) > 0$. By Taylor's expansion we have $f(\mathbf{x}^* + h) - f(\mathbf{x}^*) = \sum_{i=1}^{n} h \frac{\partial f}{\partial x_i}(\mathbf{x}^*) + \frac{1}{2!} d^2 f(\mathbf{x}^* + \theta h);\ 0 \le \theta \le 1$. The first term on the right-hand side of the Taylor's expansion dominates the second remainder term. Hence, the sign of $f(\mathbf{x}^* + h) - f(\mathbf{x}^*)$ for a given sign of $\frac{\partial f}{\partial x_i}(\mathbf{x}^*)$ changes with the sign of h, violating the requirement

Test for Positive or Negative Definiteness. It has been said in Section 2.6.1 that a matrix is positive definite if all its eigenvalues are positive, negative definite if all eigenvalues are negative, and indefinite if some eigenvalues are positive and some are negative, or if some are zero. Recall from Section 2.6.1 that one of the tests for positive (negative) definiteness of a matrix $\mathbf{A}$ is as follows: for a nonzero vector $\mathbf{x}$, $\mathbf{x}^T\mathbf{A}\mathbf{x} > 0$ (or < 0).

Another test for definiteness is based on the properties of principal minors of the determinant.[5] We state the test without proof [see Gantmacher (1950) and Chiang (1984) for proof]: For a matrix

$$\mathbf{H} = \begin{bmatrix} h_{11} & h_{12} & \cdots & h_{1n} \\ h_{21} & h_{22} & \cdots & h_{2n} \\ \vdots & \vdots & \vdots & \vdots \\ h_{n1} & h_{n2} & \cdots & h_{nn} \end{bmatrix} \tag{5.45}$$

the necessary and sufficient condition for *positive definiteness* is that the determinants of the principal minors

$$|D_1| = h_{11}, \qquad |D_2| = \begin{vmatrix} h_{11} & h_{12} \\ h_{21} & h_{22} \end{vmatrix}, \qquad \cdots |D_n| = \begin{vmatrix} h_{11} & h_{12} & \cdots & h_{1n} \\ h_{21} & h_{22} & \cdots & h_{2n} \\ \vdots & \vdots & \vdots & \vdots \\ h_{n1} & h_{n2} & \cdots & h_{nn} \end{vmatrix}$$

are all positive. The condition corresponding to the necessary and sufficient condition for *negative definiteness* is that the principal minors alternate in sign, that is,

$$|D_1| < 0, \qquad |D_2| > 0, \qquad |D_3| < 0, \ldots, \qquad \text{etc.,}$$

such that all even-numbered principal minors are positive and all odd-numbered principal minors are negative.

for a minima. The same violation of the required sign can be seen at a maxima when the sign of $\frac{\partial f}{\partial x_i}(\mathbf{x}^*)$ is < 0. Consequently, $\frac{\partial f}{\partial x_i}(\mathbf{x}^*) = 0 : \ i = 1, n$.

The argument for the sufficient condition proceeds in a similar fashion assuming $\frac{\partial f}{\partial x_i}(\mathbf{x}^*)$ to be zero. Consider the above Taylor's expansion written as $f(\mathbf{x}^* + h) - f(\mathbf{x}^*) = \sum_{i=1}^{n} h \frac{\partial f}{\partial x_i}(\mathbf{x}^*) + \frac{1}{2!}\sum_{i=1}^{n}\sum_{j=1}^{n} h_i h_j \frac{\partial^2 f}{\partial x_i \partial x_j}|_{\mathbf{x}=\mathbf{x}^*+\theta h}$; $0 \le \theta \le 1$. The first right-hand term is zero from the first-order condition, and the product $h_i h_j$ is always positive. Hence the sign of the left-hand side is determined by the partial second-derivative terms. This term can be written as $\frac{1}{2}h^T\mathbf{H}h$ (see (5.3.1)), which is positive if H is positive definite (see Section 2.6.1), indicating a minimum (and vice versa). If the Hessian is zero, a saddle point is indicated.

[5] A quadratic function is said to be positive definite if the Hessian is positive definite.

Example: The matrix

$$\begin{bmatrix} 1 & -2 & 0 \\ -1 & 8 & -2 \\ 0 & -2 & 4 \end{bmatrix}$$

is positive definite because

$$1 > 0, \qquad \begin{vmatrix} 1 & -2 \\ -1 & 8 \end{vmatrix} = 6 > 0, \qquad \begin{vmatrix} 1 & -2 & 0 \\ -1 & 8 & -2 \\ 0 & -2 & 4 \end{vmatrix} = 12 > 0.$$

However, the matrix

$$\begin{bmatrix} 2 & 3 & 2 \\ 3 & 4 & -1 \\ 0 & 4 & -4 \end{bmatrix}$$

is neither positive nor negative definite because

$$2 > 0, \qquad |D_2| = \begin{vmatrix} 2 & 3 \\ 3 & 4 \end{vmatrix} = -1 < 0,$$

which violates the condition for either positive or negative definiteness.

5.4.1 Saddle Point

If the Hessian matrix is neither positive definite nor negative definite at a point ($\overline{\mathbf{x}}$) where the first-order condition is satisfied, such a point is called the saddle point.

For example, consider the function

$$f(x_1, x_2) = 0.6x_1^2 - 0.8x_2^2$$

which is shown plotted in Figure 5.2 using MATLAB.

The first derivatives are

$$\frac{\partial f}{\partial x_1} = 1.2x_1 \quad \text{and} \quad \frac{\partial f}{\partial x_2} = -1.6x_2.$$

The first-order condition requires these to be zero, which is the case at $(\overline{x}_1, \overline{x}_2) = (0, 0)$. The Hessian matrix at $(\overline{x}_1, \overline{x}_2)$ is given by

$$H = \begin{bmatrix} 1.2 & 0 \\ 0 & -1.6 \end{bmatrix},$$

which is neither positive definite nor negative definite (since the determinant of the first minor is 1.2, and that of the second minor is 1.2 times -1.6, which equals -1.92).

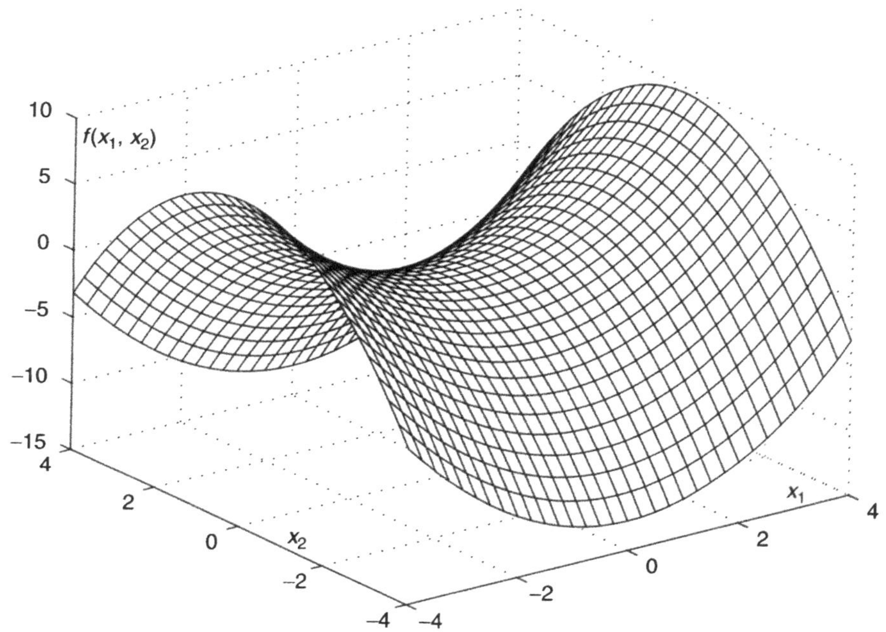

Figure 5.2. Example of saddle point.

The characteristics of the saddle point can be seen in Figure 5.2. The function $f(x_1, \overline{x}_2) = f(x_1, 0)$ has a relative minimum and the function $f(\overline{x}_1, x_2) = f(0, x_2)$ has a relative maximum at the saddle point $(\overline{x}_1, \overline{x}_2) = (0, 0)$. Another way of saying this is that we observe a relative minimum proceeding along the direction of one principal axis, and a relative maximum along the direction of another principal axis.

The principal minors test for positive definiteness is applicable to multidimensional functions as well.

Example: Find the extreme points of the function

$$f(x_1, x_2) = -5x_1^3 - 5x_1^2 + x_2^3 + 4x_2^2 + 8.$$

Solution: Figure 5.3 shows a three-dimensional plot of this function obtained using MATLAB. The first-order necessary conditions for the existence of an extremum are

$$\frac{\partial f}{\partial x_1} = -15x_1^2 - 10x_1 = -x_1(15x_1 - 10) = 0, \tag{5.46}$$

$$\frac{\partial f}{\partial x_2} = 3x_2^2 + 8x_2 = x_2(3x_2 + 8) = 0. \tag{5.47}$$

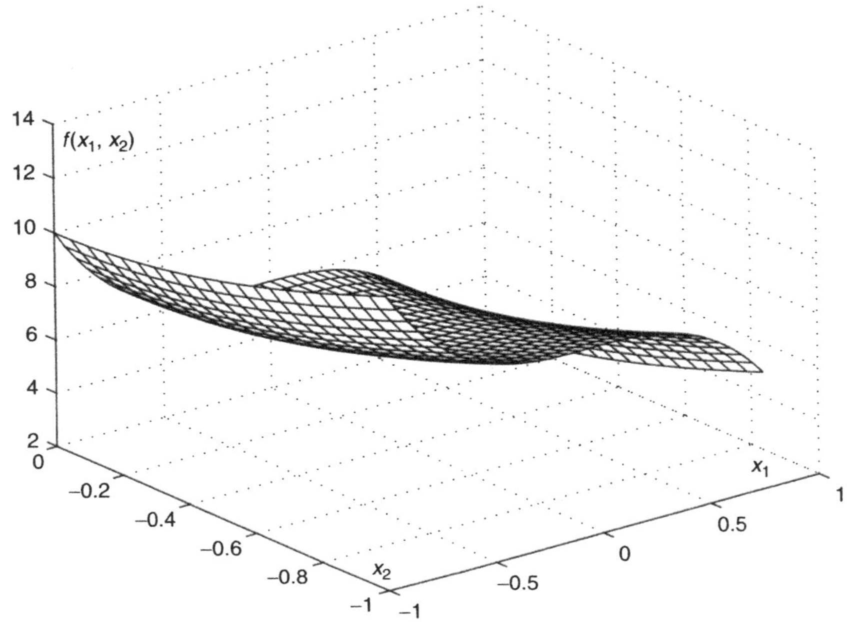

Figure 5.3. Example of multiple extreme points.

Sufficiency conditions involving the second derivatives give the following Hessian matrix:

$$H = \begin{bmatrix} \dfrac{\partial^2 f}{\partial x_1^2} & \dfrac{\partial f}{\partial x_1 \partial x_2} \\[2ex] \dfrac{\partial f}{\partial x_1 \partial x_2} & \dfrac{\partial^2 f}{\partial x_2^2} \end{bmatrix} = \begin{bmatrix} -(30x_1 + 10) & 0 \\ 0 & (6x_2 + 8) \end{bmatrix}.$$

The determinants of two principal minors to test for definiteness are

$$H_1 = |-(30x_1 + 10)| \quad \text{and} \quad H_2 = \begin{vmatrix} -(30x_1 + 10) & 0 \\ 0 & (6x_2 + 8) \end{vmatrix}.$$

In general, the first-order conditions yield a set of nonlinear equations the solution of which indicates stationary points. Such a solution can be attempted by using the Jacobian matrix to iterate toward a solution (see Section 5.3.2). We postpone our discussion of solution methods until later, but note for now that there are four possible solutions for the system of equations (5.46) and (5.47). The solution can be obtained by inspection; it has four stationary values at (0, 0), $(0, \frac{-8}{3})$, $(\frac{10}{15}, 0)$, and $(\frac{10}{15}, \frac{-8}{3})$. The table below shows the values of the determinants of two principal minors at these stationary points. A comparison of these attributes permits us to classify the extreme points as follows.

$\overline{\mathbf{x}}$	$\lvert H_1\rvert$	$\lvert H_2\rvert$	Eigenvalues of H	Nature of $\mathbf{x}$	$f(\mathbf{x})$
$(0, 0)$	-10	-80	Negative definite	Relative maximum	8
$\left(0, \frac{-8}{3}\right)$	-10	80	Indefinite	Saddle point	17.4818
$\left(\frac{10}{15}, 0\right)$	-16.66	-240	Negative definite	Relative maximum	4.2962
$\left(\frac{10}{15}, \frac{-8}{3}\right)$	-16.66	$+240$	Indefinite	Saddle point	13.777

5.5 QUADRATIC FUNCTIONS

Consider quadratic function of the form

$$f(\mathbf{x}) = c + \mathbf{d}^T\mathbf{x} + \frac{1}{2}\mathbf{x}^T\mathbf{H}\,\mathbf{x}, \tag{5.48}$$

where $\mathbf{H}$ is a symmetric matrix.

This equation is similar to the three term truncated Taylor expansion of (5.28). The constant term $\mathbf{c}$ corresponds to the term $f(\mathbf{x}^*)$ in (5.28). Our goal now is to examine the general nature of quadratic forms because most search techniques use an approximation that behaves like a quadratic function in a small neighborhood $\Delta\mathbf{x}$. As a consequence, our understanding of the behavior of quadratic functions relate directly to procedures of optimization that use a three-term approximation (up to the second derivative) of the infinite Taylor expansion.

Before we embark on such an examination, if an objective function is a quadratic function as in (5.48), the first-order condition for the extrema gives

$$f'(\mathbf{x}) = \nabla f(\mathbf{x}) = 0. \tag{5.49}$$

To compute the gradient in the above equation, we make use of the properties

$$\nabla(\mathbf{x}^T h) = \nabla(h^T\mathbf{x}) = h,$$

where h is a constant vector, and for a symmetric matrix $\mathbf{H}$

$$\nabla(\mathbf{x}^T\mathbf{H}\mathbf{x}) = \mathbf{H}\mathbf{x} + \mathbf{H}^T\mathbf{x} = 2\mathbf{H}\mathbf{x}.$$

Thus, for the gradient we obtain

$$f'(\mathbf{x}) = \nabla f(\mathbf{x}) = \mathbf{d} + \mathbf{H}\mathbf{x}, \tag{5.50}$$

from which the first-order condition gives

$$\mathbf{d} + \mathbf{Hx} = \mathbf{0}. \tag{5.51}$$

To compute the Hessian of (5.48), from the above relations we have the following for the second derivative:

$$\nabla^2 f(\mathbf{x}) = f''(\mathbf{x}) = \mathbf{H}.$$

If $\mathbf{H}$ is a constant matrix, the first-order condition is nothing but a *system of linear equations*, a solution of which gives the unconstrained extreme points. Even if constraints are present, as we shall see later, the first-order conditions represent a linear programming problem! Consequently, quadratic programming (or finding the extrema of a quadratic function) transforms itself to that of solving for the first-order conditions using a linear program.

Reverting to the examination of quadratic functions, since $\mathbf{H}$ is symmetric, it has distinct eigenvalues. From a diagonal transformation of the $\mathbf{H}$ matrix using the eigenvector matrix (see Section 2.6), it can be shown that the maximum second derivative of the function occurs in the direction of the eigenvector associated with the largest eigenvalue.[6] In fact the second derivative of f along each eigenvector will be equal to the eigenvalue associated with it. In directions other than those of the eigenvectors, the second derivative will be a weighted average of the second derivatives along the eigenvectors' directions, called the principal axes (Hagen, Demuth and Beale, 1999). The eigenvectors also define a new system of coordinates where the cross terms in the quadratic expression vanish (that is, the quadratic expression is "diagonalized"). This property is exploited in algorithms that search for the best direction of travel toward an extremum starting from an initial guess.

In a multidimensional case, the quadratic function represents a hyperellipsoid. However, in two dimensions, it represents an ellipse. We now study an example in two dimensions to reinforce the above concepts.

Example, Quadratic Function 1: Consider the quadratic form $\mathbf{x}^T\mathbf{A}\mathbf{x}$ in which $\mathbf{A}$ is given by $\begin{bmatrix} 5 & 3.5 \\ 2 & 3 \end{bmatrix}$. This translates to $f(\mathbf{x}) = 5x_1^2 + 3x_2^2 + 5.5x_1x_2$. Figure 5.4 shows the contours of this function produced by using MATLAB. Either by using MATLAB or by the procedure outlined in Sections 2.10.4, we confirm that the two eigenvalues are 6.8284 and 1.1716 with two distinct corresponding eigenvectors $[0.8863, 0.4630]^T$ and $[-0.6747, 0.7381]^T$, respectively. These eigenvectors are shown in Figure 5.4. Since the first eigenvalue is larger than the second, maximum curvature (second derivative) will occur in the direction of that eigenvector. Therefore, we will cross the contour lines more quickly in the direction of the bigger eigenvector as can be confirmed in Figure 5.4.

As a fallout from the above, it is evident that if the two eigenvectors of a Hessian matrix are equal, one obtains circular contour lines. Consequently, the change in

[6] For example, see (Hagen, Demuth and Beale, 1999) for a proof of these propositions.

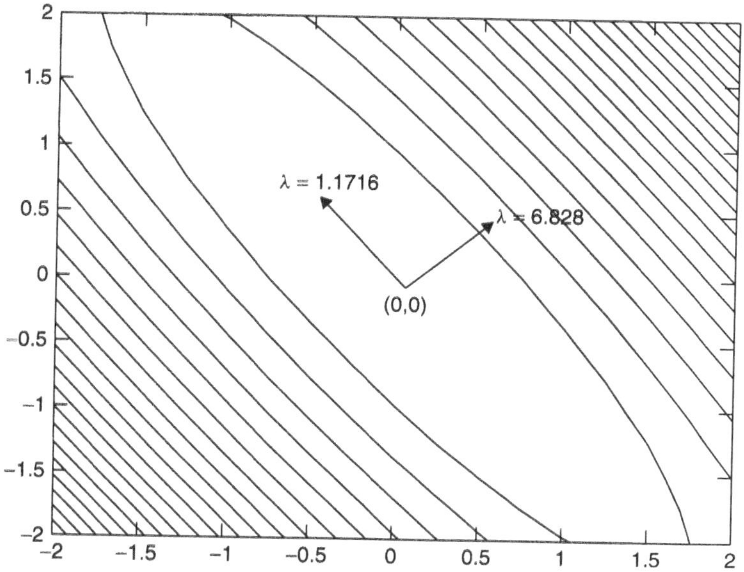

Figure 5.4. Characteristics of quadratic functions.

the curvature of the function is the same in the direction of either eigenvector. For example, such is the case for a diagonal matrix $\mathbf{H} = \begin{bmatrix} 2 & 0 \\ 0 & 2 \end{bmatrix}$ which has $\lambda_1 = \lambda_2 = 2$ with corresponding eigenvectors $[1, 0]^T$ and $[0, 1]^T$.

The above observations have assumed that the two eigenvalues are of the same sign. If the eigenvalues are of different signs, the Hessian is indefinite; therefore it represents a saddle point. Also, if one or more eigenvalues are zero, it does not represent an extreme point either; it represents a flat valley in the two-dimensional case. The change in curvature along the direction of the eigenvector corresponding to zero eigenvalue is zero, indicating flatness along the valley. The reader can find interesting examples with pictorial depictions of contours in (Hagen, Demuth and Beale, 1999).

Example, Quadratic Function 2: In later sections, the following function is used to illustrate the general nature of several minimization procedures:

$$f(x_1, x_2) = 6x_1^2 + 3x_2^2 + 4x_1x_2 + 5.5x_1 - 2x_2. \tag{5.52}$$

This function, as any other quadratic function, can be expressed in the general form as

$$f(\mathbf{x}) = \frac{1}{2}\mathbf{x}^T\mathbf{H}\mathbf{x} + \mathbf{d}\mathbf{x} + c$$

When (5.52) is written in this form, we obtain (reader to confirm the algebra)

$$f(\mathbf{x}) = \frac{1}{2}\mathbf{x}^T \begin{bmatrix} 12 & 4 \\ 4 & 6 \end{bmatrix} \mathbf{x} + [5.5, -2]^T \mathbf{x} + c, \tag{5.53}$$

where c is a constant. Evidently, we confirm that the gradient vector of (5.52) is

$$\nabla f(\mathbf{x}) = \begin{bmatrix} 12x_1 + 4x_2 + 5.5 \\ 6x_2 + 4x_1 - 2 \end{bmatrix} = \mathbf{Hx} + \mathbf{d}. \tag{5.54}$$

Note that the Hessian matrix in (5.53) is independent of x_1 and x_2. It need not be so when the function is of an order greater than two.

5.6 EXERCISE PROBLEMS

1. Obtain a second-order (quadratic) Taylor series approximation in terms of h_1, h_2, and h_3 at the point $\mathbf{x} = [1, -2, 1]$ for the function

$$f(x_1 x_2 x_3) = x_2^2 \varepsilon^{2x_3} + x_3 x_1 \varepsilon^{x_2}.$$

Compare the value of the function obtained by the above approximation at $\mathbf{x} = [1.1, -1.8, 0.9]$ with the actual value of the function.

2. Express the function

$$f(x_1, x_2, x_3, x_4) = -x_1^2 - x_2^2 + 3x_1x_2 + 2x_2x_4 - x_3^2 + x_4^2 + 5x_1x_3$$
$$-4x_3 + 2x_4 + 6$$

in matrix form as

$$f(\mathbf{x}) = \frac{1}{2}\mathbf{x}^T [\mathbf{H}] \mathbf{x} + \mathbf{dX} + \mathbf{c}.$$

Determine if [**H**] is positive definite, negative definite, or indefinite.

3. For the problem

Minimize $f(\mathbf{x}) = 3x_1^3 + 14x_2^2 - 6x_1x_2 + 15$,

subject to

$$x_1^2 + x_1x_2 - 2 = 0,$$
$$6x_1 - 2x_2^2 \leq 3,$$

(a) write equations representing the first-order conditions for minima and

(b) obtain the solution using Excel Solver and check that all first-order conditions are satisfied.

4. Consider the quadratic function

$$f(\mathbf{x}) = [-2, 2]\mathbf{x} + \frac{1}{2}\mathbf{x}^T \begin{bmatrix} 2 & 4 \\ 1 & 2 \end{bmatrix} \mathbf{x}.$$

What are the stationary values of $f(\mathbf{x})$? Support your conclusion by examining the eigenvalues and eigenvectors of the Hessian matrix.

5. A quadratic function

$$f(\mathbf{x}) = \frac{1}{2}\mathbf{x}^T \mathbf{H}\mathbf{x} + \mathbf{d}\mathbf{x} + \mathbf{c}$$

has $\mathbf{d} = [2, 2, 2]$ and $\mathbf{c} = 5$, and the Hessian matrix is

$$H = \begin{bmatrix} 4.0 & -2.0 & -2.5 \\ -2.0 & 3.0 & 3.0 \\ -2.5 & 3.0 & 4.0 \end{bmatrix}.$$

The Hessian matrix is the same as that of problem 3 in Section 2.12 with the indicated eigenvectors therein.

(a) What is the gradient of f?

(b) What is the gradient of f in the direction of (directional derivative) the vector $[1, 1, 1]^T$?

(c) What is the directional second derivative along $[-0.5617, 0.5301, 0]^T$?

(d) What is directional derivative of f along the three eigenvectors? Which directional derivative has the maximum value and why?

(e) Comment on the positive definiteness or otherwise of matrix H.

CHAPTER 6

UNCONSTRAINED NONLINEAR OPTIMIZATION

Most practical optimization problems involve constraints. Constraints arise from a myriad of reasons such as limited available capital, size of materials, amount of capacity or energy available, and other such physical reasons. Therefore, in practice, unconstrained optimization problems are less common than those with constraints. Nonetheless, there can be situations in neural network design, mathematics, and other endeavors where unconstrained optimization is of great interest. Furthermore, by examining unconstrained methods, we study procedures that are applicable to constrained optimization problem solution as well, but with appropriate modifications.

In the following and in subsequent chapters, we refer to the *minimization* of an objective. This terminology is used for convenience, not to exclude procedures that require maximization. The corresponding changes that are applicable to maximization should be obvious to the reader.

6.1 CLASSIFICATION OF MINIMIZATION METHODS

The methods for the solution of problems using digital computers can be put under two broad categories: (a) direct search methods and (b) descent or ascent methods.

Direct search methods are those where the optimal point is sought by "trial and error." Of course, the "trials" are based on logic and guidance. Techniques called *pattern search* methods identify the appropriate directions in which successive trials have to proceed. These methods require only the objective function and not its

Optimization Principles, by Narayan S. Rau
ISBN 0-471-45130-4

partial derivatives in order to find the minima. Hence such methods are often called *zeroth-order methods* or *nongradient methods*. In general, direct search methods are not as efficient as descent methods and are most suitable for problems that are simple and involve a relatively small number of variables.

On the other hand, descent methods utilize, in addition to the function value, the first, and in some cases second, derivative of the objective function. Therefore, these methods are more efficient than direct search methods because more information about the function (in the way of derivatives) is used in the minimization procedure. Descent methods are also called *gradient methods*. Methods that use the gradient and the second derivative are called *second-order methods*, while those that use only the first derivative are called the *first-order* methods.

A list of popular direct search methods is

- Random search method
- Univariate method
- Grid search method
- Simplex method
- Hook–Jeeves method
- Powell's method

A list of popular descent methods is

- Steepest descent method
- Fletcher–Reeves method
- Newton's method
- Marquard method
- Davidon–Power–Fletcher method
- Broyden–Fletcher–Goldfarb-Shanno method

Of these, the last two procedures are classified as quasi-Newton methods.

In addition to the above, there are other important methods for unconstrained optimization as well as constrained optimization that are not listed. An example of a very powerful procedure for constrained optimization is that of penalty function method. Since our goal in this book is to render an understanding of practical methods, a mathematical description of several such important methods is omitted. Thus, we discuss only some of the above methods that impart an understanding of optimization processes.

6.2 RANDOM SEARCH METHODS

These are methods in which the function value is evaluated at several points and the minimum is chosen. In order to ensure that all points are searched, one has to have a general idea of the space in which the solution lies. Otherwise, the solution may lie outside the space in which a search was conducted.

In two-dimensional space, for example, the space can be divided into a fine grid and a search can be conducted by evaluating the function value at all grid points. The same procedure can be followed in multidimensional space as well. However, such a procedure is crude and is computationally burdensome. Consequently, one embarks upon some refinements. The refinements can fall in the category of following some behavior pattern of the function value, which directs us to certain more probable grid points. There are several other possibilities to conduct an intelligent search.

We will not dwell on the details of search methods because, except for simple problems, other methods to be discussed in the following sections are infinitely superior and more efficient.

6.3 UNIVARIATE METHOD

In this method, only one variable is changed at a time to seek an improvement in the function value to progress toward the minimum point.

Starting from an arbitrary point $\mathbf{x}_1$ in the first iteration, a search direction $\mathbf{s}_1$ is found by perturbing the first variable by a small amount as $\mathbf{x} = \mathbf{x}_1 + \mathbf{s}_1$, where $\mathbf{s}_1 = (\mp\varepsilon, 0, 0, \ldots, 0)$, where ε is a small quantity. The sign of ε to be used is established by comparing the values of the function with a positive sign and negative sign. Next, the optimum distance γ to move from the starting point along $\mathbf{s}_1$ is determined.

In the second iteration, using the same procedure of iteration 1, the optimal distance γ to travel in the direction $\mathbf{s}_2 = (0, \mp\varepsilon, 0, \ldots 0)$ is determined. After a travel in all n directions is completed, the procedure is repeated for the second time for all directions. The process is continued until the change in the function value is within a tolerance value.

An example in two dimensions will elucidate the procedure.

Example: Minimize

$$f(x_1, x_2) = 6x_1^2 + 3x_2^2 + 4x_1x_2 + 5.5x_1 - 2x_2, \tag{6.1}$$

starting from $\mathbf{x}_0 = (0, 0)$.

Solution[1]***:*** Figure 6.1 shows a contour of the function and the trajectory of the minimization procedure. The value of ε is chosen as 0.01.

Iteration $i = 1$, Step 1.
$\mathbf{x}_0 = [0, 0]^T$

[1]The solution to this illustrative problem can be obtained analytically. The first-order necessary conditions give the following two equations:

continued

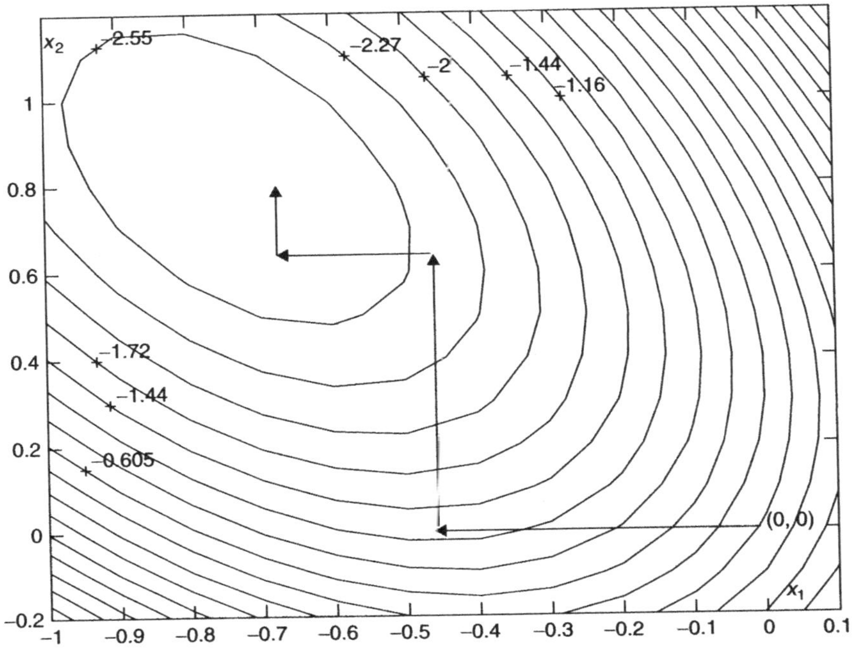

Figure 6.1. Univariate method of minimizing functions.

We have $f(\mathbf{x}_0) = 0$.

Choose search direction as $\mathbf{s}_1 = [1, 0]^T$. For function values, we have $f^+ = f(\mathbf{x}_0 + \varepsilon\mathbf{s}_1) = f(0.01, 0) = 0.0556$, and $f^- = f(\mathbf{x}_0 - \varepsilon\mathbf{s}_1) = f(-0.01, 0) = -0.0544$. Therefore, the correct direction to proceed in order to minimize the function is $-\mathbf{s}_1 = [-1, 0]^T$. The question then is how far to proceed in this direction. As can be seen from Figure 6.1, if we proceed too far, say to $(-1, 0)$, contour lines show that the value of the function increases. Therefore, the optimum distance γ to proceed is determined as follows.

$$\frac{\partial f}{\partial x_1} = 12x_1 + 4x_2 + 5.5 = 0,$$

$$\frac{\partial f}{\partial x_2} = 4x_1 + 6x_2 - 2 = 0,$$

the solution of which is $\bar{\mathbf{x}} = [-0.73214, 0.821428]^T$, with $f(\bar{\mathbf{x}}) = -2.8348$. If the problem is more complex in nature involving powers higher than two, the first-order conditions give a system of nonlinear equations, which has to be solved. That may not be a trivial task. Nevertheless, obtaining a solution by solving a set of nonlinear equations may not be more computationally efficient than iterative search schemes. However, solution of nonlinear equations may not be applicable to many problems with constraints. In this example, we shall use our knowledge of the optimum obtained by the analytical scheme only to serve as an illustrative metric to evaluate the proposed iterative schemes.

We want to determine γ such that the function

$$f(\mathbf{x}_0 - \gamma \mathbf{s}_1) = f(-\gamma, 0)$$

is a minimum. A substitution of $\mathbf{x} = \mathbf{x}_0 - \gamma \mathbf{s}_1 = (-\gamma, 0)$ into (6.1) gives

$$f(\mathbf{x}_0 - \gamma \mathbf{s}_1) = f(-\gamma, 0) = 6\gamma^2 - 5.5\gamma.$$

Hence, we want

$$\frac{\partial}{\partial \gamma} f(\mathbf{x}_0 - \gamma \mathbf{s}_1) = \frac{\partial}{\partial \gamma} \left[6\gamma^2 - 5.5\gamma\right] = 0.$$

The above yields $12\gamma - 5.5 = 0$, giving a value of $\frac{5.5}{12} = 0.4583$ for γ. We have $\mathbf{x}_1 = \mathbf{x}_0 - \gamma \mathbf{s}_1 = (-0.4583, 0)$, $f(\mathbf{x}_1) = f(-0.4583, 0) = -1.2604$.

Iteration $i = 1$, Step 2.
Choose search direction $\mathbf{s}_2 = [0, 1]^T$. Then,

$$f(\mathbf{x}_1 \mp \varepsilon \mathbf{s}_2) = f(-0.4583, \mp 0.01).$$

Trying both positive and negative values for ε, we get $f^+ = -1.2984$, and $f^- = -1.2218$. Therefore, $\mathbf{s}_2 = [0, +1]^T$ is the right direction to proceed.

As in the previous step, if we proceed too far in this direction, we might go to higher function values (see Figure 6.1). We have to find the optimum distance, γ, to proceed in this direction. By substituting $\mathbf{x} = (-0.4583, \gamma)$ into (6.1), we have

$$f(\mathbf{x}_1 + \gamma \mathbf{s}_2) = 6\ (-0.4583)^2 + 3\gamma^2 - 4\ (0.4583)\ \gamma + 5.5\ (-0.4583) - 2\gamma,$$

which has a minimum value when

$$\frac{\partial f}{\partial \gamma} = 0 = \frac{\partial}{\partial \gamma}[3\gamma^2 - 3.83\gamma - 3.78065].$$

Hence,

$$6\gamma - 3.83 = 0,$$

which gives a value of 0.6388 for γ, resulting in $\mathbf{x}_2 = (-0.4583, 0.6388)$. Then $f(\mathbf{x}_2) = -2.4849$ that is less than $f(\mathbf{x}_0)$.

Iteration $i = 2$, Step 1.
Start with $\mathbf{x}_2 = (-0.4583, 0.6383)^T$ obtained previously.

We are back to the direction $\mathbf{s}_1$. We note that $f^+ = f\begin{pmatrix} -0.4583 + 0.01 \\ 0.6383 \end{pmatrix} = -2.4849$, and $f^- = f\begin{pmatrix} -0.4583 - 0.01 \\ 0.6383 \end{pmatrix} = -2.50108$. Hence, proceeding in the direction $\mathbf{s}_1 = [-1, 0]^T$ reduces the value of the function.

As in previous steps, the optimum length γ to proceed in this direction is obtained by substituting $(\mathbf{x}_2 + \gamma\mathbf{s}_1) = (-0.4583 - \gamma, 0.6383)$ into (6.1), which results in

$$\frac{\partial f(\mathbf{x}_2 + \gamma\mathbf{s}_1)}{\partial\gamma} = \frac{\partial}{\partial\gamma}[6\gamma^2 - 2.55\gamma - 2.4872] = 0 = 12\gamma - 2.55.$$

Hence we obtain $\gamma = 0.21376$, which gives $\mathbf{x}_3 = \mathbf{x}_2 + \gamma\mathbf{s}_1 = (-0.6720, 0.6388)$, and $f(\mathbf{x}_3) = -2.7566$.

Iteration $i = 2$, Step 2.
Proceeding now in the direction $\mathbf{s}_2 = [0, 1]^T$, we find that $f^+ = f\begin{pmatrix} -0.6720 \\ 0.6388 + 0.01 \end{pmatrix} = -2.7648$, and $f^- = f\begin{pmatrix} -0.6720 \\ 0.6388 - 0.01 \end{pmatrix} = -2.7477$. Therefore, the correct direction to proceed is $\mathbf{s}_2 = [0, +1]^T$.

As in the previous steps, the optimal length γ to be traveled in this direction can be obtained as (work out details) 0.142. Thus, at the end of second iteration, we have $\mathbf{x}_4 = (-0.6720, 0.7813)$ and $f(\mathbf{x}_4) = -2.8175$.

We can continue this process of iterations further, but it is evident that the process is converging toward the desired solution of $\bar{\mathbf{x}} = (-0.732, 0.82146)$ where $f(\mathbf{x}) = -2.8348$.

6.4 ADVANCED METHODS INVOLVING PATTERN DIRECTIONS

In the univariate method, the search for the minimum proceeded along directions parallel to the axes as shown in Figure 6.1. It is possible that this method may not converge in some cases, particularly if the minima are in a deep valley. Furthermore, as can be seen from the example, its convergence is slow as we approach the optimum point. This problem can be avoided by altering the search directions instead of directions parallel to the coordinate axis. For example, in Figure 6.1, after two moves, we come to the point $(-0.458, 0.6583)$ from the starting point $(0, 0)$. We note that diagonal lines joining the alternate points of a univariate search point generally in the direction of the minimum. This diagonal direction starting from $(0, 0)$ is in the direction $(-0.458, 0.6583)$. Such a direction is called *pattern direction*. Guided by this fact, it is logical to examine the outcome if our subsequent search directions were the diagonal lines. In other words, the search directions would be those of the lines joining the alternate points such as 1–3, 2–4, 3–5, and so on. Is there a guarantee that such lines pass through the minimum?[2] Unfortunately, while such lines may pass through the minimum for some cases of two variable problems, such is not the case for multivariable functions, even if they are quadratic functions. Nevertheless, this idea of changing direction of search, particularly near the optimum point, speeds convergence and is used in two well-known methods: Powell's method and Hooke–Jeeves method. We will discuss the former in some detail. We refer the reader to Rao (1996), Gill et al. (1981), and other references for a more detailed description of this and other such *pattern search methods*.

6.5 CONJUGATE DIRECTIONS

Before a discussion of Powell's method, we digress to discuss conjugate directions. It can be proved that Powell's method searches along conjugate directions. A general nonlinear function can be approximated to a reasonable accuracy by a quadratic function, particularly near its minimum (by considering only three terms of the Taylor expansion). Using this approximation, searches along conjugate directions is expected to speed the convergence of nonlinear programs.

Definition: It is shown in Section 2.2.4 that any two vectors $\mathbf{s}_i$ and $\mathbf{s}_j$ are said to be orthogonal if their dot product

$$\mathbf{s}_i^T.\mathbf{s}_j = 0, \qquad i \neq j.$$

Extending this to include the relationship of all n orthogonal vectors (of directions of search), we express these n vectors that are orthogonal as

$$\mathbf{s}_i^T \mathbf{I}\, \mathbf{s}_j = 0 \quad \text{for all } i \neq j, i = 1, 2, \ldots n, \ j = 1, 2 \ldots, n,$$

where $\mathbf{I}$ is an identity matrix.

As a further extension of this to a symmetric matrix $\mathbf{A}$, a set of directions (or vectors) are said to be $\mathbf{A}$-conjugate (or conjugate) if

$$\mathbf{s}_i^T \mathbf{A}\, \mathbf{s}_j = 0 \text{ for all } i = 1, 2; \ j = 1, 2 \ldots, n.$$

Discussion of Conjugate Directions: Consider a quadratic function as in (5.48)

$$f(\mathbf{x}) = c + \mathbf{d}^T\mathbf{x} + \frac{1}{2}\mathbf{x}^T\mathbf{H}\mathbf{x}. \tag{6.2}$$

[2] In the above example, we have determined that the minimum point is (−0.732, 0.8216). Hence, the directions given by joining the starting point (0, 0) and the third point (−0.645833, 0.6388), or the line joining the second and the fourth points (−0.6388, 0), and (−0.6720, 0.6388) do not pass through the minimum. Had it been that any of these directions passed through the minimum, one search in that direction for an optimal γ would have taken the solution to the minimum!

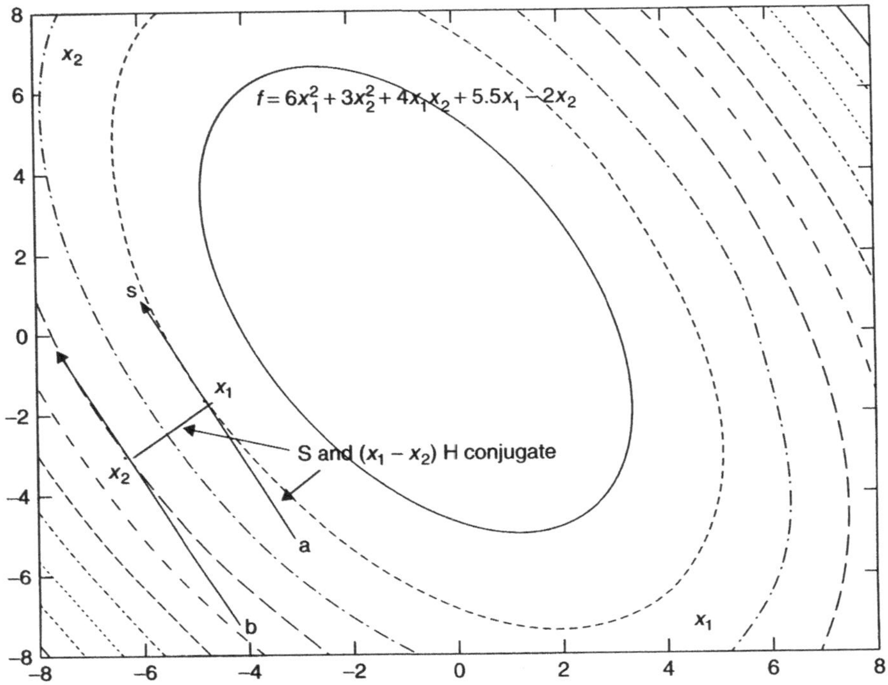

Figure 6.2. Conjugate directions for a quadratic function.

The gradient of f is given by

$$\nabla f(\mathbf{x}) = \mathbf{H}\mathbf{x} + \mathbf{d}. \tag{6.3}$$

Now consider two separate points **a** and **b**. We proceed from these points in parallel directions to find the minimum of a function along this direction. Figure 6.2 depicts this situation in which the contours of the function

$$f(\mathbf{x}) = \frac{1}{2}\mathbf{x}^T \begin{bmatrix} 12 & 4 \\ 4 & 6 \end{bmatrix} \mathbf{x} + [5.5, -2]^T \mathbf{x} \tag{6.4}$$

are shown[3] along with two starting points **a** and **b** and the parallel direction **s**. The minimum of the function along these directions for the two starting points are indicated as $\mathbf{x}_1$ and $\mathbf{x}_2$.

We shall now demonstrate that the line joining $\mathbf{x}_1$ and $\mathbf{x}_2$ is *conjugate* with the direction **s**.

[3]It should be obvious that this expresses the relationship $f(x_1, x_2) = 6x_1^2 + 3x_2^2 + 4x_1x_2 + 5.5x_1 - 2x_2$.

The gradient of (6.4) is given by

$$\nabla f(\mathbf{x}) = \mathbf{Hx} + \mathbf{d}, \tag{6.5}$$

and hence at $\mathbf{x}_1$ and $\mathbf{x}_2$ we have

$$\nabla f(\mathbf{x}_1) = \mathbf{Hx}_1 + \mathbf{d}, \tag{6.6}$$

$$\nabla f(\mathbf{x}_2) = \mathbf{Hx}_2 + \mathbf{d}. \tag{6.7}$$

Consider two parallel vectors semanating from a and b. In Figure 6.2, points $\mathbf{x}_1$ and $\mathbf{x}_2$ are the minimum values of the function along these directions. The vector $\mathbf{s}$ is orthogonal to the gradients $\nabla f(\mathbf{x}_1)$, $\nabla f(\mathbf{x}_2)$. Thus

$$\mathbf{s}^T \nabla f(\mathbf{x}_1) = \mathbf{s}^T \mathbf{H}\,\mathbf{x}_1 + \mathbf{s}^T \mathbf{d} = 0, \tag{6.8}$$

$$\mathbf{s}^T \nabla f(\mathbf{x}_2) = \mathbf{s}^T \mathbf{H}\,\mathbf{x}_2 + \mathbf{s}^T \mathbf{d} = 0. \tag{6.9}$$

By subtracting (6.9) from (6.8), we obtain

$$\mathbf{s}^T \mathbf{H}(\mathbf{x}_1 - \mathbf{x}_2) = 0. \tag{6.10}$$

Hence, $\mathbf{s}$ and $(\mathbf{x}_1 - \mathbf{x}_2)$ are conjugate.

Use of Conjugate Directions in Algorithms. Many algorithms use the efficiency of searching for the minima in conjugate directions. The following theorem illustrates the advantage.

Theorem: If a quadratic function of order n

$$f(\mathbf{x}) = c + \mathbf{d}^T \mathbf{x} + \frac{1}{2}\mathbf{x}^T \mathbf{H}\,\mathbf{x} \tag{6.11}$$

is sequentially minimized, once along each direction of a set of n directions that are mutually conjugate, irrespective of the starting point, the minimum will be found at or before the nth iteration.

For a proof of this theorem, refer to Rao (1996) and Bazaraa et al. (1979).

Example: Find a vector $\mathbf{s}_2$ that is conjugate to the vector $\mathbf{s}_1 = [2, -3, 6]^T$ with respect to the matrix

$$[\mathbf{A}] = \begin{bmatrix} 1 & 2 & 3 \\ 2 & 5 & 6 \\ 3 & 6 & 7 \end{bmatrix}$$

Solution: We want $\mathbf{s}_1^T \mathbf{A}\,\mathbf{s}_2$ to be zero. Since $\mathbf{s}_1^T \mathbf{A} = [14, 25, 30]$, we need $14s_2^1 + 25s_2^2 + 30s_2^3 = 0$. We arbitrarily select the three components of vector $\mathbf{s}_2$ as $s_2^1 = 0$, $s_2^2 = \frac{1}{25}$, and $s_2^3 = -\frac{1}{30}$ in order to satisfy this relationship. Normalizing the magnitude of $\mathbf{s}_2$ to unity, we obtain $\mathbf{s}_2 = [0,\ 0.76822,\ 0.64018]^T$.

6.6 PATTERN SEARCH METHOD—POWELL'S METHOD

Consider the earlier example:

Example: Minimize

$$f(x_1, x_2) = 6x_1^2 + 3x_2^2 + 4x_1x_2 + 5.5x_1 - 2x_2, \tag{6.12}$$

starting from $\mathbf{x}_0 = (0, 0)$.

Solution:

Univariate Cycle 1: We first perform a univariate search for the minimum. From the previous example, we recall that the results of the first two iterations were

$$\mathbf{x}_0 = \begin{pmatrix} 0 \\ 0 \end{pmatrix}, \qquad \mathbf{x}_1 = \begin{pmatrix} -0.4583 \\ 0 \end{pmatrix}, \qquad \mathbf{x}_2 = \begin{pmatrix} -0.4583 \\ 0.6388 \end{pmatrix},$$

$$\mathbf{x}_3 = \begin{pmatrix} -0.6720 \\ 0.6388 \end{pmatrix}, \qquad \mathbf{x}_4 = \begin{pmatrix} -0.6720 \\ 0.7813 \end{pmatrix},$$

and the corresponding function values were

$$f(\mathbf{x}_0) = 0,\ f(\mathbf{x}_1) = -1.2604, \qquad f(\mathbf{x}_2) = -2.4848,$$

$$f(\mathbf{x}_3) = -2.7566, \quad \text{and} \quad f(\mathbf{x}_4) = -2.8145.$$

Pattern Search, Cycle 1: We can choose the search direction to be either $\mathbf{x}_2 - \mathbf{x}_0$, $\mathbf{x}_3 - \mathbf{x}_1$, or $\mathbf{x}_4 - \mathbf{x}_2$.

We start a pattern search from $\mathbf{x}_2 = [-0.4583, 0.6383]^T$ with $f(\mathbf{x}_2) = -2.4848$ with a search direction as $\mathbf{s}_{p1} = (\mathbf{x}_2 - \mathbf{x}_0) = [-0.4583,\ 0.6383]^T$.

As before, using a value of $\varepsilon = 0.01$, we obtain $f^+ = f(\mathbf{x}_2 + \varepsilon\mathbf{s}_{p1}) = -2.4965$, and $f^- = f(\mathbf{x}_2 - \varepsilon\mathbf{s}_{p1}) = -2.4730$. Consequently, $+\mathbf{s}_{p1}$ is the right direction to proceed.

The optimal length γ to proceed along $\mathbf{s}_{p1}$ is determined as follows. We want

$$f(-0.4583 - 0.4583\gamma,\ 0.6383 + 0.6383\gamma)$$

to be a minimum. From a substitution of these coordinates for $\mathbf{x}$ into (6.12), along with a simplification, one obtains[4]

$$f[\mathbf{x}_2 + \gamma\mathbf{s}_{p1}] = 1.3123\gamma^2 - 1.1728\gamma - 2.4836.$$

For a minimum, $\partial f/\partial\gamma = 0$. Hence one obtains $\gamma = 0.4468$, which gives $\mathbf{x}_3 = \mathbf{x}_2 + \gamma\mathbf{s}_{p1} = [-0.6630,\ 0.92349]^T$ with a corresponding $f(\mathbf{x}_3) = -2.7466$.

[4]In Section 6.7.2, (6.17) shows a more elegant and easier way of calculating the optimum distance γ than by the quadratic equations procedure shown in this example.

Univariate Cycle 2:
STEP 1. Reverting to univariate cycle at $\mathbf{x}_3$ obtained above, we examine a search direction $\mathbf{s}_1 = [1, 0]^T$ to see if $+\mathbf{s}_1$ or $-\mathbf{s}_1$ is the proper direction to proceed. A test in the two direction adding $\varepsilon = 0.01$ reveals $f^- = -2.75851$ and $f^+ = -2.7336$. Consequently, $-\mathbf{s}_1$ is the correct direction to proceed.

In order to determine the optimal length to proceed in this direction, a substitution of $\mathbf{x} = \mathbf{x}_3 - \gamma\mathbf{s}_1$ into (6.12) obtains

$$6\gamma^2 - 1.2373\gamma - 2.55913,$$

which has a minimum value when $\gamma = -0.1031$. Hence $\mathbf{x}_4 = \mathbf{x}_3 + \gamma\mathbf{s}_1 = [-0.76615,\ 0.92349]^T$ with corresponding $f(\mathbf{x}_4) = -2.81053$.
STEP 2. Search for optimum length along $\mathbf{s}_2 = [0, 1]^T$ is done in a manner identical to the above step. We determine that $-\mathbf{s}_2$ is the right direction to proceed with a corresponding γ of 0.079. Thus we obtain $\mathbf{x}_5 = (\mathbf{x}_4 - \gamma\mathbf{s}_2) = [-0.76615,\ 0.84414]^T$ with corresponding $f(\mathbf{x}_5) = -2.8294$.

Pattern Search, Cycle 2: Starting now from $\mathbf{x}_5 = [-0.76615,\ 0.84414]^T$, the search direction $\mathbf{s}_{p2}$ now is $(\mathbf{x}_3 - \mathbf{x}_5) = [-0.76615,\ 0.84414]^T - [-0.663,\ 0.92349]^T = [-0.103,\ -0.0792]^T$. By adding $\mp\varepsilon\mathbf{s}_{p2}$ to $\mathbf{x}_5$, we find that $f^+ = -2.82909$ and $f^- = -2.82974$. Since $f(\mathbf{x}_5) = 2.8294$, the appropriate search direction is $-\mathbf{s}_{p2}$, which is equal to $[0.103,\ 0.0792]^T$.

Substituting $x_1 = (-0.76615 + 0.103\gamma)$ and $x_2 = (0.84414 + 0.0792\gamma)$, we obtain (algebra omitted)

$$f(\mathbf{x}_5 + \gamma\mathbf{s}_{p2}) = 0.2599\gamma^2 - 0.03229\gamma - 0.89,$$

giving an optimum value for $\gamma = 0.0621$. Hence we obtain $\mathbf{x}_6 = [-0.76615 + 0.621 \times 0.103,\ \ 0.84414 + 0.621 \times 0.0792]^T = [-0.7597,\ \ 0.84905]^T$. Thus we have $f(\mathbf{x}_6) = -2.83102$.

At this stage, we note that the function is continuously decreasing, but has not reached the theoretical solution of $\overline{\mathbf{x}} = [-0.73214,\ 0.82142]^T$ where $f(\overline{\mathbf{x}}) = -2.8348$. However, at $\mathbf{x}_6$ the value of the function is only $(-2, 8348\ -\ 2.83102) = 0.00378$ away from the theoretical solution. In general, since we do not always know the theoretical minimum, we measure the change in function value from iteration to iteration, and if the absolute magnitude of change is less than a small tolerance value τ, we terminate iterations. In this case, the change in function value from $\mathbf{x}_5$ to $\mathbf{x}_6$ is $|(2.8294 - 2.83102)| = 0.00162$. Accepting this to be within a tolerance of 0.002, we terminate iterations.[5]

Figure 6.3 traces the path of iterations. Compare the path of iterations by the Powell method of pattern search with that of the univariate method in Figure 6.1.

[5]Since we want the gradient to be zero, an alternative is to examine the norm of the gradient vector and to terminate iterations when the norm is a very small value.

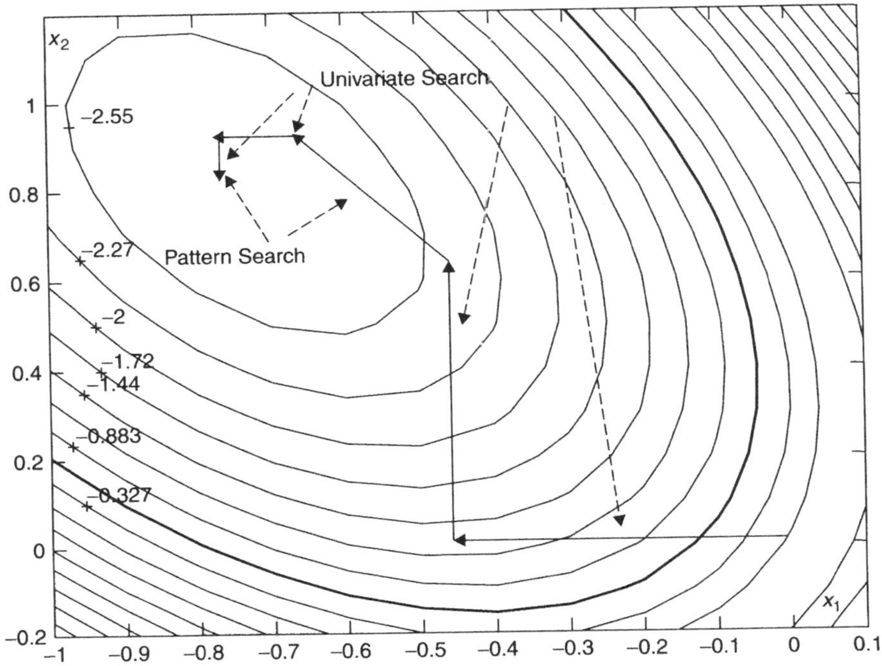

Figure 6.3. Powell's method of pattern search to minimize functions.

6.7 DESCENT METHODS

Descent methods are "indirect methods," which rely on the gradient of a function to find the extreme points. Such methods that use only information about the gradient are called the *first-order* methods while those that use the Hessian matrix are called the *second-order* methods. Because of the use of gradients, these methods are also called search techniques.

6.7.1 Steepest Descent Method

Descent methods use the gradient information in a manner similar to that of a blind person who finds her way to the lowest point in a terrain by using a stick[6] to obtain the gradient information at a given location **x**. Clearly, the gradient produces only local information. It does not reveal a possible barrier of a higher elevation that may lie between the point and the minima. As a consequence, first-order methods can be inefficient search procedures and can lead to a breakdown of the minimization procedure for some complex problems.

For problems whose analytical expression is known (differentiable in the region of interest), the gradient can be computed in closed form by a formula. For some

[6]Descent methods are also called *hill climbing* methods.

problems, the evaluation of the resulting formulae might require large computational time. In those circumstances, the gradient can be numerically computed by using either the following forward or the central finite difference formulae by evaluating the function a small distance away from a point of interest.

The components $i, i = 1 \cdots n$, of the gradient vector at $\mathbf{x}_p$ obtained by forward finite difference approximation as

$$\left|\frac{\partial f}{\partial \mathbf{x}_i}\right|_{x_p} \simeq \frac{f(\mathbf{x}_p + \Delta x_i \mathbf{u}_i) - f(\mathbf{x}_p)}{\Delta x_i},$$

or by the central difference approximation as

$$\left|\frac{\partial f}{\partial \mathbf{x}_i}\right|_{x_p} \simeq \frac{f(\mathbf{x}_p + \Delta x_i \mathbf{u}_i) - f(\mathbf{x}_p - \Delta x_i \mathbf{u}_i)}{2\Delta x_i},$$

where Δx_i is a small scalar quantity, and $\mathbf{u}_i$ is a unit vector in the direction x_i.

In the above, that the central difference formulation requires two additional function evaluations is clear, but one obtains better results from such a procedure. For functions that are discontinuous, such as at a sharp point or a kink, derivatives do not exist. Therefore, any numerical computation of the derivative is bound to be hopelessly inaccurate. For such problems, the random search methods of Section 6.2 appear to be the only recourse.

Algorithm: The algorithm is a straightforward implementation of the first two terms of (5.28) as

$$f(\mathbf{x}^* + h) = f(\mathbf{x}^*) + h\nabla f(\mathbf{x}^*). \tag{6.13}$$

The procedure on the digital computer is as follows:

1. Start with the initial point $\mathbf{x}_1$. Set iteration count $i = 1$.
2. Set the search direction $\mathbf{s}_i$ for minimization as $\mathbf{s}_i = -\nabla f_i$ which is also expressed as $-\nabla f(\mathbf{x}_i)$
3. Compute optimum length γ_i in the direction $\mathbf{s}_i$ and set

$$\mathbf{x}_{i+1} = \mathbf{x}_i + \gamma_i \mathbf{s}_i = \mathbf{x}_i - \gamma_i \nabla f_i.$$

   This expression is the iterative representation of (6.13) above.
4. The test for convergence is made by examining if the gradient is zero or by observing if the change in function value in successive iterations is less than or equal to a tolerance. If convergence is reached, stop the process. Otherwise, set iteration counter $i = i + 1$ and go to step 2.

Example: Minimize using the descent method

$$f(x_1, x_2) = 6x_1^2 + 3x_2^2 + 4x_1x_2 + 5.5x_1 - 2x_2, \tag{6.14}$$

starting from $\mathbf{x}_0 = (0, 0)$.

Solution: The gradient of the function is given by

$$\nabla f(\mathbf{x}) = \begin{bmatrix} \dfrac{\partial f}{\partial x_1} \\ \\ \dfrac{\partial f}{\partial x_2} \end{bmatrix} = \begin{bmatrix} 12x_1 + 4x_2 + 5.5 \\ \\ 6x_2 + 4x_1 - 2 \end{bmatrix}. \tag{6.15}$$

Iteration 1: $\mathbf{x}_1 = (0,\ 0)$, the value of function $f(\mathbf{x}_1) = 0$, and $\nabla f(\mathbf{x})|_{(0,\ 0)} = \begin{bmatrix} 5.5 \\ -2 \end{bmatrix}$. Therefore, we have search direction $\mathbf{s}_1 = \begin{bmatrix} -5.5 \\ +2 \end{bmatrix}$. In order to search for the optimal distance γ_1 in this direction, substitute $x_1 = 0 - 5.5\gamma_1$ and $x_2 = 0 + 2\gamma_1$ in (6.14) to obtain, upon simplification,

$$f(\mathbf{x}_1 + \gamma_1\mathbf{s}_1) = 149\gamma^2 - 34.25\gamma + \kappa,$$

where κ represents a constant term. To minimize the function in this direction, we require $\partial f/\partial \gamma_1 = 0$. Hence, we obtain from a differentiation of the above equation $\gamma_1 = 0.1145$. Therefore, the corrected value of $\mathbf{x}$ for the next iteration is $x_2 = \begin{bmatrix} 0 - 5.5\gamma_1 \\ 0 + 2\gamma_1 \end{bmatrix} = \begin{bmatrix} -0.625 \\ 0.225 \end{bmatrix}$, resulting in $f(\mathbf{x}_2) = -1.95438$.

Iteration 2: At $\mathbf{x}_2$, (6.15) gives $\nabla f(\mathbf{x}_2) = \begin{pmatrix} -1.1 \\ -3.15 \end{pmatrix}$. Hence, $\mathbf{s}_2 = -\nabla f(\mathbf{x}_2) = \begin{pmatrix} 1.1 \\ 3.15 \end{pmatrix}$. As in the previous iteration, a substitution of $x_1 = -0.625 + 1.1\gamma_2$, $x_2 = 0.225 + 3.15\gamma_2$ into (6.14) will give

$$f(\gamma) = 50.38\gamma_2^2 - 11.1275\gamma_2 + \kappa.$$

To find optimal γ_2, a partial differentiation of the above with respect to γ_2 and setting the result to zero gives $\gamma_2 = 0.1104$, $\mathbf{x}_2 = [-0.5035, 0.5728]^T$, and $f(\mathbf{x}_2) = -2.5631$. Furthermore, we note that $\nabla f(\mathbf{x}_2) = \begin{pmatrix} 1.7482 \\ -0.5770 \end{pmatrix} \neq 0$; therefore we go to the next iteration.

Iteration 3: From the above $\nabla f(\mathbf{x}_2)$, we set $\mathbf{s}_3 = \begin{pmatrix} -1.7482 \\ 0.5770 \end{pmatrix}$. As in the previous iterations, a substitution of $\mathbf{x}_2 + \gamma_3\mathbf{s}_3$ into (6.14) will give the optimum value of $\gamma_3 = 0.1105$. The resulting $\mathbf{x}_3 = [-0.6967,\ 0.63658]^T$, $f(\mathbf{x}_3) = -2.7509$, and $\nabla f(\mathbf{x}_3) = \begin{pmatrix} -0.31408 \\ -0.96732 \end{pmatrix} \neq 0$. Therefore, we again proceed to the next iteration.

Iteration 4: We set $\mathbf{s}_4 = \begin{pmatrix} 0.31408 \\ 0.96732 \end{pmatrix}$. As in the previous iterations, a substitution of $\mathbf{x}_3 + \gamma_4 \mathbf{s}_4$ into (6.14) will give the optimum value of $\gamma_4 = 0.11213$. The resulting $\mathbf{x}_4 = [-0.66148,\ 0.74504]^T$, $f(x_4) = -2.80895$, and $\nabla f(\mathbf{x}_4) = \begin{pmatrix} 0.5424 \\ -0.17568 \end{pmatrix} \neq 0$. Therefore, we embark on another iteration.

Iteration 5: We set $\mathbf{s}_5 = \begin{pmatrix} -0.5424 \\ 0.17568 \end{pmatrix}$. A substitution of $\mathbf{x}_4 + \gamma_5 \mathbf{s}_5$ into (6.14) will give the optimum value of $\gamma_5 = 0.110714$. The resulting $\mathbf{x}_5 = [-0.72153,\ 0.76449]^T$, $f(\mathbf{x}_5) = -2.82684$, and $\nabla f(\mathbf{x}_5) = \begin{pmatrix} -0.1004 \\ -0.29918 \end{pmatrix} \neq 0$. This nonzero gradient indicates that stationary point is not yet reached. However, note that the norm of the gradient is constantly diminishing.

Further iterations are left as an exercise to the reader until the gradient vector attains a norm that is less than a chosen tolerance.

6.7.2 Minimizing along a Direction

In the above examples, at each iteration, a quadratic equation has been solved to find the optimum multiplier γ_i to be traversed along the direction $\mathbf{s}_i$ (the negative direction of the gradient vector). This can be done formally and in a compact form in the following manner which is amenable to programming, including Excel spreadsheets, as later solved problems will show.

The object of the exercise is to optimize the distance $\gamma \mathbf{s}_i$ to be traversed from a point $\mathbf{x}_i$ in order to minimize $f(\mathbf{x}_i + \gamma_i \mathbf{s}_i)$.

If f is a quadratic function as in (5.48), it can be shown from a Taylor expansion that

$$\frac{d}{d\gamma_i} f(\mathbf{x}_i + \gamma_i \mathbf{s}_i) = \nabla f(\mathbf{x}_i)^T|_{\mathbf{x}=\mathbf{x}_i} \mathbf{s}_i + \gamma_i \mathbf{s}_i^T \mathbf{H}|_{\mathbf{x}=\mathbf{x}_i} \mathbf{s}_i. \tag{6.16}$$

Since the left-hand side is equal to zero at the optimum distance γ_i, we have

$$\gamma_i = \frac{-\mathbf{s}_i^T \nabla f(\mathbf{x}_i)}{\mathbf{s}_i^T \mathbf{H}_{\mathbf{x}_i} \mathbf{s}_i}, \tag{6.17}$$

where $\nabla f(\mathbf{x}_i)$ is the gradient at $\mathbf{x}_i$, and $\mathbf{H}_{\mathbf{x}_i}$ is the Hessian evaluated at $\mathbf{x}_i$.

That this formula indeed gives the optimum distance to be traversed is supported by the following confirmation of the third iteration in the above descent method example.

Confirmation by Example: The earlier example expressed by (6.14) can be written in quadratic form as

$$f(\mathbf{x}) = \frac{1}{2}\mathbf{x}^T \begin{bmatrix} 12 & 4 \\ 4 & 6 \end{bmatrix} \mathbf{x} + [5.5, -2]^T \mathbf{x}, \tag{6.18}$$

which has a gradient given by (6.15), and the Hessian matrix is indicated in (6.18). These are substituted into (6.17) to obtain γ_i in the following.

At the end of the third iteration we had arrived at $\mathbf{x}_i = [-0.6967, \ 0.63658]^T$, and the negative of the gradient vector as a search direction was computed to be $\mathbf{s}_3^T = [0.31408, \ 0.96732]$. Hence, for the numerator of (6.17) we obtain 1.03435 (confirm by multiplying in an Excel spreadsheet setup or otherwise); and the denominator obtains a value of 9.228, giving us a value of 0.1121 for γ, which is the same as that obtained by the longhand procedure in the fourth iteration of the above example.

In future examples of optimization, we will use the relation of (6.17) to compute the optimum distance to be traveled.

Remarks: Figure 6.4 illustrates the trajectory of five iterations. Note that the path traversed is more efficient than that of the univariate method or the pattern search method. Furthermore, the alternate directions are perpendicular to each other. The reason for this is apparent from the discussion on conjugate search directions. In addition, it is important to observe that several iterations are needed for the problem to converge to the stationary point, particularly as one gets closer to the optimum.

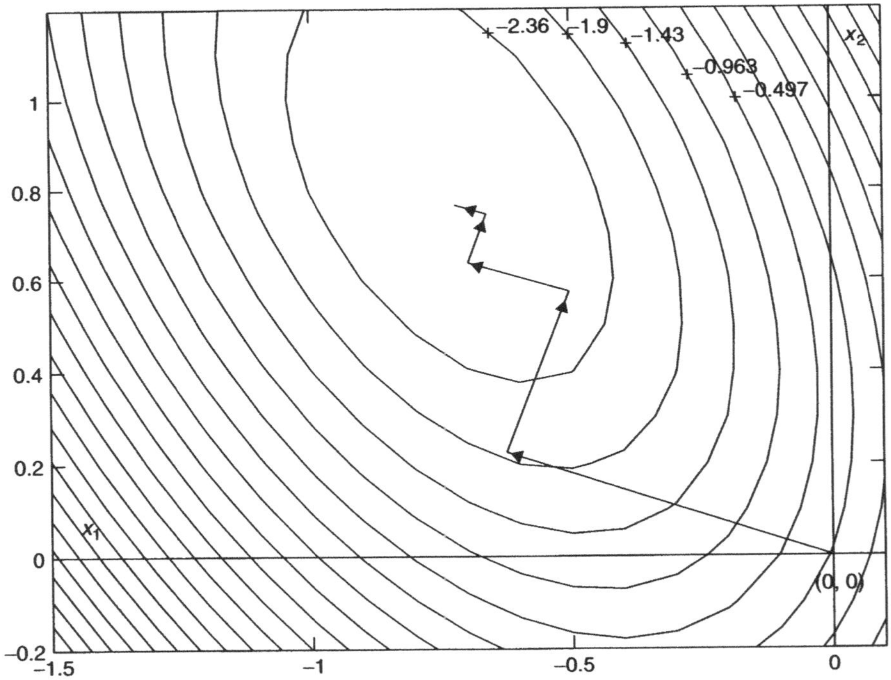

Figure 6.4. Method of descent.

6.8 NEWTON'S METHOD

Consider the second-order Taylor expansion of a multivariate function $f(\mathbf{x}_i)$ at a point $\mathbf{x}_i$ to approximate the function at $\mathbf{x}_{i+1}$ as in (5.28):

$$f(\mathbf{x}_{i+1}) = f(\mathbf{x}_i + \Delta\mathbf{x}) = f(\mathbf{x}_i) + \nabla f(\mathbf{x}_i)^T \Delta\mathbf{x} + \frac{1}{2}\Delta\mathbf{x}_i^T \mathbf{H}_i \, \Delta\mathbf{x}, \tag{6.19}$$

in which $\mathbf{H}_i$ is the Hessian matrix of the function, and $\nabla f(\mathbf{x}_i)$ is the gradient of $f(\mathbf{x})$, both of which are evaluated[7] at $\mathbf{x}_i$.

Location of a stationary point requires that the gradient is zero. Therefore, we take the gradient of (6.19) with respect to $\Delta\mathbf{x}$ and set it equal to zero to obtain[8]

$$\nabla f(\mathbf{x}_i) + \mathbf{H}_i \Delta\mathbf{x} = 0. \tag{6.20}$$

Solving for $\Delta\mathbf{x}$ yields

$$\Delta\mathbf{x} = -\mathbf{H}_i^{-1} \nabla f(\mathbf{x}). \tag{6.21}$$

We use this relation to suggest a search for improved points $\mathbf{x}_i + \Delta\mathbf{x}$ in a search procedure for stationary points starting from $\mathbf{x}_i$. Hence, *Newton's method* is defined as

$$\mathbf{x}_{i+1} = \mathbf{x}_i - \mathbf{H}_i^{-1} \nabla f(\mathbf{x}_i). \tag{6.22}$$

Note that unlike the procedure of descent method, which uses only the gradient, Newton's method algorithm of (6.20) uses the gradient as well as the Hessian to improve the search for a stationary point. Therefore, Newton's method is a *second order method*.

The search direction $-\mathbf{H}_i^{-1}\nabla f(\mathbf{x}_i)$ in (6.22) for updates to $\mathbf{x}_i$ is called the Newton's direction.

Example: As an illustration, we apply Newton's method to the previous example of (6.14):

Minimize

$$f(\mathbf{x}) = \frac{1}{2}\mathbf{x}^T \begin{bmatrix} 12 & 4 \\ 4 & 6 \end{bmatrix} \mathbf{x} + [5.5, -2]^T \mathbf{x} + \mathbf{c}. \tag{6.23}$$

Starting from an initial guess of $\mathbf{x}_0 = \begin{pmatrix} 0 \\ 0 \end{pmatrix}$, the gradient from (6.15) is

$$\nabla f(\mathbf{x}) = \begin{bmatrix} 12x_1 + 4x_2 + 5.5 \\ 6x_2 + 4x_1 - 2 \end{bmatrix}, \tag{6.24}$$

[7] For a quadratic function, the Hessian will contain only constant matrix elements. However, for higher-order functions, the elements of the Hessian matrix contain expressions involving variables.

[8] See (5.50) where it is shown that the gradient ∇f of a quadratic function is of the form $\mathbf{Hx} + \mathbf{d}$.

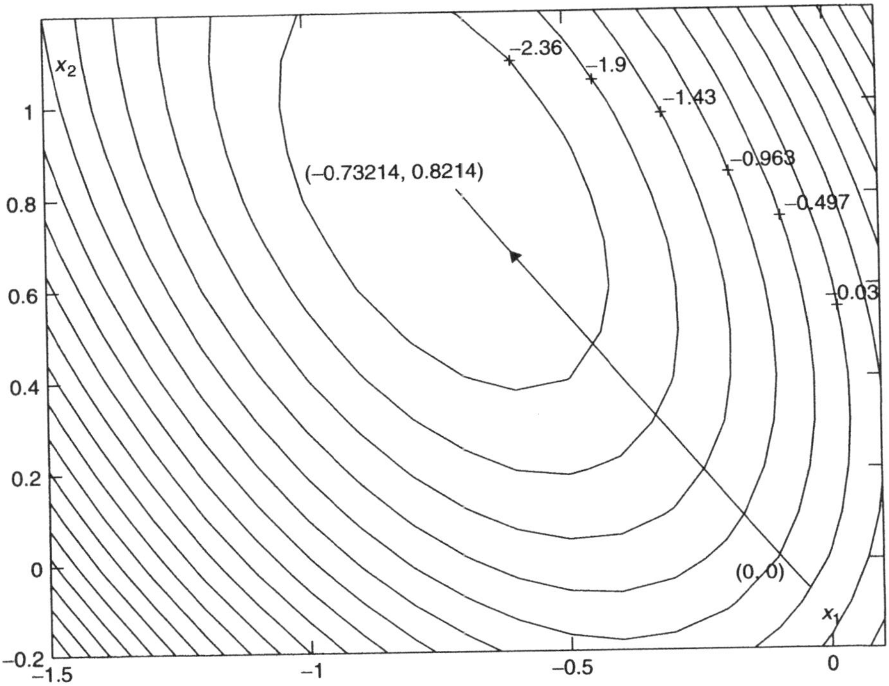

Figure 6.5. Trajectory of Newton's method to minimize quadratic functions.

$$\mathbf{H} = \begin{bmatrix} 12 & 4 \\ 4 & 6 \end{bmatrix}, \qquad \mathbf{H}^{-1} = \begin{bmatrix} 0.1071 & -0.0714 \\ -0.0714 & 0.2143 \end{bmatrix}. \tag{6.25}$$

From (6.24), at $\mathbf{x} = \mathbf{x}_0$, $\nabla f(\mathbf{x}_0) = \begin{bmatrix} 5.5 \\ -2 \end{bmatrix}$. Therefore, the second term of the right-hand side of (6.22) becomes

$$-\mathbf{H}_0^{-1} \nabla f(\mathbf{x}_0) = \begin{bmatrix} -0.73214 \\ +0.82142 \end{bmatrix}. \tag{6.26}$$

Hence, from (6.22) we get $\mathbf{x}_1 = [-0.73214, \; 0.82142]^T$!

At this value of $\mathbf{x}_1$, $\nabla f(\mathbf{x}_1) = 0$.

The above indicates that in one iteration we have obtained the stationary point, which happens to be the global minimum.[9] Figure 6.5 Shows the trajectory of the process. In fact, for a quadratic function, irrespective of the initial starting point,

[9]That this is a stationary point can be confirmed by evaluating the gradient at this point which is a null vector.

the stationary point can be reached in one iteration. We expect the reader to confirm this by using any other arbitrary starting point $\mathbf{x}_0$ to work out this example.

Example 2: Apply Newton's method and examine the convergence properties of the problem: Minimize

$$f(\mathbf{x}) = 6x_1^2 + 3x_2^2 + 3.5x_1^2x_2 + 4x_1x_2 + 5.5x_1 - 2x_2. \tag{6.27}$$

Solution: This problem is the same as that of (6.14) except for the additional term $3.5x_1^2x_2$ which has rendered $f(\mathbf{x})$ a third-order function. Hence it is no longer a simple quadratic function.

We have

$$\nabla f(\mathbf{x}) = \begin{bmatrix} 12x_1 + 7x_1x_2 + 4x_2 + 5.5 \\ 6x_2 + 3.5x_1^2 + 4x_1 - 2 \end{bmatrix}, \tag{6.28}$$

and

$$\mathbf{H} = \begin{bmatrix} 12 + 7x_2 & 7x_1 + 4 \\ 7x_1 + 4 & 6 \end{bmatrix}. \tag{6.29}$$

Because $f(\mathbf{x})$ is of third order, the Hessian is not a matrix with constant terms. We first set the tolerance for the convergence of the gradient at the extreme point at 0.005. If we start with $\mathbf{x}_0 = \begin{pmatrix} 0 \\ 0 \end{pmatrix}$, evaluating the above expressions we get

Iteration 1: At $\mathbf{x}_0$, we have $f(\mathbf{x}_0) = 0$. From (6.28) and (6.29), for $\nabla f(\mathbf{x}_0)$ and $\mathbf{H}_0$ we get

$$\nabla f(\mathbf{x}_0) = \begin{pmatrix} 5.5 \\ -2 \end{pmatrix}, \quad \mathbf{H}_0 = \begin{bmatrix} 12 & 4 \\ 4 & 6 \end{bmatrix}, \quad \mathbf{H}_0^{-1}\nabla f(\mathbf{x}_0) = \begin{pmatrix} 0.7321 \\ -0.8214 \end{pmatrix},$$

giving[10]

$$\mathbf{x}_1 = (\mathbf{x}_0 - \mathbf{H}_0^{-1}\nabla f(\mathbf{x}_0)) = \begin{pmatrix} -0.7321 \\ 0.8214 \end{pmatrix}.$$

and

$$f(\mathbf{x}_1) = -1.29373.$$

Iteration 2: Using the above value of $\mathbf{x}_1$ and evaluating $\nabla f(\mathbf{x}_1)$ and $\mathbf{H}_1$, we get

$$\nabla f(\mathbf{x}_1) = \begin{pmatrix} 0.71875 \\ 1.8761 \end{pmatrix}, \quad \mathbf{H}_1 = \begin{bmatrix} 17.75 & -1.125 \\ -1.125 & 6 \end{bmatrix},$$

$$\mathbf{H}_1^{-1}\nabla f(\mathbf{x}_1) = \begin{pmatrix} 0.061036 \\ 0.32413 \end{pmatrix},$$

[10]The reader may set up an Excel spreadsheet or a use MATLAB to check the algebra.

resulting in

$$\mathbf{x}_2 = (\mathbf{x}_1 - \mathbf{H}_1^{-1}\nabla f(\mathbf{x}_1)) = \begin{pmatrix} -0.79318 \\ 0.49729 \end{pmatrix}$$

and

$$f(\mathbf{x}_2) = -1.3231.$$

Iteration 3: With the above $\mathbf{x}_2$, we have

$$\nabla f(\mathbf{x}_2) = \begin{pmatrix} -1.8063 \\ 0.0130 \end{pmatrix}, \quad \mathbf{H}_2 = \begin{bmatrix} 15.849 & -1.5526 \\ -1.5526 & 6 \end{bmatrix},$$

$$\mathbf{H}_2^{-1}\nabla f(\mathbf{x}_2) = \begin{pmatrix} -0.11956 \\ -0.02876 \end{pmatrix}.$$

Hence we get

$$\mathbf{x}_3 = (\mathbf{x}_2 - \mathbf{H}_2^{-1}\nabla f(\mathbf{x}_2)) = \begin{pmatrix} -0.6732 \\ 0.5260 \end{pmatrix}$$

and

$$f(\mathbf{x}_3) = -1.7862.$$

Iteration 4: Above values of $\mathbf{x}_3$ give

$$\nabla f(\mathbf{x}_3) = \begin{pmatrix} 0.19661 \\ 0.050032 \end{pmatrix}, \quad \mathbf{H}_3 = \begin{bmatrix} 15.5624 & -0.7153 \\ -0.7153 & 6 \end{bmatrix},$$

$$\mathbf{H}_3^{-1}\nabla f(\mathbf{x}_3) = \begin{pmatrix} 0.01298 \\ 0.00988 \end{pmatrix},$$

and thus

$$\mathbf{x}_4 = (\mathbf{x}_3 - \mathbf{H}_3^{-1}\nabla f(\mathbf{x}_3)) = \begin{pmatrix} -0.68661 \\ 0.51616 \end{pmatrix},$$

and

$$f(\mathbf{x}_4) = -1.7465.$$

We note that at this value of $\mathbf{x}_4$, $\nabla f(\mathbf{x}_4) = [0.05842,\ 0.0059]^T$. Carrying one more iteration, the gradient becomes $\nabla f(\mathbf{x}_5) = [-0.00245,\ 0.000049]^T$ at $\mathbf{x}_5 = [-0.683,\ 0.5165]^T$ with $f(\mathbf{x}) = -1.75852$. Since the gradient is within the specified tolerance from zero, we terminate the iterations and accept $\mathbf{x}_5$ as the solution.

The function, not being a purely quadratic function, did not converge in one iteration. However, a quadratic approximation was used in each iteration by first three terms of the Taylor series. The function has been minimized in four iterations or three corrections by searching along Newton's directions from the starting value of $\mathbf{x}_0$. This represents *quadratic convergence* as discussed later in Section 7.16.

6.8.1 Effect of Starting Point

In the above example we started out the iterations with $\mathbf{x}_0 = [0, 0]^T$. We leave it to the reader to work out the consequences of starting with $\mathbf{x}_0 = \begin{pmatrix} 1.5 \\ -1.75 \end{pmatrix}$. The result diverges and does not converge.

The reason for this is apparent from Figure 6.6. The direction of the gradient vector g from this starting point is shown in the figure. Successive iterations from this starting point lead to a part of the contour in the southeast corner of the figure where the function value is continuously decreasing. Figure 6.7 is a three-dimensional plot of the function which also indicates this situation vividly. However, for multivariable problems, it is not possible a priori to determine if the problem has several minima by a graphical approach. The only way of analytically determining if a problem has multiple minima is by examining the convexity or concavity of the function, and by examining the positive definiteness of the Hessian matrix. A practical way is to start the iterations from different points. For the function in (6.27), the Hessian matrix would be positive definite at the local minima of $\bar{\mathbf{x}} = [-0.683,\ 0.5165]^T$ and would be indefinite at the saddle point of Figure 6.6. It is expected of the reader to confirm this.

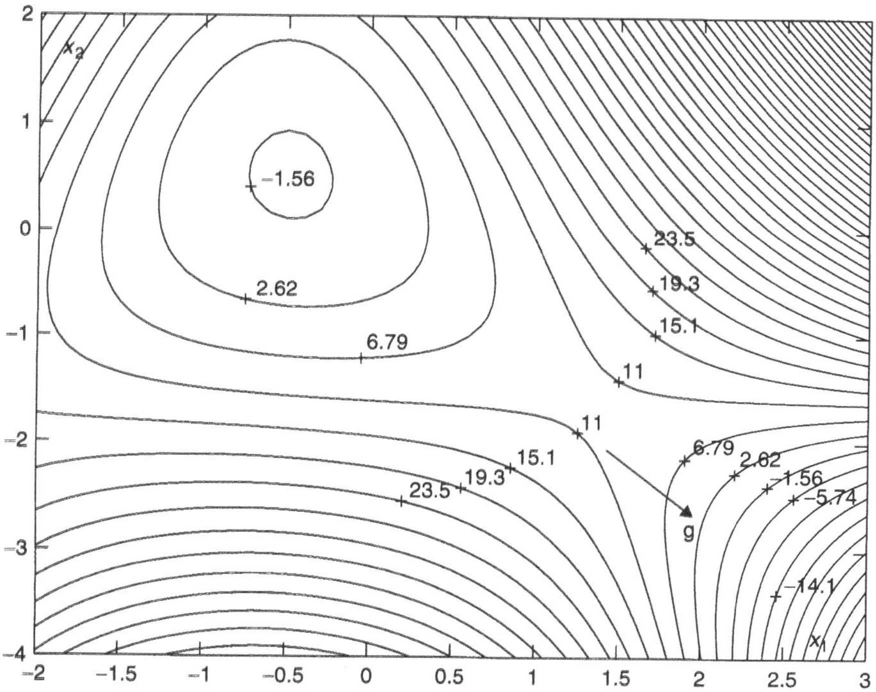

Figure 6.6. Contour plot of example problem.

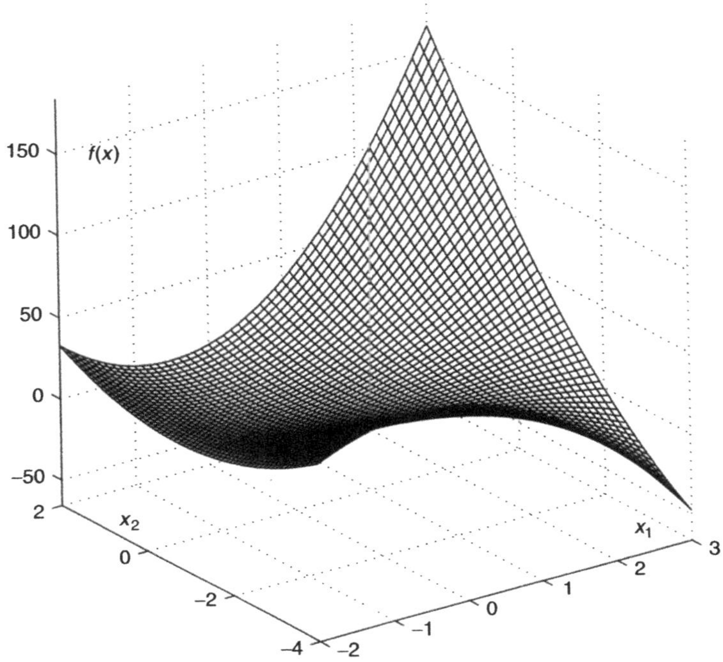

Figure 6.7. Three-dimensional plot of example problem.

6.9 SOLUTION OF NONLINEAR EQUATIONS

In Section 5.3.2, we have discussed a method to solve a system of nonlinear equations iteratively by using the Jacobian matrix. Here we show via an example that a system of nonlinear equations can be also be solved by using Newton's method or any other method by structuring the problem as an unconstrained nonlinear optimization problem.

Example: Solve the system of nonlinear equations,

$$f_1(x, y) = x^2 + y^2 = 2 \tag{6.30}$$

$$f_2(x, y) = 10x^2 - 10y - 5x = -1. \tag{6.31}$$

We solve this in three different ways. First we develop and illustrate the method of solution as an unconstrained optimization problem. Then we show the method of solution using the Jacobian matrix and by using the Excel Solver.

1. *Unconstrained Optimization Problem.* The idea is to guess and obtain successive iterates for x and y in (6.30) and (6.31) such that the sum of squares in the errors (for the right-hand side) is minimized. For example, we want $x^2 + y^2 - 2$ to be equal to zero. However, for the guessed values of x' and y', the right-hand side

may have a value of ε_1. Similarly, let the value of $10x^2 - 10y - 5x + 1$ for the assumed values x' and y' be ε_2. Our goal is to minimize the sum of errors given by $(\varepsilon_1 - 0)^2 + (\varepsilon_2 - 0)^2$.

The mathematical statement of this is as follows:
Minimize

$$f = (x^2 + y^2 - 2)^2 + (10x^2 - 10y - 5x + 1)^2. \tag{6.32}$$

Note that there are no constraints whatsoever and the assumption is that the solution exists.

In order to minimize the above equation, we first obtain the gradient and the Hessian matrix as

$$\nabla f = \begin{bmatrix} 404x^3 - 300x^2 + 4xy^2 - 400xy + 82x + 100y - 10 \\ 4x^2y + 4y^3 - 200x^2 + 192y + 100x - 20 \end{bmatrix}, \tag{6.33}$$

$$\mathbf{H} = \begin{bmatrix} 1212x^2 + 4y^2 - 600x - 400y + 82 & 8xy - 400x + 100 \\ 8xy - 400x + 100 & 4x^2 + 12y^2 + 192 \end{bmatrix}. \tag{6.34}$$

Using the Newton's method, from (6.22) the relation for successive iterates of x and y is

$$\begin{bmatrix} x^{(i+1)} \\ y^{(i+1)} \end{bmatrix} = \begin{bmatrix} x^{(i)} \\ y^{(i)} \end{bmatrix} - \mathbf{H}_i^{-1} \nabla f(x, y). \tag{6.35}$$

Open the spreadsheet "NL.eqn.System.Xls." Which can be found at ftp://ftp.wiley.com/public/sci_tech_med/electric_power/. Assume a starting value of, say, $x = 1$ and $y = 0$ in cells A6:B6. The gradient and the Hessian matrix representing (6.33) and (6.34) are entered in cells B7:B9 and A11:B12, respectively. The updated values for x and y are obtained according to (6.35) in cells F14:G14. Paste the value of these cells in A6:B6. Continue the procedure a few times to obtain the final solution as indicated in Figure 6.8

NL.eqn.System.Xls

2. *Jacobian Matrix Method.* The Jacobian matrix of the equation system (6.30), (6.31) is

$$J = \begin{bmatrix} 2x & 2y \\ 20x - 5 & -10 \end{bmatrix}, \tag{6.36}$$

which is in cells A20:B21. The values of (6.30) and (6.31) resulting from x, y values of cells A6:B6 are in cells E20:E21. The error in the solution, labeled $\Delta\mathbf{x}$, is computed in cell G20:G21. The corrections obtained by using (5.42) are in cells C24:C25. The updated values for x and y are in cells D24:D24, which, upon successive pasting in cells A6:B6, yields the same solution as obtained by the above procedure.

3. *Solution Using Excel Solver.* Cell F23 shows the sum of squares of errors of the right-hand side of (6.30) and (6.31) for assumed value of x and y. Open the

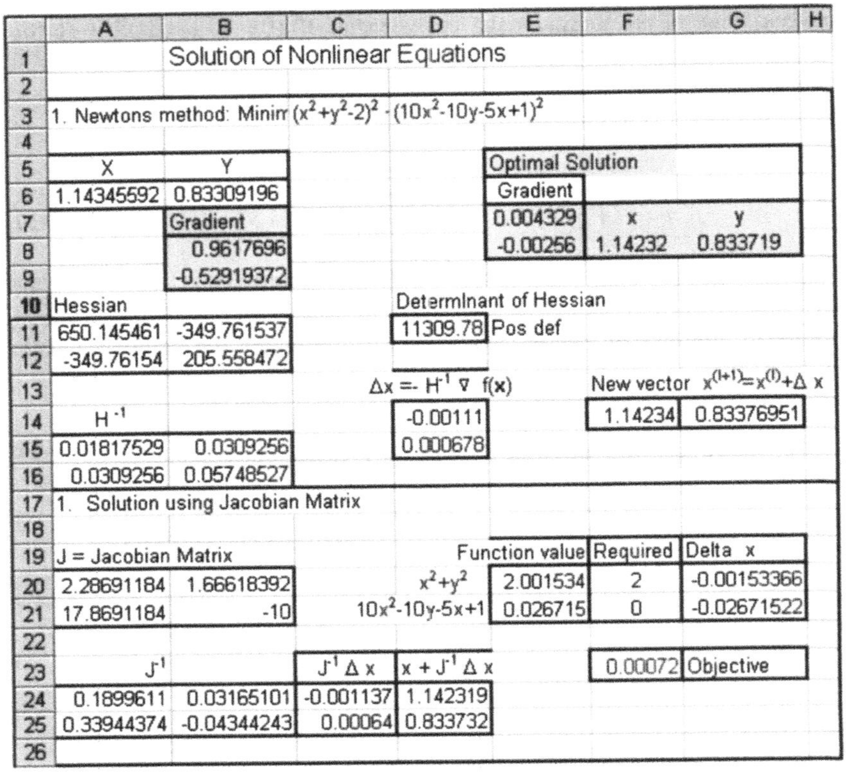

	A	B	C	D	E	F	G	H
1		Solution of Nonlinear Equations						
2								
3	1. Newtons method: Minim $(x^2+y^2-2)^2 - (10x^2-10y-5x+1)^2$							
4								
5	X	Y			Optimal Solution			
6	1.14345592	0.83309196			Gradient			
7		Gradient			0.004329	x	y	
8		0.9617696			-0.00256	1.14232	0.833719	
9		-0.52919372						
10	Hessian			Determinant of Hessian				
11	650.145461	-349.761537		11309.78	Pos def			
12	-349.76154	205.558472						
13				$\Delta x = -H^{-1} \nabla f(x)$		New vector $x^{(l+1)} = x^{(l)} + \Delta x$		
14	H^{-1}			-0.00111		1.14234	0.83376951	
15	0.01817529	0.0309256		0.000678				
16	0.0309256	0.05748527						
17	1. Solution using Jacobian Matrix							
18								
19	J = Jacobian Matrix			Function value		Required	Delta x	
20	2.28691184	1.66618392		x^2+y^2	2.001534	2	-0.00153366	
21	17.8691184	-10		$10x^2-10y-5x+1$	0.026715	0	-0.02671522	
22								
23		J^{-1}	$J^{-1} \Delta x$	$x + J^{-1} \Delta x$		0.00072	Objective	
24	0.1899611	0.03165101	-0.001137	1.142319				
25	0.33944374	-0.04344243	0.00064	0.833732				
26								

Figure 6.8. Example illustrating solution of a system of nonlinear equations.

Solver screen, which is programmed to minimize cell F23 by selecting values for x and y in cells A6:B6. There are no constraints. The optimal solution obtained is shown in cells F8:G8. Importantly, observe that the gradient of the expression approaches a value of zero at the minima.

One of the tests for the iterative procedures described above is that of testing the norm of the gradient to be within a small tolerance value.

Remarks: Newton's method is a second-order method while the use of Jacobian matrix is a first-order method. One expects the second-order method to result in faster convergence. However, during the use of the above spreadsheet to obtain solution, one finds that the Newton's method is slightly slower in converging to the final solution than the Jacobian matrix method. This is because Newton's method is being used to find the minima of a fourth-order nonlinear equation system (because of squaring the error) while the Jacobian matrix solves a system of second-order nonlinear equation system. This goes to show that one has to be careful in choosing an appropriate method to solve a problem at hand.

6.10 EXERCISE PROBLEMS

1. Solve the system of nonlinear equations

$$10x_1^3 - 10x_1 - x_2 = 2,$$
$$2x_2^3 - 10x_2 - x_1 = 3$$

(a) using the Jacobian matrix approach as in Section 6.9 and
(b) as an unconstrained minimization problem (two iterations of any method you choose);
(c) confirm your answer or procedure by using Excel Solver.

2. Figure 6.9 represents a four-node network. The load demand at nodes, branch impedances, and the required generator output at node #1 are indicated in the figure.

The following equations representing the net power injection into the network at four nodes have errors.

$$S_1 = |Y_{11}||V_1^2|\angle\theta_{11} - |Y_{12}||V_1||V_2|\angle(\theta_{12} + \delta_2 - \delta_1)$$
$$- Y_{13}|V_1||V_3|\angle(\theta_{13} + \delta_3 - \delta_1) - |Y_{14}||V_1||V_4|\angle(\theta_{14} + \delta_4 - \delta_1),$$
$$S_2 = |Y_{22}||V_2^2|\angle\theta_{22} - |Y_{21}||V_2||V_1|\angle(\theta_{21} + \delta_1 - \delta_2)$$
$$- Y_{23}|V_2||V_3|\angle(\theta_{23} + \delta_3 - \delta_2) - |Y_{14}||V_1||V_4|\angle(\theta_{14} + \delta_4 - \delta_1),$$
$$S_1 = |Y_{33}||V_3^2|\angle\theta_{33} - |Y_{32}||V_3||V_1|\angle(\theta_{31} + \delta_1 - \delta_3)$$
$$- Y_{32}|V_3||V_2|\angle(\theta_{32} + \delta_2 - \delta_3) - |Y_{34}||V_3||V_4|\angle(\theta_{34} + \delta_4 - \delta_1),$$

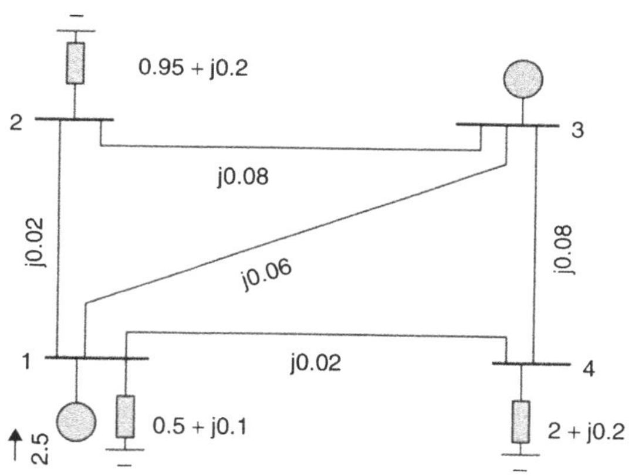

Figure 6.9. Four-node network.

and

$$S_4 = |Y_{44}||V_4^2|\angle\theta_{44} - |Y_{41}||V_4||V_1|\angle(\theta_{41} + \delta_1 - \delta_4)$$
$$- Y_{42}|V_4||V_2|\angle(\theta_{41} + \delta_4 - \delta_1).$$

(a) Correct the errors in the equations.

(b) Formulate nonlinear equations representing real and reactive power injections at nodes. Solve the nonlinear equations by employing the Jacobian matrix method of Section 6.9 to compute the unknown network parameters. Assume node #3 to be the slack bus with reference angle $\delta_3 = 0$, and assume that the voltage at the generator node #1 is maintained at 1.02 pu.

3. In order to minimize $f(\mathbf{x}) = x_1 - x_2 + 2x_1^2 + 2x_1x_2 + x_2^2$, consider a search direction $\mathbf{s}_1 = [-1, 1]^T$. Find a conjugate direction for the minimization.

4. Minimize $f(\mathbf{x}) = 4x_1^2 + 6x_2^2 - 6x_1x_2 - 4x_1$ starting from the point $(0, 0)$ by using:

(a) univariate method (two iterations),

(b) steepest descent method(two iterations),

(c) Newton's method.

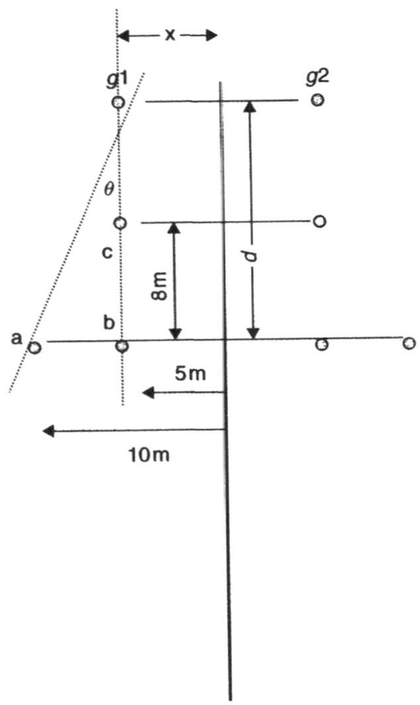

Figure 6.10. Contemplated transmission tower design.

5. Figure 6.10 indicates the contemplated design of a double circuit transmission tower. The object is to determine the optimal height for the placement of two ground wires to effectively shield the power lines from lightning strokes.

   The effectiveness of shielding is proportional to cosine of the angle θ. Additionally, if the ground wires are too far away from the conductors, the effectiveness of shielding reduces. The effectiveness of shielding as a function of d is given by the expression $10[\varepsilon^{-0.1d} - \varepsilon^{-0.11d}]$. The overall shielding performance is expressed by an index, which is a product of the above two effectiveness parameters.

   (a) Unconstrained problem: Obtain the optimum distance d for the ground wires to maximize overall shielding performance by using Newton's method.Check your answer with that obtained by using Excel Solver.

   (b) Constrained Problem: Refer to problem 4 in Section 8.11.

CHAPTER 7

CONSTRAINED NONLINEAR OPTIMIZATION

This chapter addresses solutions to constrained problems where the goal is to find a m vector of choice variables $\mathbf{x}$ which minimizes $f(\mathbf{x})$ subject to

$$g_i(\mathbf{x}) \leq 0, \qquad i = 1, 2, \ldots, n, \tag{7.1}$$

$$h_j(\mathbf{x}) = 0, \qquad j = 1, 2, \ldots, p. \tag{7.2}$$

In the above equations, we have n inequality constraints and p equality constraints. Understandably, either n or p can be zero, and when both are equal to zero, the problem degenerates to that of unconstrained optimization.

Nonlinear constrained optimization is referred to as nonlinear programming.

7.1 EFFECT OF CONSTRAINTS ON THE SOLUTION SPACE OF NONLINEAR PROGRAMS

Unlike linear optimization where the optimum point is at the vertex of a polytope or polyhedra (see Section 3.4), nonlinear programs have different characteristics. The necessary and sufficient conditions at the minimum point $\overline{\mathbf{x}}$ of a constrained minimum problem are

$$\nabla f|_{\mathbf{x}=\overline{\mathbf{x}}} = 0, \tag{7.3}$$

Optimization Principles, by Narayan S. Rau
ISBN 0-471-45130-4

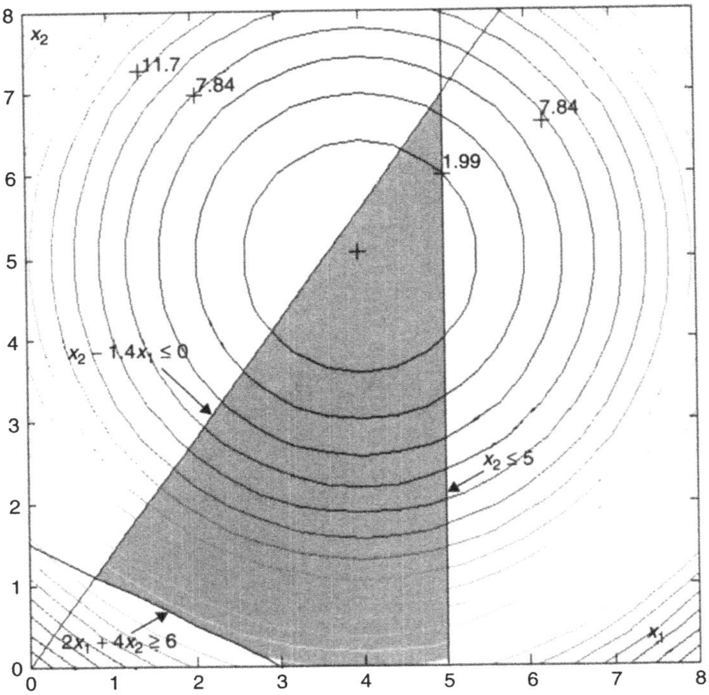

Figure 7.1. Optimization where constraints do not affect the solution.

and

$$\mathbf{H}|_{\mathbf{x}=\bar{\mathbf{x}}} = \text{Positive definite.} \tag{7.4}$$

However, constraints may affect the solution in the following way.

- Constraints of a nonlinear program may have no effect on the optimum point. Consider the problem of selecting x_1 and x_2 to minimize

$$(x_1 - 4)^2 + (x_2 - 5)^2$$

the contours of which are shown in Figure 7.1. Clearly, the optimum of the unconstrained problem is $[\bar{x}_1, \bar{x}_2] = [4, 5]$, which is shown in the figure. Now consider constraints to this problem as

$$x_2 - 1.4x_1 \leq 0,$$

$$2x_1 + 4x_2 \geq 6,$$

$$0 \leq x_2 \leq 6, \qquad \text{and} \qquad x_1 \geq 0.$$

The boundaries of constraints are shown in Figure 7.1. The shaded region is the permissible region,and it includes the global minimum. Hence constraints do not affect the solution of the unconstrained problem.

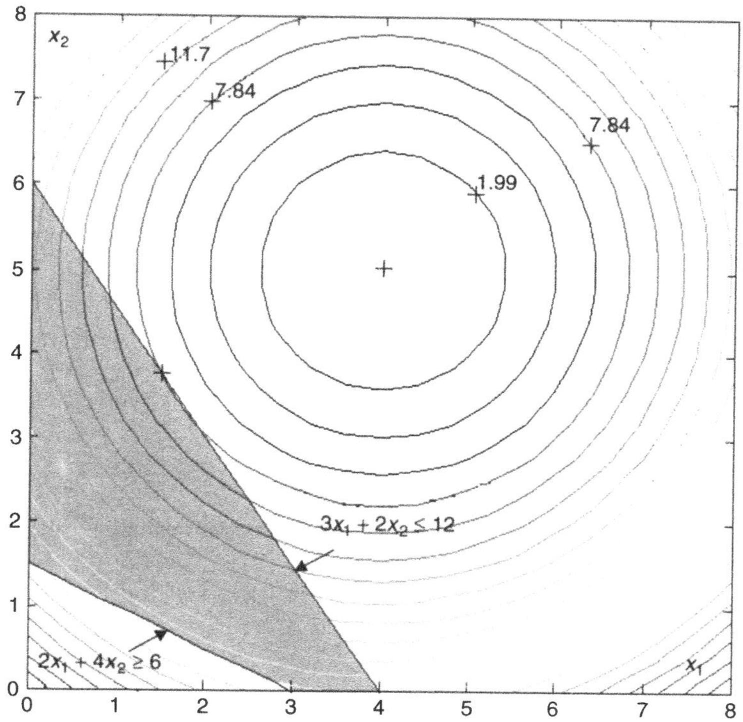

Figure 7.2. Depiction of constraints affecting the solution.

- *Constraints affect the solution.* Consider the above function which had an unconstrained minima at $[\bar{x}_1, \bar{x}_2] = [4, 5]$, but now with the following constraints:

$$3x_1 + 2x_2 \leq 12,$$

$$2x_1 + 4x_2 \geq 6,$$

 and $x_1, x_2 \geq 0$.

 Figure 7.2 depicts the shaded permissible region of the solution space. Evidently, the solution is where the contour line (circle in this example) is tangential to the constraint $3x_1 + 2x_2 \leq 12$. Unlike linear optimization in which the solution was at a vertex of the bounding region, this example illustrates that even when constraints are linear, the solution to nonlinear programs need not be at a vertex of the bounding polytope.

 In this example with linear constraints, the optimum is at a constraint boundary. Even if the constraint were to be nonlinear, the optimum can be on a constraint boundary. The fact that the optimum is at a constraint boundary can be detected by Kuhn–Tucker conditions to be described later.

- The solution space may be nonconvex; that is, it is disjointed.

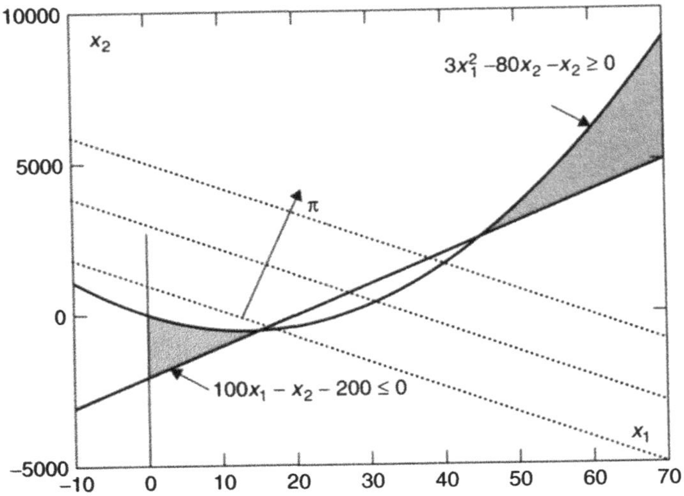

Figure 7.3. Maximization problem with nonconvex regions.

Consider the following problem:

Maximize

$$\pi = 85x_1 + x_2,$$

subject to

$$3x_1^2 - 80x_1 - x_2 \geq 0,$$

$$100x_1 - x_2 - 200 \leq 0, \text{‘}$$

and

$$x_1, x_2 \geq 0.$$

The domain of the problem and its parameters are shown in Figure 7.3. The permissible regions are shown shaded. The objective function π and direction of its increase are also indicated in the figure. The two regions are not even connected, indicating that the solution space is nonconvex. Therefore, if we start from the southwest region, we would stop at the first intersection point of the two curves as the solution to the maximum value of π. However, if we continue to go to the next region in the northeast, the value of the function can be further increased!

Furthermore, we note as well that the global maximum is unbounded. If there were an additional constraint (for example, $0 \leq x_1 \leq 70$), the constrained problem would have a bounded solution. However, even in that case, we would not know where to begin searching for the solution. Clearly, in multi-dimensional space, this can be a complex task because there may exist several disjointed regions.

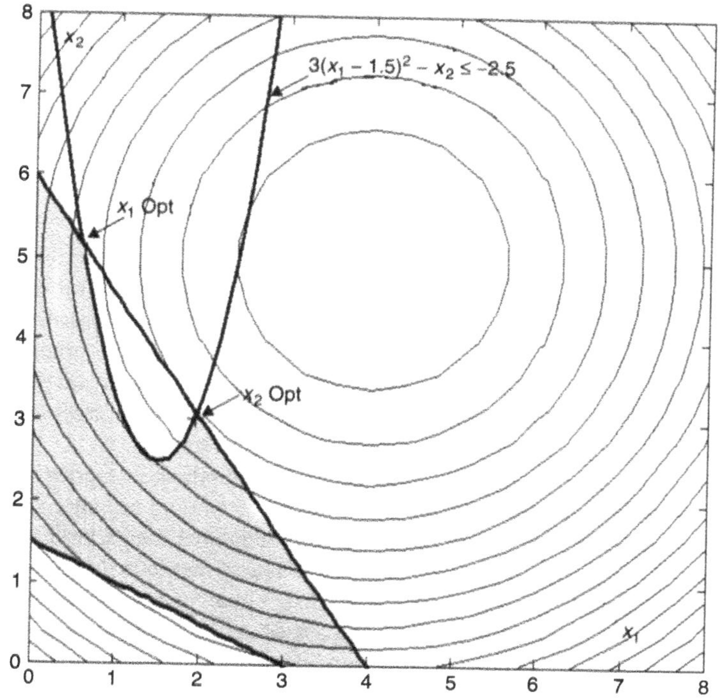

Figure 7.4. Introduction of relative minima by constraints.

- *Constraints Introduce Multiple Minima.* In some cases, although a function that is minimized may have a single global minimum, constraints may introduce multiple local relative minima. Consider the earlier example into which we introduce another constraint as follows:

  Minimize

$$(x_1 - 4)^2 + (x_2 - 5)^2,$$

  subject to

$$3x_1 + 2x_2 \le 12,$$

$$2x_1 + 4x_2 \ge 6,$$

$$3(x_1 - 1.5)^2 - x_2 + 2.5 \le 0,$$

  and $x_1, x_2 \ge 0$.

  The permissible region for this problem is shown in Figure 7.4, which is the same as Figure 7.2 but with an additional constraint. It can be seen that there are two relative minima, $x_1opt.$ and $x_2opt.$, for the problem.

- *Multiple Minima and Saddle Points.* It has been shown in Section 5.4.1 and in Figures 5.3 and 6.6 that the objective function itself may contain local and multiple minima.

It is evident from above examples that nonlinear programming can be complex. In contrast with the solution of linear programs being at the vertex of a polyhedra, the behavior of the solution space can be complex and may not be conceived *a priori* during the search for a solution.

7.2 METHODS FOR SOLVING NONLINEAR PROGRAMS

There are several methods for the solution of nonlinear programs. Indeed, even within a particular method, variations exist which make an algorithm more efficient, or particularly suited, for a class of problems. It is not possible to describe all the methods and their variations in a book where the emphasis is on the practical use of algorithms and software. Therefore, suffice it to say that methods are classified into two broad categories: direct methods and indirect methods. Indirect methods, by themselves, are many, including the penalty function methods.

In the following, some major and popular direct methods are described. Under the category of direct methods, we omit a description of random search and heuristic search methods because these are somewhat covered under the earlier chapter of unconstrained nonlinear optimization. Direct substitution methods and the method of constrained variations are briefly described. Gradient projection methods, in particular the reduced gradient method, are discussed at some length. A somewhat detailed description with examples is reserved for sequential linear programming and sequential quadratic programming methods because of their popularity. These methods are based on Lagrange multipliers discussed in Section 7.5 and are called Lagrange multiplier methods.

7.3 DIRECT SUBSTITUTION METHOD

Consider an optimization problem with equality constraints given by the following: Select an m vector $\mathbf{x}$ in order to minimize

$$f(\mathbf{x}), \tag{7.5}$$

subject to constraints

$$h_j = 0, \qquad j = 1, 2, \ldots, p. \tag{7.6}$$

In theory, problems with only equality constraints can be solved by the method of constrained variation. It is possible to solve simultaneously for p variables in the above in terms of $(m - p)$ remaining variables. Then by substituting the expressions for p variables in the original objective function, a new objective function not subject to any constraint will result. The optimum of such an unconstrained function can be computed by any of the techniques of the previous chapter.

Example: Consider the following problem:

Choose x_1, x_2, and x_3 to minimize

$$f(\mathbf{x}) = f(x_1, x_2, x_3) = (6x_1 + 12x_2)x_3, \tag{7.7}$$

subject to the nonlinear relations

$$2x_1^2 + 3x_2^2 + x_3 = 8,$$
$$4x_1^2 - 5x_2^2 - x_3 = 9.$$

These constraint equations can be written as

$$\begin{bmatrix} 2 & 3 \\ 4 & -5 \end{bmatrix} \begin{bmatrix} x_1^2 \\ x_2^2 \end{bmatrix} + \begin{bmatrix} 1 & 0 \\ 0 & 1 \end{bmatrix} \begin{bmatrix} x_3 \\ x_3 \end{bmatrix} = \begin{bmatrix} 8 \\ 9 \end{bmatrix}.$$

Since

$$\begin{bmatrix} 2 & 3 \\ 4 & -5 \end{bmatrix}^{-1} = \begin{bmatrix} 0.227 & -0.136 \\ 0.1818 & 0.0909 \end{bmatrix},$$

we get

$$\begin{bmatrix} x_1^2 \\ x_2^2 \end{bmatrix} + \begin{bmatrix} 0.227 & -0.136 \\ 0.1818 & 0.0909 \end{bmatrix} \begin{bmatrix} 1 & 0 \\ 0 & 1 \end{bmatrix} \begin{bmatrix} x_3 \\ x_3 \end{bmatrix}$$
$$= \begin{bmatrix} 0.227 & -0.136 \\ 0.1818 & 0.0909 \end{bmatrix} \begin{bmatrix} 8 \\ 9 \end{bmatrix},$$

which gives

$$x_1 = (-0.227\ x_3 + 0.5909)^{\frac{1}{2}} \tag{7.8}$$

and

$$x_2 = (-0.0909\ x_3 + 2.2727)^{\frac{1}{2}}. \tag{7.9}$$

A substitution of these values for x_1 and x_2 in (7.7) will transform the expression for $f(x_1, x_2, x_3)$ to a nonlinear function only in terms of x_3 which can be solved for its unconstrained optimum by setting the necessary condition $\partial f(\mathbf{x})/\partial x_3 = 0$. A substitution of the value of x_3 so obtained back into (7.8) and (7.9) will give the values for x_1 and x_2.

This procedure of direct substitution for solving problems with equality constraints may appear straightforward. In fact, it is of no practical use because the resulting nonlinear equations may be very complex, and even impossible, to solve. Therefore, this direct procedure is of use only for small and simple problems.

7.4 METHOD OF CONSTRAINED VARIATIONS

In the following, only basic principles of this method are addressed. However, we suggest that the advanced reader consult Rao (1996) and Pierre (1969) for details regarding the extension of the method to general problems. Such extensions offer additional information about the development of Lagrange multipliers.

Consider the following simple problem:

Minimize

$$f(\mathbf{x}) = f(x_1, x_2),$$

subject to the single equality constraint

$$h(\mathbf{x}) = h_1(x_1, x_2) = 0.$$

The necessary condition at optimum $\overline{\mathbf{x}}$ is that the total differential at $\overline{\mathbf{x}}$ is equal to zero; that is,

$$df = \frac{\partial f}{\partial x_1} dx_1 + \frac{\partial f}{\partial x_2} dx_2 = 0. \tag{7.10}$$

However, in order to honor the above equality constraint, we invoke the necessary relationship between changes dx_1 and dx_2. The variations of dx_1 and dx_2 cannot be arbitrary—any changes to x_1 and x_2 should keep $dh = 0$. From Taylor expansion, for dh we have

$$dh = \nabla h(\overline{x}_1, \overline{x}_2) + \frac{\partial h(\overline{x}_1, \overline{x}_2)}{\partial x_1} dx_1 + \frac{\partial h(\overline{x}_1, \overline{x}_2)}{\partial x_2} dx_2. \tag{7.11}$$

In order to have $dh = 0$, since $\nabla h(\overline{x}_1, \overline{x}_2) = 0$ at the minimum point, from the above we get

$$dx_2 = \frac{\partial h/\partial x_1}{\partial h/\partial x_2}. \tag{7.12}$$

Substitution of this in (7.10) gives

$$df = \left| \left(\frac{\partial f}{\partial x_1} - \frac{\partial h/\partial x_1}{\partial h/\partial x_2} \frac{\partial f}{\partial x_2} \right) \right|_{\mathbf{x}=\overline{\mathbf{x}}} dx_1 = 0. \tag{7.13}$$

To extend the above procedure to a general case of m variables and $n + p$ equality constraints (n inequalities converted to equality by adding dummy variables as shown later) is a straightforward procedure. Rao (1996) and Chiang (1984) show the procedure. It entails the solution of a set of homogeneous equations formed from the objective function and constraints using the Jacobian matrix. Such a task may be prohibitive for problems involving more than three constraints. Therefore, the procedure is very difficult to apply to general problems. The method of Lagrange multipliers discussed in the next section is preferred and is more commonly used to solve multivariate problems.

7.5 LAGRANGE MULTIPLIERS: KUHN–TUCKER CONDITIONS

Consider the following minimization problem:

Select an m vector $\mathbf{x}$ to minimize

$$f(\mathbf{x}), \tag{7.14}$$

subject to n inequality conditions

$$g_i(\mathbf{x}) \leq k_i, \qquad i = 1, 2, \ldots, n, \tag{7.15}$$

where k_i is a constant, $-\infty \leq k_i \leq \infty$, $i = 1, \cdots, n$, and p equality conditions[1]

$$h_j(\mathbf{x}) = 0, \qquad j = 1, 2, \ldots, p. \tag{7.16}$$

The inequality constraints (7.15) can be transformed to

$$g_i(\mathbf{x}) \leq 0, \qquad i = 1, 2, \ldots, n, \tag{7.17}$$

by transferring k to the left-hand side.

Clearly, the inequality condition of (7.17) can handle both $\leq$ and $\geq$ conditions. For example, any $\leq$ can be converted to $\geq$ by changing the sign of the left-hand side of (7.17). Similarly, the equality constraints of (7.16) can be converted to two inequalities, one with $\geq$ and the other with $\leq$, both with the same scalar on the right-hand side.

In the following theoretical development, we convert all constraints to equalities. In order to do so, we add an appropriate nonnegative slack variable[2] z_i^2, the value of which is yet unknown, to the ith $\leq$ constraint of (7.17).

This transforms (7.17) to

$$g_i(\mathbf{x}, \mathbf{z}) = g_i(\mathbf{x}) + z_i^2 = 0, \qquad i = 1, 2, \ldots, n \tag{7.18}$$

[1] We have used the convention of representing inequalities as $g_i(\mathbf{x})$ and equalities as $h_j(\mathbf{x})$. However, since inequalities are converted to equalities by adding slack variables, when we refer to $g(x)$ in a general sense, it implies all constraints, equalities as well as inequalities. It is only when we want to draw particular attention to equality constraints, or when we want to distinguish them from inequalities, that we use $h(\mathbf{x})$.

[2] For example, consider an inequality $g_i(\mathbf{x}) \leq -7$. We transform this to $g_i(\mathbf{x}) + 7 + z_i^2 = 0$. If the solution to the optimization results in $g_i(\mathbf{x}) = -8$, we obtain $z_i^2 = -(-8 + 7) = 1$. On the other hand, if we consider the example of an inequality constraint as $g_i(\mathbf{x}) \leq 10$, we transform it to $g_i(\mathbf{x}) - 10 + z_i^2 = 0$. If the solution to the problem gives $g_i = +4$, the value of z_i^2 is equal to +6. Two important observations are that z_i^2 is nonnegative and that if the value of z_i^2 is known, the value of the constraint equation can be computed (and vice versa). The above procedure can convert any inequality into equality.

which can be written in a vector form as

$$g(\mathbf{x}, \mathbf{z}) = g(\mathbf{x}) + \mathbf{z}^2 = 0, \tag{7.19}$$

where $\mathbf{z}^2$ is a n vector associated with n inequalities given by

$$\mathbf{z} = \begin{pmatrix} z_1^2 \\ z_2^2 \\ \vdots \\ z_n^2 \end{pmatrix}. \tag{7.20}$$

Now, in order to minimize $f(\mathbf{x})$, we minimize the *Lagrange function* (or the *Lagrangian*) defined as

$$L(\mathbf{x}, \mathbf{z}, \boldsymbol{\lambda}) = f(\mathbf{x}) + \sum_{i=1}^{n+p} \lambda_i g_i(\mathbf{x}, \mathbf{z}), \tag{7.21}$$

where $\boldsymbol{\lambda}$ is a vector of *Lagrange multipliers* which is given by $[\lambda_1, \lambda_2, \cdots \lambda_{n+p}]^T$.

In (7.21), all inequalities and equalities are combined. Some prefer to separate inequalities and equalities with their own Lagrange multipliers and express the relation as

$$L(\mathbf{x}, \mathbf{z}, \boldsymbol{\lambda}) = f(\mathbf{x}) + \sum_{i=1}^{n} \lambda_i g_i(\mathbf{x}, \mathbf{z}) + \sum_{k=1}^{p} \beta_k h_k(\mathbf{x}, \mathbf{z}), \tag{7.22}$$

in which the Lagrange multipliers for inequalities are $\boldsymbol{\lambda}$, and those for the equalities are $\boldsymbol{\beta}$.

When we minimize $L(\mathbf{x}, \mathbf{z}, \boldsymbol{\lambda})$ in (7.21), if we ensure that the second part on the right-hand side is zero, then it is the same as minimizing $f(\mathbf{x})$.

Necessary conditions for minimizing (7.21) require that

$$\frac{\partial L(\mathbf{x}, \mathbf{z}, \boldsymbol{\lambda})}{\partial x_j} = 0 = \frac{\partial f(\mathbf{x})}{\partial x_j} + \sum_{i=1}^{n+p} \lambda_i \frac{\partial g_i(\mathbf{x}, \mathbf{z})}{\partial x_j}, \quad j = 1, 2, \cdots, m \tag{7.23}$$

(m equations),

$$\frac{\partial L(\mathbf{x}, \mathbf{z}, \boldsymbol{\lambda})}{\partial \lambda_i} = 0 = \frac{\partial f(\mathbf{x})}{\partial \lambda_i} + g_i + z_i^2, \quad i = 1, 2 \cdots, (n+p) \tag{7.24}$$

($n + p$ equations),

and

$$\frac{\partial L(\mathbf{x}, \mathbf{z}, \boldsymbol{\lambda})}{\partial z_i} = 0 = 2 z_i \lambda_i, \quad i = 1, 2, \cdots, n \tag{7.25}$$

(n equations).

The above represents $m + p + 2n$ equations for $m + p + 2n$ unknowns, namely the m vector $\mathbf{x}$, the $(n + p)$ vector $\boldsymbol{\lambda}$, and the n vector of slack variables $\mathbf{z}^2$. Equation (7.24) is nothing but a restatement of constraints in (7.19) because the first term on the right-hand side, $\partial f(\mathbf{x})/\partial \lambda_i$, is zero since f is not a function of $\boldsymbol{\lambda}$.

Equations (7.25) is satisfied when either z_i is zero, or if λ_i is zero.[3] These relations are called *transversality conditions*.

If λ_i is zero, it means that z_i is nonzero, and hence the value of g_i is less than k_i in (7.15). Consequently the constraint is said to be *nonbinding* or *inactive*. On the other hand, if $z_i = 0$, the constraint is *active*[4] and *binding* since we observe from (7.18) that $g_i(\mathbf{x}) = 0$.

Of the n inequality constraints of (7.17), let a represent the number of active constraints.[5]

Since $z_i = 0$ when constraints are active, we ignore them in binding constraints and write the constraint set as

$$g_k(\mathbf{x}) = 0, \qquad k = 1, \ldots, a \tag{7.26}$$

and

$$g_r(\mathbf{x}) + z_r^2 = 0, \qquad r = a + 1, \ldots, n. \tag{7.27}$$

Clearly, since $\lambda_i \neq 0$ for active constraints, (7.23) can be written for a active constraints as

$$\frac{\partial L(\mathbf{x}, \mathbf{z}, \boldsymbol{\lambda})}{\partial x_j} = 0 = \frac{\partial f(\mathbf{x})}{\partial x_j} + \sum_{i=1}^{a} \lambda_i \frac{\partial g_i(\mathbf{x})}{\partial x_j}, \qquad j = 1, 2, \ldots, m \tag{7.28}$$

which in expanded form gives

$$\frac{-\partial f}{\partial x_1} = \lambda_1 \frac{\partial g_1}{\partial x_1} + \lambda_2 \frac{\partial g_2}{\partial x_1} + \cdots + \lambda_a \frac{\partial g_a}{\partial x_1} \tag{7.29}$$

$$\frac{-\partial f}{\partial x_2} = \lambda_1 \frac{\partial g_1}{\partial x_2} + \lambda_2 \frac{\partial g_2}{\partial x_2} + \cdots + \lambda_a \frac{\partial g_a}{\partial x_2} \tag{7.30}$$

$$\vdots = \vdots \qquad \vdots \qquad \vdots \tag{7.31}$$

$$\frac{-\partial f}{\partial x_m} = \lambda_1 \frac{\partial g_1}{\partial x_m} + \lambda_2 \frac{\partial g_2}{\partial x_m} + \cdots + \lambda_a \frac{\partial g_a}{\partial x_m}. \tag{7.32}$$

[3] This means that if one of either z_i or λ_i is nonzero, the other has to be necessarily zero. The second right-hand term in (7.21) will be zero if λ_i is zero, or if z_i is zero (which indicates a binding constraint). In either case, since the second right-hand term in (7.21) is zero, minimizing L is the same as minimizing $f(\mathbf{x})$.

[4] By active or binding constraint, we mean that the constraint (7.17), and of course (7.16), are satisfied with an equality sign, that is, $g_i(\mathbf{x}) = 0$.

[5] Since (7.21) includes p equalities and since all equalities are binding at the optimal solution, a has to be at least equal to or greater than p.

The above equation can be written as

$$-\nabla f(\mathbf{x}) = \lambda_1 \nabla g_1(\mathbf{x}) + \lambda_2 \nabla g_2(\mathbf{x}) + \cdots + \lambda_a \nabla g_a(\mathbf{x}), \tag{7.33}$$

in which $\nabla f(\mathbf{x})$ is the gradient of the objective function, and $\nabla g_a(\mathbf{x})$ is the gradient of the ath binding constraint.

Equation (7.33) indicates that *the negative gradient of the objective function can be expressed as a linear combination of the gradients of active constraints.*[6]

7.5.1 Kuhn–Tucker Conditions

Equation (7.28) for a binding constraints is

$$\frac{\partial L(\mathbf{x}, \mathbf{z}, \boldsymbol{\lambda})}{\partial x_j} = 0 = \frac{\partial f(\mathbf{x})}{\partial x_j} + \sum_{i=1}^{a} \lambda_i \frac{\partial g_i(\mathbf{x})}{\partial x_j}, \qquad j = 1, 2, \ldots, m \tag{7.34}$$

$$\lambda_i > 0, \qquad i = 1, 2, \ldots, a. \tag{7.35}$$

However, since active or binding constraints are not known *a priori*, a general way of writing the above conditions, called the Kuhn–Tucker conditions named after the mathematicians who derived them, is as follows.

If all n inequalities are transformed to equalities as in (7.18) and included with p equalities of the problem, the Kuhn–Tucker conditions are

$$\frac{\partial L(\mathbf{x}, \mathbf{z}, \boldsymbol{\lambda})}{\partial x_j} = 0 = \frac{\partial f(\mathbf{x})}{\partial x_j} + \sum_{i=1}^{n+p} \lambda_i \frac{\partial g_i(\mathbf{x})}{\partial x_j}, \qquad j = 1, 2, \ldots, m, \tag{7.36}$$

$$\lambda_i g_i = 0, \qquad i = 1, 2 \cdots, n + p, \tag{7.37}$$

$$\lambda_i \geq 0, \qquad i = 1, 2 \cdots, n + p. \tag{7.38}$$

$$g_i \leq 0, \qquad i = 1, 2 \cdots, n + p, \tag{7.39}$$

The above equations, without loss of generality, are applicable to equality as well as inequalities as transformed by (7.27). In order to clearly recognize that the Lagrange multiplier associated with equality constraints at the optimum is nonzero, and to explicitly show inequalities and equalities in the Lagrangian (7.36), the above conditions can be written in the following form.

[6] If there is only one constraint (line or surface), it can be seen from (7.33) that the gradient of the objective and that of the constraint are collinear. That means that the constraining line or surface is tangential to the function contour or surface.

For the problem choose x_i, $i = 1, m$, in order to minimize

$$f(\mathbf{x}),$$

subject to

$$g_i(\mathbf{x}) \leq 0, \qquad i = 1, 2 \cdots, n,$$
$$h_n(\mathbf{x}) = 0, \qquad n = 1, 2 \cdots, p.$$

The Kuhn–Tucker conditions become

$$\nabla f + \sum_{i=1}^{n} \lambda_i \nabla g_i + \sum_{n=1}^{p} \beta_n h_n, = 0 \tag{7.40}$$
$$\lambda_i g_i = 0, \qquad i = 1, 2, \ldots, n, \tag{7.41}$$
$$g_i \leq 0, \qquad i = 1, 2, \ldots, n, \tag{7.42}$$
$$\lambda_i \geq 0, \qquad i = 1, 2, \ldots, n, \tag{7.43}$$
$$h_n = 0, \qquad n = 1, 2, \ldots, p, \tag{7.44}$$
$$\beta_n \neq 0, \qquad n = 1, 2, \ldots, p, \tag{7.45}$$

where λ_i, and β_n denote Lagrange multipliers associated with constraints $g_i \leq 0$ and $h_n = 0$, respectively. Unlike the sign of $\boldsymbol{\lambda}$, the sign of β_n can be either positive or negative.

Kuhn–Tucker conditions are necessary conditions to ensure a relative minimum (or maximum). However, they are not sufficient conditions to ensure a relative minimum for general problems. Only for a class of problems called *convex programming problems* [see Rao (1996), Bazaraa et al. (1979), and Gill et al. (1981)], Kuhn–Tucker conditions are necessary and sufficient conditions to ensure a relative extremum.

We see from Kuhn–Tucker conditions (7.41), if λ_i is nonzero, $g_i = 0$. If a constraint has a upper and lower bound, if the optimal solution requires that the solution is at its upper or lower limit, then the value of g_i is zero. Correspondingly, if $g \neq 0$, then $\lambda_i = 0$.

For the conventions adopted, we observe that the Lagrange multiplier is negative if the constraint is at its lower limit, positive if the constraint is at its upper limit.[7] If the constraint is unbounded (free), since g_i is nonzero, the Lagrange multiplier is zero because of (7.41).

For example, if the constraint g_i for a single variable x_1 is $0 \leq (x_1 - 5) \leq 10$, if the optimum solution to minimize some objective function requires $x_1 = 15$, then λ_i associated with this constraint is positive, and $g_i = x_1 - 5 = 10$, satisfying

[7]See Rao (1996) and Bazaraa et al. (1979) for a detailed proof of these propositions.

the upper bound. Similarly, if the optimum solution requires $x_1 = 5$ satisfying the lower bound, the Lagrange multiplier will be negative because $g_i = 0$. However, if the solution requires $x_1 = 7$, then the associated Lagrange multiplier will be zero because neither $g_i(7) = 0$ nor $g_i(7) = 10$.

Cautionary Note: We note that the signs of Lagrange multipliers depend on the conventions used and the type of constraint. If the constraint g_i were to be of the $\geq$ type, the sign of the multipliers would be the opposite of above. Similarly, the signs would be the opposite of those indicated above if the problem is of maximization type. Furthermore, we have conventionally transformed the inequality constraints of the type $g_i \leq k$ to $g_i - k \leq 0$ or to $g_i - k + z^2 = 0$. However, if we were to transform them as $k - g_i \geq 0$, the signs of the Lagrange multipliers will change.

For minimization problems using Excel Solver, λ_i is positive at the lower bound of a constraint i and is negative at the upper bound.

7.6 INTERPRETATION OF LAGRANGE MULTIPLIERS

Consider the objective function with two choice variables and one inequality constraint as follows:

Minimize

$$f(x_1, x_2), \tag{7.46}$$

subject to

$$g(x_1, x_2) + z = 0 \tag{7.47}$$

where the inequality constraint has been expressed as an equality by the addition of a nonnegative slack variable z.

The Lagrangian function is given by

$$L = f(x_1, x_2) + \lambda[g(x_1, x_2) + z]. \tag{7.48}$$

At the optimum value $\overline{x}_1$ and $\overline{x}_2$, the variation of the Lagrangian with respect to z is

$$\begin{aligned} \frac{\partial L}{\partial z} &= \frac{\partial f}{\partial x_1}\frac{\partial x_1}{\partial z} + \frac{\partial f}{\partial x_2}\frac{\partial x_2}{\partial z} + \frac{\partial \lambda}{\partial z}[g(x_1, x_2) + z] \\ &\quad + \lambda\left[\frac{\partial g}{\partial x_1}\frac{\partial x_1}{\partial z} + \frac{\partial g}{\partial x_2}\frac{\partial x_2}{\partial z} + 1\right]. \end{aligned} \tag{7.49}$$

A rearrangement of terms of this equation gives

$$\frac{\partial L}{\partial z} = \left[\frac{\partial f}{\partial x_1} + \lambda\frac{\partial g}{\partial x_1}\right]\frac{\partial x_1}{\partial z} + \left[\frac{\partial f}{\partial x_2} + \lambda\frac{\partial g}{\partial x_2}\right]\frac{\partial x_2}{\partial z} + \frac{\partial \lambda}{\partial z}\left[g(x_1, x_2) + z\right] + \lambda. \tag{7.50}$$

The first three right-hand terms in (7.50) are zero because of (7.28) and (7.47). Hence from the above we get

$$\frac{\partial L}{\partial z} = \lambda, \tag{7.51}$$

indicating that *the sensitivity of the Lagrangian function (and hence the objective function) to changes in the constraint is given by the Lagrange multiplier.*

We now extend the above problem to one with two binding constraints as follows:

Minimize

$$f(x_1, x_2), \tag{7.52}$$

subject to

$$g_1(x_1, x_2) + z_1 = 0 \tag{7.53}$$

and

$$g_2(x_1, x_2) + z_2 = 0. \tag{7.54}$$

Performing partial differentiation as in (7.49), we obtain (reader to confirm algebra)

$$\frac{\partial L}{\partial z_1} = \left[\frac{\partial f}{\partial x_1} + \lambda_1\frac{\partial g_1}{\partial x_1} + \lambda_2\frac{\partial g_2}{\partial x_1}\right]\frac{\partial x_1}{\partial z_1} + \left[\frac{\partial f}{\partial x_2} + \lambda_1\frac{\partial g_1}{\partial x_2} + \lambda_2\frac{\partial g_2}{\partial x_2}\right]\frac{\partial x_2}{\partial z_1} + \frac{\partial \lambda_1}{\partial z_1}\left[g_1(x_1, x_2) + z_1\right] + \frac{\partial \lambda_2}{\partial z_1}\left[g_2(x_1, x_2) + z_2\right] + \lambda_1, \tag{7.55}$$

where λ_1 and λ_2 are nonzero Lagrange multipliers associated with the two binding constraints g_1 and g_2.

As before, all the right-hand-side terms with the exception of λ_1 are zero because of (7.28) and (7.47). Thus the Lagrange multiplier λ_1 indicates the change in the objective function with changes to the constraint g_1. Similarly, we can obtain an expression for $\partial L/\partial z_2$ in order to confirm that λ_2 is the sensitivity of the objective function to changes in the constraint g_2.

7.7 CONSTRAINT QUALIFICATION

In view of (7.22), which explicitly expresses equality and inequality constraints, consider equations (7.40) to (7.45).

If $\bar{\mathbf{x}}$ is a feasible solution to the problem, then for a solution to exist such that $\bar{\mathbf{x}}$, $\boldsymbol{\lambda}$, and $\boldsymbol{\beta}$ satisfy these equations, $\nabla g_i(\bar{\mathbf{x}}),\ i = 1, 2, \ldots, n$ should be linearly independent, and $\nabla h_k(\bar{\mathbf{x}}),\ k = 1, 2, \ldots, p$ should be linearly independent. For a proof of this see Bazaraa et al. (1979).

The above condition that ∇g_i and ∇h_k should be linearly independent is called *constraint qualification*. If the constraint qualification is not satisfied at the optimum point, the problem may not have a solution.

It is difficult to apply this qualification test in practice because $\bar{\mathbf{x}}$ should be known *a priori* to verify this test. However, it can be proved (Bazaraa et al., 1979) that problems with the following characteristics always satisfy the qualification test:

- All equality and inequality constraints are linear.
- All equality constraints are linear, all inequality constraints are convex, and there exists at least one vector $\mathbf{w} = (x_1, x_2, \ldots, x_m)$ that lies strictly within the feasible region satisfying the relations

$$g_i(\mathbf{w}) < 0,\ i = 1, 2, \ldots, n, \quad \text{and} \quad h_k(\mathbf{w}) = 0,\ k = 1, 2, \ldots, p.$$

Example: Consider the problem:
Minimize $f(\mathbf{x}) = x_1^2 + x_2^2$, subject to constraints

$$g_1 = x_2 + x_1^2 - 2, \leq 0,$$

and

$$g_2 = x_2 - x_1^2 - 2 \geq 0.$$

Figure 7.5 shows a contour plot of the function and constraints. The feasible regions are shown shaded. From an examination of this figure, we see that the solution lies at the intersection of the two feasible regions, namely, $\mathbf{x}^* = (0, 2)$. The gradients of the two constraints are shown in the figure, and they are collinear as confirmed by the following:

$$\nabla g_1 = \begin{bmatrix} 2x_1 \\ 1 \end{bmatrix}_{(0,2)} = \begin{bmatrix} 0 \\ 1 \end{bmatrix},$$

and

$$\nabla g_2 = \begin{bmatrix} -2x_1 \\ 1 \end{bmatrix}_{(0,2)} = \begin{bmatrix} 0 \\ 1 \end{bmatrix},$$

which indicates that the two gradients are not linearly independent.

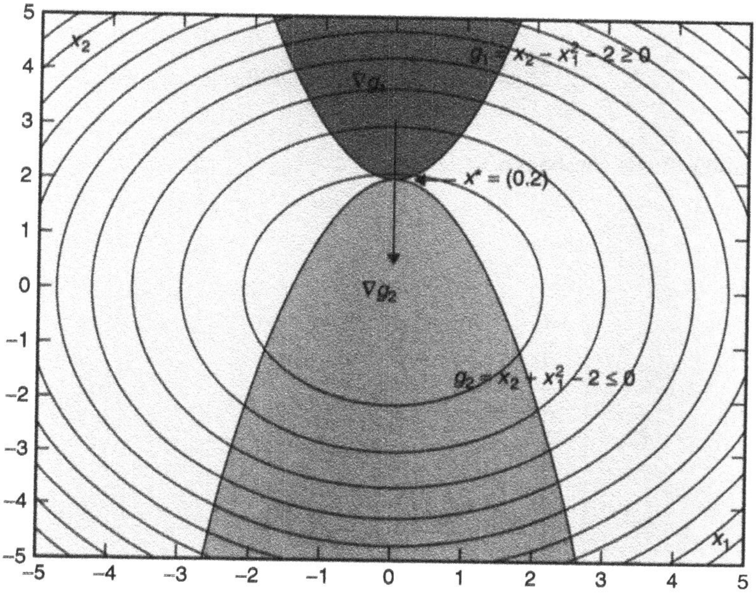

Figure 7.5. Illustration of constraint qualification.

Since

$$\nabla f(\bar{\mathbf{x}}) = \begin{bmatrix} 2x_1 \\ 2x_2 \end{bmatrix}_{(0.2)} = \begin{bmatrix} 0 \\ 4 \end{bmatrix},$$

Substituting ∇g_1 and ∇g_2 in the above, we get Kuhn–Tucker conditions of (7.34) and (7.35) as

$$0 + \lambda_1(0) + \lambda_2(0) = 0, \tag{7.56}$$

$$4 + \lambda_1(1) + \lambda_2(1) = 0, \tag{7.57}$$

$$\lambda_1 > 0, \tag{7.58}$$

$$\lambda_2 > 0. \tag{7.59}$$

Clearly, since (7.58), and (7.59) require λ_1 and λ_2 to be positive, (7.57) cannot be satisfied. Hence, Kuhn–Tucker conditions are not satisfied at the optimum point.

7.8 TESTING KUHN–TUCKER CONDITIONS

The necessary condition that a point $\mathbf{x}$ is an optimum is that the Kuhn–Tucker conditions (7.34) and (7.35) should be satisfied. For the active constraints, these conditions can be written as

$$\sum_{i=1}^{a} -\lambda_i \frac{\partial g_i(\mathbf{x})}{\partial x_j} = -\frac{\partial f(\mathbf{x})}{\partial x_j}, \qquad j = 1, 2, \ldots, m. \tag{7.60}$$

Equation (7.60) can be written as

$$\mathbf{G}\boldsymbol{\lambda} = F, \tag{7.61}$$

in which

$$\mathbf{G} = \begin{bmatrix} \frac{\partial g_1}{\partial x_1} & \frac{\partial g_2}{\partial x_1} & \cdots & \frac{\partial g_a}{\partial x_1} \\ \frac{\partial g_1}{\partial x_2} & \frac{\partial g_2}{\partial x_2} & \cdots & \frac{\partial g_a}{\partial x_2} \\ \vdots & \vdots & \vdots & \vdots \\ \frac{\partial g_1}{\partial x_m} & \frac{\partial g_2}{\partial x_m} & \cdots & \frac{\partial g_a}{\partial x_m} \end{bmatrix}, \tag{7.62}$$

$$\boldsymbol{\lambda} = \begin{bmatrix} -\lambda_1 \\ -\lambda_2 \\ \vdots \\ -\lambda_a \end{bmatrix}, \quad \text{and} \quad \mathbf{F} = \begin{bmatrix} \frac{\partial f(\mathbf{x})}{\partial x_1} \\ \frac{\partial f(\mathbf{x})}{\partial x_2} \\ \vdots \\ \frac{\partial f(\mathbf{x})}{\partial x_m} \end{bmatrix}. \tag{7.63}$$

From the above, we can write the following for $\boldsymbol{\lambda}$:

$$\boldsymbol{\lambda} = [\mathbf{G}^T\mathbf{G}]^{-1}\mathbf{G}^T\mathbf{F}. \tag{7.64}$$

In this expression, we require all the components of the vector $\boldsymbol{\lambda}$ to be positive in order to satisfy the Kuhn–Tucker conditions.

We now prove that if point $\mathbf{x}_t$ is the optimum point, the Lagrange multipliers associated with the binding constraints must be positive (for minimization problems).

Consider a situation when there are two binding constraints. Then from (7.33) we have

$$-\nabla f(\mathbf{x}) = \lambda_1 \nabla g_1(\mathbf{x}) + \lambda_2 \nabla g_2(\mathbf{x}). \tag{7.65}$$

If there is a feasible direction s to reduce f, a move along that direction requires the constraints g_1 and g_2 to be honored, thus requiring $\mathbf{s}^T g$ to be < 0. Multiplying (7.65) by s, we get

$$-\mathbf{s}^T \nabla f(\mathbf{x}) = \lambda_1 \mathbf{s}^T \nabla g_1(\mathbf{x}) + \lambda_2 \mathbf{s}^T \nabla g_2(\mathbf{x}). \tag{7.66}$$

If $\lambda_1 > 0$ and $\lambda_2 > 0$, since $\mathbf{s}^T g$ has to be < 0, the term $\mathbf{s}^T \nabla f(\mathbf{x})$ in (7.66) should be positive. Since ∇f indicates the direction along which the function value increases at a maximum rate, $\mathbf{s}^T \nabla f$ is the component of increase in f along the direction $\mathbf{s}$. If $\mathbf{s}^T \nabla f(\mathbf{x}) > 0$, the function value increases as we move along $\mathbf{s}$. Hence, if λ_1 and λ_2 are positive, it is not possible to find a direction in the feasible domain along which the function value can be further decreased. Since there is no further feasible direction to reduce f, our assumption that we are already at an optimal point is confirmed. Consequently, $\mathbf{x}_t$ is the optimum, and it is only possible when λ_1 and λ_2 are positive. This completes the proof.

In order to apply the above mentioned test for Kuhn–Tucker conditions, the difficulty is in knowing *a priori* which are the active constraints. During computation, no constraint will be *exactly* equal to zero at the optimum point $\mathbf{x}$. Therefore, we take a constraint to be active whenever

$$\|g_i(\mathbf{x})\| \leq \varepsilon,$$

where ε is a small tolerance value of the order of 10^{-4} or less.

The application of the above test is made clearer by examples in Section 7.15.

7.9 NATURE OF SOLUTION OF NONLINEAR PROGRAMS

In the following, we examine the general procedure to obtain the solution to a constrained nonlinear minimization problem using Kuhn–Tucker conditions.

Consider the problem of finding the m vector $\mathbf{x}$ which minimizes $f(\mathbf{x})$, subject to p inequalities and n equalities expressed as

$$h(\mathbf{x}), \qquad k = 1, 2, \ldots, n + p, \tag{7.67}$$

where p inequalities have been converted to equalities by adding variables $\mathbf{z}$ as in (7.18).

The Lagrange function corresponding to this problem is given by

$$L = f(\mathbf{x}) + \sum_{k=1}^{n+p} \lambda_k h_k(\mathbf{x}), \tag{7.68}$$

where λ_k is the Lagrange multiplier of the kth constraint.

The Kuhn–Tucker necessary conditions give

$$\nabla L = \nabla f(\mathbf{x}) + \sum_{k=1}^{n+p} \lambda_k \nabla h_k(\mathbf{x}) = 0 \tag{7.69}$$

and

$$h_k(\mathbf{x}) = 0, \qquad k = 1, \ldots, n + p. \tag{7.70}$$

Equation (7.69) can be written as

$$\nabla f + [\mathbf{G}]^T \boldsymbol{\lambda} = 0, \tag{7.71}$$

where $\mathbf{G}$ is a $m \times (n+p)$ matrix whose ith column denotes the gradient of the ith constraint h_i.

Equations (7.71) and (7.70) represent $m+n+p$ nonlinear equations in $m+n+p$ unknowns (x_i, $i = 1, 2, \ldots, m$ and λ_k, $k = 1, 2, \ldots, n+p$). These nonlinear equations can be solved using the Jacobian matrix method as shown in Section 5.3.2. Equations (7.69) and (7.70) can be written in compact form as

$$\mathbf{F}(\mathbf{z}) = \mathbf{0}, \tag{7.72}$$

where

$$\mathbf{F} = \left\{ \begin{matrix} \nabla L \\ h \end{matrix} \right\}_{(m+n+p)}, \qquad \mathbf{z} = \left\{ \begin{matrix} \mathbf{x} \\ \boldsymbol{\lambda} \end{matrix} \right\}_{(m+n+p)}, \qquad \mathbf{0} = \left\{ \begin{matrix} \mathbf{0} \\ \mathbf{0} \end{matrix} \right\}_{(m+n+p)}, \tag{7.73}$$

with the dimensions of vectors indicated by subscripts.

Starting from an assumed value of $\mathbf{z}^j$ in iteration j, corrections $\Delta\mathbf{z}^{(j)}$ to obtain $\mathbf{z}^{(j+1)}$ are obtained by using the Jacobian matrix method as follows. The corrections $\Delta\mathbf{z}^{(j)}$ in

$$\mathbf{z}^{(j+1)} = \mathbf{z}^{(j)} + \Delta\mathbf{z}^{(j)} \tag{7.74}$$

are given by (see (5.41))

$$\mathbf{J}\Delta\mathbf{z}^{(j)} = \text{error} = \mathbf{0} - F(\mathbf{z}^{(j)}). \tag{7.75}$$

Since $\mathbf{J} = [\nabla F]^T$, from the above we get

$$[\nabla F]_j^T \, \Delta\mathbf{z}^{(j)} = -F(\mathbf{z}^{(j)}) \tag{7.76}$$

where $[\nabla F] = \left[\nabla F(\mathbf{z}^{(j)})\right]$ is the Jacobian matrix of the nonlinear equations whose ith columns represent the gradient of the ith nonlinear equation $F_i(\mathbf{z})$ with respect to the $(m+n+p)$ vector $\mathbf{z}$.

A substitution of (7.72) and (7.73) into (7.76) results in

$$\begin{bmatrix} [\nabla^2 L] & [\mathbf{G}] \\ [\mathbf{G}]^T & [\mathbf{0}] \end{bmatrix}^{(j)} \begin{bmatrix} \Delta\mathbf{x}^{(j)} \\ \Delta\boldsymbol{\lambda}^{(j)} \end{bmatrix} = -\begin{bmatrix} \nabla L \\ h \end{bmatrix}^{(j)}, \tag{7.77}$$

where $[\nabla^2 L]_{m+n}$ is the Hessian matrix of the Lagrange function with m design variables and n variables z_n^2 arising from n inequalities of (7.18). The algorithm to obtain the successive iterates of $\mathbf{x}$ and the Lagrange multipliers $\boldsymbol{\lambda}$ is

$$\mathbf{x}^{(j+1)} = \mathbf{x}^{(j)} + \Delta\mathbf{x}^{(j)}, \tag{7.78}$$

$$\boldsymbol{\lambda}^{(j+1)} = \boldsymbol{\lambda}^{(j)} + \Delta\boldsymbol{\lambda}^{(j)}. \tag{7.79}$$

The following example will clarify the procedure.

7.9.1 Examples

Example 1: Select x_1, x_2, and x_3 to minimize

$$f(\mathbf{x}) = f(x_1, x_2, x_3) = 4x_1^2 + 3x_2^2 + 5x_3^2 - 6x_1x_2 + x_1x_3 - 3x_1 - 2x_2 + 15, \tag{7.80}$$

subject to one inequality ($n = 1$) and two equalities ($p = 2$) as follows:

$$g(\mathbf{x}) = x_1^2 + x_3^2 - 30 \leq 0, \tag{7.81}$$

$$h_2(\mathbf{x}) = 2x_1 + 3x_2^3 + x_3 - 8 = 0, \tag{7.82}$$

$$h_3(x) = 4x_1 + 5x_2 - x_3^3 - 9 = 0. \tag{7.83}$$

We first add a nonnegative slack variable, x_4^2 to the inequality of (7.81) to convert it to an equality constraint $h_1(\mathbf{x})$ as

$$h_1(\mathbf{x}) = x_1^2 + x_3^2 - 30 + x_4^2 = 0. \tag{7.84}$$

To simplify matters, we have not included any range constraints for choice variables. Such constraints can be easily treated by adding positive variables as in the case of above inequality.[8]

The Lagrangian is given by

$$\begin{aligned} L = &\left[4x_1^2 + 3x_2^2 + 5x_3^2 - 6x_1x_2 + x_1x_3 - 3x_1 - 2x_2 + 15\right] \\ &+ \lambda_1\left[x_1^2 + x_3^2 - 30 + x_4^2\right] + \lambda_2\left[2x_1 + 3x_2^3 + x_3 - 8\right] \\ &+ \lambda_3\left[4x_1 + 5x_2 - x_3^3 - 9\right]. \end{aligned} \tag{7.85}$$

[8]If the m design variables x are required to be ≥ 0, surplus variables w_m^2 are introduced. The relation $x_m \geq 0$ now becomes the equality $-x_j + w_j^2 = 0,\ j = 1, \ldots, m$, which is handled in exactly the same way as other equality relationships arising from the addition of slack variables to inequalities.

The Kuhn–Tucker necessary condition requires that the gradient of the Lagrangian with respect to four choice variables $x_1, \ldots, x_4$ is zero. That is,

$$\nabla L = \begin{bmatrix} 8x_1 - 6x_2 + x_3 - 3 + 2\lambda_1 x_1 + 2\lambda_2 + 4\lambda_3 \\ 6x_2 - 6x_1 - 2 + 9\lambda_2 x_2^2 + 5\lambda_3 \\ 10x_3 + x_1 + 2\lambda_1 x_3 + \lambda_2 - 3\lambda_3 x_3^2 \\ 2\lambda_1 x_4 \end{bmatrix} = \mathbf{0}. \tag{7.86}$$

Corresponding to (7.71), the above can be written as

$$\nabla f + [\mathbf{G}]\lambda = 0, \tag{7.87}$$

where

$$\nabla f = \begin{bmatrix} 8x_1 - 6x_2 + x_3 - 3 \\ 6x_2 - 6x_1 - 2 \\ 10x_3 + x_1 \end{bmatrix} \tag{7.88}$$

and

$$\mathbf{G} = \begin{bmatrix} \frac{\partial h_1}{\partial x_1} & \frac{\partial h_2}{\partial x_1} & \frac{\partial h_3}{\partial x_1} \\ \frac{\partial h_1}{\partial x_2} & \frac{\partial h_2}{\partial x_2} & \frac{\partial h_3}{\partial x_2} \\ \frac{\partial h_1}{\partial x_3} & \frac{\partial h_2}{\partial x_3} & \frac{\partial h_3}{\partial x_3} \\ \frac{\partial h_1}{\partial x_4} & \frac{\partial h_2}{\partial x_4} & \frac{\partial h_3}{\partial x_4} \end{bmatrix} = \begin{bmatrix} 2x_1 & 2 & 4 \\ 0 & 9x_2^2 & 5 \\ 2x_3 & 1 & -3x_3^2 \\ 2x_4 & 0 & 0 \end{bmatrix}. \tag{7.89}$$

Equation (7.87) along with constraint equations (7.82), (7.83), and (7.84) represent an $m + p + (n + p)$ nonlinear equation system with $3 + 1 + 3$ variables (x_1, x_2, x_3, x_4, and $\lambda_1, \lambda_2, \lambda_3$). In these, the unknowns are choice variables or design variables x_1, x_2, x_3, and x_4, and three Lagrange multipliers λ_1, λ_2, and λ_3 correspond to the sum of the number of inequalities and equalities, $(p + n)$. Hence, the total number of unknowns are $m + p + p + n$. In order to solve these equations ((7.86) and constraints), we make use of the Jacobian matrix as shown by the procedure in Section 5.3.2 as follows.

Assuming $\mathbf{z}^{(i)}$ as the solution in iteration i, the necessary changes $\Delta\mathbf{z}^{(i)}$ to generate an improved solution for the next iteration is given by

$$\mathbf{z}^{(i+1)} = \mathbf{z}^{(i)} + \Delta\mathbf{z}^{(i)}, \tag{7.90}$$

where $\mathbf{z}$ represents $[x_1, x_2, x_3, x_4, \lambda_1, \lambda_2, \lambda_3]^T$.

The corrections $\Delta\mathbf{z}^{(i)}$ necessary to improve the solution are obtained by

$$\Delta\mathbf{z}^{(i)} = -\mathbf{J}^{-1}\left\{\begin{array}{c}\nabla L(\mathbf{z}^{(i)}) - \mathbf{0}\\ h(\mathbf{z}^{(i)}) - \mathbf{0}\end{array}\right\} = -\mathbf{J}^{-1}\left\{\begin{array}{c}\nabla L(\mathbf{z}^{(i)})\\ \hline h(\mathbf{z}^{(i)})\end{array}\right\}, \tag{7.91}$$

in which the jth column of the Jacobian matrix denotes the gradient of the system of equations comprising (7.86) and constraints as the block matrix

$$\mathbf{J} = \left[\begin{array}{cccc|ccc}\dfrac{\partial\nabla L}{\partial x_1} & \dfrac{\partial\nabla L}{\partial x_2} & \dfrac{\partial\nabla L}{\partial x_3} & \dfrac{\partial\nabla L}{\partial x_4} & \dfrac{\partial\nabla L}{\partial\lambda_1} & \dfrac{\partial\nabla L}{\partial\lambda_2} & \dfrac{\partial\nabla L}{\partial\lambda_3}\\ \hline \dfrac{\partial h}{\partial x_1} & \dfrac{\partial h}{\partial x_2} & \dfrac{\partial h}{\partial x_3} & \dfrac{\partial h}{\partial x_4} & \dfrac{\partial h}{\partial\lambda_1} & \dfrac{\partial h}{\partial\lambda_2} & \dfrac{\partial h}{\partial\lambda_3}\end{array}\right], \tag{7.92}$$

where h signifies the three constraints h_1, h_2, and h_3 of (7.84), (7.82), and (7.83).

From (7.82), (7.83), (7.84), and (7.86), we compute $\mathbf{J}$ corresponding to (7.92) as

$$\mathbf{J} = \left[\begin{array}{cccc|ccc}8+2\lambda_1 & -6 & 1 & 0 & 2x_1 & 2 & 4\\ -6 & 6+18\lambda_2 x_2 & 0 & 0 & 0 & 9x_2^2 & 5\\ 1 & 0 & 10+2\lambda_1+6\lambda_3 x_3 & 0 & 2x_3 & 1 & -3x_3^2\\ 0 & 0 & 0 & 2\lambda_1 & 2x_4 & 0 & 0\\ \hline 2x_1 & 0 & 2x_3 & 2x_4 & 0 & 0 & 0\\ 2 & 9x_2^2 & 1 & 0 & 0 & 0 & 0\\ 4 & 5 & -3x_3^2 & 0 & 0 & 0 & 0\end{array}\right]. \tag{7.93}$$

Note that $\mathbf{J}$ in (7.93) is of the form shown in (7.77), namely,

$$\mathbf{J} = \begin{bmatrix}[\nabla^2 L] & [\mathbf{G}]\\ [\mathbf{G}]^T & [\mathbf{0}]\end{bmatrix}, \tag{7.94}$$

where $\nabla^2 L$ is the Hessian of L with respect to the four choice variables x_1, x_2, x_3, and x_4. It is expected of the reader to confirm this by performing the necessary differentiation of the Lagrangian in (7.85) to obtain the Hessian matrix. The partitioned matrices in (7.93) correspond to the elements of $\mathbf{J}$ in (7.94).

It can be shown by some algebraic manipulation (Rao, 1996) that (7.77) can be transformed to

$$\begin{bmatrix}[\nabla^2 L] & [\mathbf{G}]\\ [\mathbf{G}]^T & [\mathbf{0}]\end{bmatrix}^{(j)}\begin{bmatrix}\Delta\mathbf{x}^{(j)}\\ \boldsymbol{\lambda}^{(j+1)}\end{bmatrix} = -\begin{bmatrix}\nabla f\\ h\end{bmatrix}^{(j)}. \tag{7.95}$$

By comparing the above with (7.77), carefully observe the difference between the two procedures. A subsequent example will employ both procedures to show

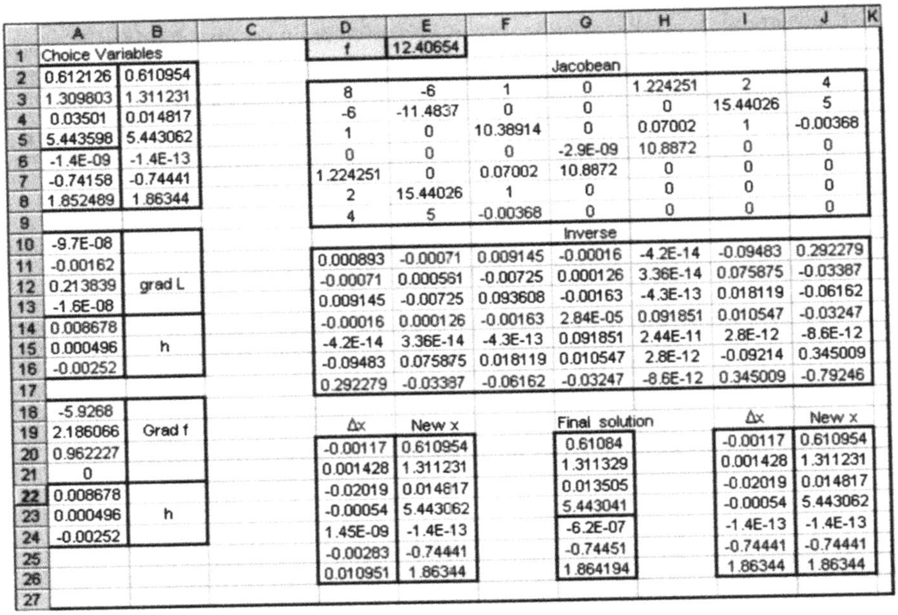

	A	B	C	D	E	F	G	H	I	J	K
1	Choice Variables			f	12.40654						
2	0.612126	0.610954					Jacobean				
3	1.309803	1.311231		8	-6	1	0	1.224251	2	4	
4	0.03501	0.014817		-6	-11.4837	0	0	0	15.44026	5	
5	5.443598	5.443062		1	0	10.38914	0	0.07002	1	-0.00368	
6	-1.4E-09	-1.4E-13		0	0	0	-2.9E-09	10.8872	0	0	
7	-0.74158	-0.74441		1.224251	0	0.07002	10.8872	0	0	0	
8	1.852489	1.86344		2	15.44026	1	0	0	0	0	
9				4	5	-0.00368	0	0	0	0	
10	-9.7E-08						Inverse				
11	-0.00162			0.000893	-0.00071	0.009145	-0.00016	-4.2E-14	-0.09483	0.292279	
12	0.213839	grad L		-0.00071	0.000561	-0.00725	0.000126	3.36E-14	0.075875	-0.03387	
13	-1.6E-08			0.009145	-0.00725	0.093608	-0.00163	-4.3E-13	0.018119	-0.06162	
14	0.008678			-0.00016	0.000126	-0.00163	2.84E-05	0.091851	0.010547	-0.03247	
15	0.000496	h		-4.2E-14	3.36E-14	-4.3E-13	0.091851	2.44E-11	2.8E-12	-8.6E-12	
16	-0.00252			-0.09483	0.075875	0.018119	0.010547	2.8E-12	-0.09214	0.345009	
17				0.292279	-0.03387	-0.06162	-0.03247	-8.6E-12	0.345009	-0.79246	
18	-5.9268										
19	2.186066	Grad f		Δx	New x		Final solution		Δx	New x	
20	0.962227			-0.00117	0.610954		0.61084		-0.00117	0.610954	
21	0			0.001428	1.311231		1.311329		0.001428	1.311231	
22	0.008678			-0.02019	0.014817		0.013505		-0.02019	0.014817	
23	0.000496	h		-0.00054	5.443062		5.443041		-0.00054	5.443062	
24	-0.00252			1.45E-09	-1.4E-13		-6.2E-07		-1.4E-13	-1.4E-13	
25				-0.00283	-0.74441		-0.74451		-0.74441	-0.74441	
26				0.010951	1.86344		1.864194		1.86344	1.86344	
27											

Figure 7.6. Sequential solution of (7.80) using the inverse of a Jacobian matrix.

that they do indeed converge to the same solution. In addition, we will later use (7.95) to solve sequential quadratic programs.

From the above equation, in jth iteration, knowing the Hessian of the Lagrange function ($\nabla^2 L$ matrix) and the gradient of the objective function and the constraint, changes to the values of $\mathbf{x}$ of the jth iteration and the values of Lagrange multipliers $\boldsymbol{\lambda}^{(j+1)}$ for the next iteration can be computed. Obviously, the values of $\mathbf{x}$ in the $(j+1)$th iteration is given by $\mathbf{x}^{(j+1)} = \mathbf{x}^{(j)} + \Delta\mathbf{x}^{(j)}$.

NLP.Xls *Solution:* Figure 7.6 shows the Excel spreadsheet "NLP.Xls" in which cells A2:A8 are choice variables. The Jacobian matrix per (7.93) is entered in cells D3:J9, and its inverse is obtained in cells D11:J17. The right-hand-side vector corresponding to (7.77) is evaluated in cells A10:A16, and that corresponding to (7.95) is evaluated in cells A18:A24. All Excel spreadsheets can be downloaded from Wiley's ftp site at ftp://ftp.wiley.com/public/sci_tech_med/electric_power/.

To obtain the solution using (7.77), enter a starting value of one in cells A2:A8. The change vector $\Delta\mathbf{x}$ is indicated in cells D20:D26, and the new vector of choice variables for the next iteration is computed in cells E20:E26. For the next iteration, paste the value of these cells in cells A2:A8. The new vector is now displayed in cells E20:E26. Again, paste the values of these in cells A2:A8. By continuing this process, after a few iterations, we obtain the final solution shown in cells G20:G26.

Similarly, to use the relation of (7.95), we start with a value of one entered in cells A2:A8. We obtain new iterates for the choice variable vector in cells J20:J26.

By repeating the process of pasting these values in cells A2:A8, we obtain the final solution after a few iterations, which is identical to the previous solution. The minimum value of the function subject to constraints of the problem is 12.40654, which is in cell E1.

As a confirmation, we now use the Excel Solver routine to minimize cell D1 subject to constraints of the problem. The Solver solution will be identical to the previously obtained result, as it should be. We leave it to the reader as an exercise to confirm this.

Example 2: We take the problem from Wood and Wollenberg (1984) where Kuhn–Tucker conditions have been used to solve a constrained minimization problem. However, in the following we apply the procedure indicated in (7.91) to obtain a solution.

Problem:
Minimize

$$f(x_1, x_2) = 0.25x_1^2 + x_2^2, \tag{7.96}$$

subject to

$$h_1(x_1, x_2) = 5 - x_1 - x_2 = 0, \tag{7.97}$$

and

$$g(x_1, x_2) = x_1 + 0.2x_2 - 3 \leq 0. \tag{7.98}$$

We transform all constraints to equalities by adding slack variable x_3^2 as

$$h_1(x_1, x_2) = 5 - x_1 - x_2 = 0, \tag{7.99}$$

and

$$h_2(x_1, x_2) = g(x_1, x_2) = x_1 + 0.2x_2 - 3 + x_3^2 = 0. \tag{7.100}$$

We now solve for x_1, x_2, x_3, λ_1, and λ_2 as follows:
Since

$$L = 0.25x_1^2 + x_2^2 + \lambda_1 [5 - x_1 - x_2] + \lambda_2 \left[x_1 + 0.2x_2 - 3 + x_3^2\right], \tag{7.101}$$

we have

$$\nabla L = \begin{bmatrix} 0.5x_1 - \lambda_1 + \lambda_2 \\ 2x_2 - \lambda_1 + 0.2\lambda_2 \\ 2\lambda_2 x_3 \end{bmatrix}, \tag{7.102}$$

$$\nabla^2 L = \begin{bmatrix} 0.5 & 0 & 0 \\ 0 & 2 & 0 \\ 0 & 0 & 2\lambda_2 \end{bmatrix}, \tag{7.103}$$

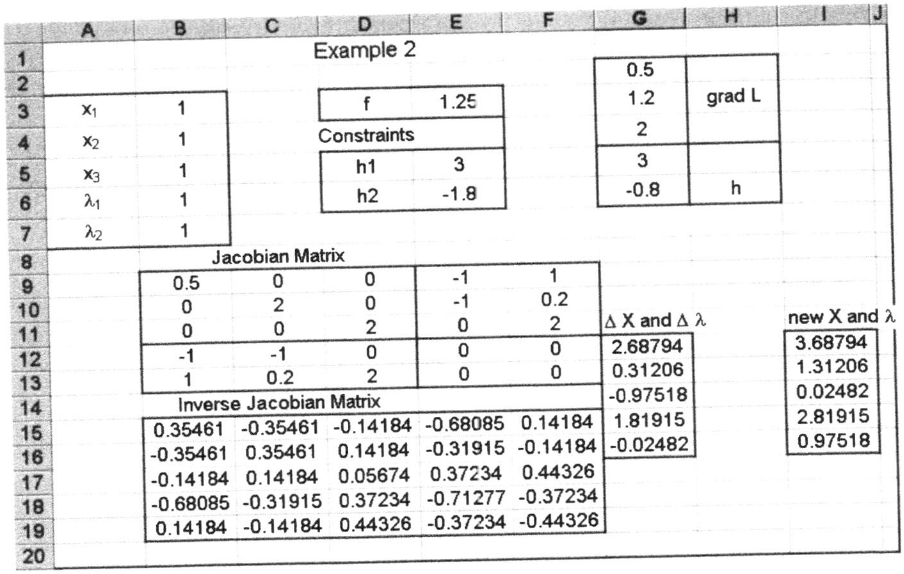

Figure 7.7. Sequential solution of (7.96) using the inverse of a Jacobian matrix.

and **G** according to (7.89) as

$$\mathbf{G} = \begin{bmatrix} -1 & 1 \\ -1 & 0.2 \\ 0 & 2x_3 \end{bmatrix}. \tag{7.104}$$

Second Eg.Xls

The spreadsheet "Second Eg.Xls" shows the procedure to obtain a solution to this problem. A snapshot of the spreadsheet is shown in Figure 7.7.

Variables of the problem x_1, x_2, x_3, λ_1, and λ_2 are in cell B3:B7. The Jacobian matrix **J** with component matrices $\nabla^2 L$ and **G** according to (7.94) is in cells B9:F13. Its inverse is in cells B15:F19. The constraint h and ∇L, using the values in cells B3:B9, is computed in cells G5:G6 and G2:G4, respectively.

We start out by assuming a value of one for all variables. The corrections $\Delta\mathbf{z}$ for variables according to (7.91) are computed in cells G12:G16, and the new vector of variables is obtained in cells I12:I16. Paste the values (values only) in cells B3:B7. Continue this process until the following values are obtained after four iterations.

$$x_1 = 2.5,\ x_2 = 2.5,\ x_3 = 2.03 \times 10^{-06},\ \lambda_1 = 5.9375, \text{ and } \lambda_2 = 4.6875.$$

A solution to the problem by using the Excel Solver (programmed into the spreadsheet) obtains the same solution. It is expected of the reader to study the Lagrange multipliers in the sensitivity sheet.

7.10 QUADRATIC PROGRAMS

In somewhat of a digression from the main theme, we now examine the nature of a quadratic program. The purpose of the digression is twofold. The first is that a procedure similar to that of the previous section can be used for its solution. The second purpose is to show that solving a corresponding linear program with linearized constraint equations can in turn solve any quadratic program. The importance of this observation is that the solution to a quadratic program can be obtained by solving a linear program derived from it. In Section 7.13 we show that a quadratic program can approximate a nonlinear program. Such a solution procedure is called Sequential Quadratic Programs (SQP). Therefore, the solution to some nonlinear programs can be obtained by solving a linear program derived from it, through the intermediary quadratic program.

7.10.1 Example of Quadratic Program

In order to examine the nature of solution to the minimization of quadratic functions, consider the problem of Section 6.7.1:

Minimize

$$f(\mathbf{x}) = 6x_1^2 + 3x_2^2 + 4x_1x_2 + 5.5x_1 - 2x_2, \tag{7.105}$$

subject to

$$g_1(\mathbf{x}) = (x_1 + 0.4)^2 - x_2 \leq 0, \tag{7.106}$$

$$g_2(\mathbf{x}) = -x1 + x_2 \leq 0.8. \tag{7.107}$$

We will use Kuhn–Tucker conditions and the procedure of Section 5.1 to solve this problem.

The inequalities are first converted to equalities by adding a nonnegative variable as

$$h_1(\mathbf{x}) = (x_1 + 0.4)^2 - x_2 + x_3^2 = 0, \tag{7.108}$$

and

$$h_2(\mathbf{x}) = -x_1 + x_2 - 0.8 + x_4^2 = 0. \tag{7.109}$$

Equation (7.105) can be written as

$$f(\mathbf{x}) = \mathbf{d}^T\mathbf{x} + \frac{1}{2}\mathbf{x}^T\mathbf{H}\mathbf{x}, \tag{7.110}$$

in which $\mathbf{d}^T = [5.5,\ 2]$ and $\mathbf{H} = \begin{bmatrix} 12 & 4 \\ 4 & 6 \end{bmatrix}$.

The Lagrangian function is

$$\begin{aligned} L = {} & [6x_1^2 + 3x_2^2 + 4x_1x_2 + 5.5x_1 - 2x_2] \\ & +\lambda_1[(x_1 + 0.4)^2 - x_2 + 0.5 + x_3^2] \\ & +\lambda_2[-x_1 + x_2 - 0.8 + x_4^2]. \end{aligned} \tag{7.111}$$

The Kuhn–Tucker conditions give

$$\nabla L = 0, \tag{7.112}$$

$$h_1(\mathbf{x}) = 0, \tag{7.113}$$

$$h_2(\mathbf{x}) = 0, \tag{7.114}$$

and

$$\lambda_1 x_3^2 = 0, \tag{7.115}$$

$$\lambda_2 x_4^2 = 0. \tag{7.116}$$

In (7.112), if the gradient of the Lagrangian function is computed with respect to all variables (design variables x_1, x_2, slack variables x_3, x_4, and the Lagrange multipliers λ_1, λ_2 associated with constraints), we get

$$\begin{aligned} & 5.5 + 12x_1 + 4x_2 + 2(x_1 + 0.4)\lambda_1 - \lambda_2 = 0 \\ & -2 + 4x_1 + 6x_2 - \lambda_1 + \lambda_2 = 0 \\ & 2x_3\lambda_1 = 0 \\ & 2x_4\lambda_2 = 0 \\ & (x_1 + 0.4)^2 - x_2 + x_3^2 = 0 \\ & -x_1 + x_2 + x_4^2 - 0.8 = 0. \end{aligned} \tag{7.117}$$

The third and fourth equations in (7.117) are nothing but an alternative form of equations (7.115) and (7.116), while the last two equations are nothing but constraints of the problem expressed in (7.113), (7.114).

The equations in (7.117) can also be written as

$$\nabla L = \mathbf{d} + \mathbf{Hx} + \mathbf{G}\lambda = 0, \tag{7.118}$$

$$2[x_3, x_4]\begin{bmatrix} \lambda_1 \\ \lambda_2 \end{bmatrix} = 0, \tag{7.119}$$

$$(x_1 + 0.4)^2 - x_2 + x_3^2 = 0, \tag{7.120}$$

and

$$-x_1 + x_2 + x_4^2 - 0.8 = 0, \tag{7.121}$$

where **d** and **H** have been defined earlier, and **G** represents the matrix of partial derivatives of constraints:

$$\mathbf{G}^T = \begin{bmatrix} \dfrac{\partial h_1}{\partial x_1} & \dfrac{\partial h_1}{\partial x_2} & \dfrac{\partial h_1}{\partial x_3} & \dfrac{\partial h_1}{\partial x_4} \\ \dfrac{\partial h_2}{\partial x_1} & \dfrac{\partial h_2}{\partial x_2} & \dfrac{\partial h_2}{\partial x_3} & \dfrac{\partial h_2}{\partial x_4} \end{bmatrix}. \tag{7.122}$$

As in the previous example, we can solve the above equations for variables. The equations are nonlinear because the constraint in (7.108) is nonlinear. One option is to linearize constraints. It is interesting to note that if all constraints are linear or linearized, the set of equations in (7.117), with the exception of (7.115) and (7.116), would be linear equations. Therefore, an important observation is that a solution of such linear equations (when all constraints are linear relationships, or are linearized) that satisfies (7.115) and (7.116) minimizes $f(\mathbf{x})$.

For the problem at hand, we retain the nonlinearity of the constraint in (7.108). The Jacobian matrix,as in the previous section, per (7.93), (7.94), is given by

$$\mathbf{J} = \left[\begin{array}{cccc|cc} 12 & 4 & 0 & 0 & 2(x_1+0.4) & -1 \\ 4 & 6 & 0 & 0 & -1 & 1 \\ 0 & 0 & 2\lambda_1 & 0 & 2x_3 & 0 \\ 0 & 0 & 0 & 2\lambda_2 & 0 & 2x_4 \\ \hline 2(x_1+0.4) & -1 & 2x_3 & 0 & 0 & 0 \\ -1 & 1 & 0 & 2x_4 & 0 & 0 \end{array}\right]. \tag{7.123}$$

In the attached Excel spreadsheet "Quadnc.Xls" of Figure 7.8, the computed values for the design variables x_1, x_2 and variables $x_3, x_4, \lambda_1, \lambda_2$ are entered in cells A4:F4. We start out by entering an initial guess of one in these cells. The Jacobian matrix and its inverse are in cells A7:F12 and A15:F20. The correction to initial guess is given in cells J18:J23, and the new value for cells A4:F4 are given in cells A23:F23. Paste the values of this update in cells A4:F4. Continuing this process several times, we find that the program converges to the following final solution: **Quadnc.Xls**

x_1	x_2	x_3	x_4	λ_1	λ_2
−0.44238	0.35776	0.59716	0.06057	−0.00034	1.62257

The fact that the process has converged can be confirmed by observing the value of ∇L in cells J2:J7 which attains a value of zero (within a tolerance).

Figure 7.9 shows the solution space and the domain of constraints. Because of the closed domain of constraints, regardless of the initial guess, the solution should converge into this domain.

In Figure 7.8, the value of the function is in cell H2. The value of cell H2 is minimized by using Excel Solver subject to constraints in cells H6:H7. As expected, the final solution is the same as the above-mentioned solution.

In Section 7.18, the reader will be asked to repeat the procedure of this spreadsheet to solve a problem with multiple minima.

	A	B	C	D	E	F	G	H	I	J
1	QUADRATIC PROGRAM with Non Linear Constraints							f		Grad L
2			Choice variables					-2.22327		0.00053
3	x1	x2	x3	x4	l1	l2		Function Value		-0.00016
4	-0.44231	0.35769	0.59717	0.00057	-0.00034	1.62258				-0.00041
5								Constraints		0.00185
6			Jacobean					-0.35590		0.00071
7	12.00000	4.00000	0.00000	0.00000	-0.08462	-1.00000		0.80000		0.00000
8	4.00000	6.00000	0.00000	0.00000	-1.00000	1.00000				
9	0.00000	0.00000	-0.00069	0.00000	1.19433	0.00000				
10	0.00000	0.00000	0.00000	3.24516	0.00000	0.00114				
11	-0.08462	-1.00000	1.19433	0.00000	0.00000	0.00000				
12	-1.00000	1.00000	0.00000	0.00114	0.00000	0.00000				
13										
14			Inverse Jacobean							
15	0.03846	0.03846	0.03493	0.00014	0.00002	-0.38460				
16	0.03846	0.03846	0.03493	-0.00022	0.00002	0.61540				
17	0.03493	0.03493	0.03172	-0.00017	0.83730	0.48801		x		Δ x=J⁻¹(0-f)
18	0.00014	-0.00022	-0.00017	0.30815	0.00000	0.00076		-0.44231		0.00000
19	0.00002	0.00002	0.83730	0.00000	0.00048	0.00028		0.35769		0.00000
20	-0.38460	0.61540	0.48801	0.00076	0.00028	-2.15368		0.59717	+	-0.00059
21								0.00057		-0.00057
22			New X					-0.00034		0.00034
23	-0.44231	0.35769	0.59658	0.00000	0.00000	1.62308		1.62258		0.00050
24										

Figure 7.8. Quadratic program with nonlinear constraint.

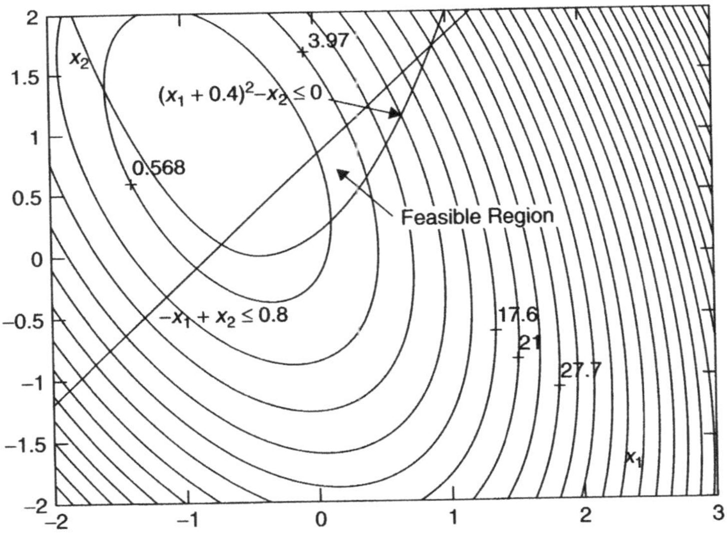

Figure 7.9. Contour plot of the constrained quadratic minimization problem.

7.10.2 Solution of Quadratic Programs via Linear Programs

We now address the method of solution of quadratic programs by solving the resulting LP. The attraction of this procedure is that it offers an efficient method for the solution of general constrained nonlinear optimization problem by forming sequential quadratic programs which in turn can be solved by a corresponding LP.

Toward that end we continue with the example of previous section and write the Lagrangian relations in terms of $\Delta\mathbf{x}$, and changes to a guessed values of $\mathbf{x}$, as follows.

The minimization problem is as follows:

Minimize

$$f(\mathbf{x}) = 6x_1^2 + 3x_2^2 + 4x_1x_2 + 5.5x_1 - 2x_2, \tag{7.124}$$

subject to

$$g_1(\mathbf{x}) = (x_1 + 0.4)^2 - x_2 + 0.5 \leq 0. \tag{7.125}$$

$$g_2(\mathbf{x}) = -x_1 + x_2 \leq 0.8. \tag{7.126}$$

To start with, we linearize the constraint, which is nonlinear as

$$g_1(\mathbf{x}) = g_1(\mathbf{x}^{(0)}) + 2(x_1^{(0)} + 0.4)\Delta x_1 - \Delta x_2 + 0.5 \leq 0, \tag{7.127}$$

and keep the second constraint that is a linear relationship as it is.

In the above, zero superscript indicates initial guesses for $\mathbf{x}$ or their value in the previous iteration depending on the context.

These linearizing relationships can be expressed in a general manner as

$$g_i(\mathbf{x}) = g_i(\mathbf{x}^{(0)}) + [\nabla g_i(\mathbf{x}^{(0)})]^T \Delta\mathbf{x}, \qquad i = 1, \ldots, n \tag{7.128}$$

and correspondingly

$$h_j(\mathbf{x}) = h_j(\mathbf{x}^{(0)}) + [\nabla h_j(\mathbf{x}^{(0)})]^T \Delta\mathbf{x}, \qquad j = 1, \ldots, p, \tag{7.129}$$

where it is implicitly understood that the gradients of g_i and h_j are computed using the values of $\mathbf{x}$ in the previous iteration as signified by the superscript of zero.

The Lagrangian function in terms of $\Delta\mathbf{x}$ is

$$\begin{aligned} L = {} & [6(x_1^{(0)} + \Delta x_1)^2 + 3(x_2^{(0)} + \Delta x_2)^2 \\ & + 4(x_1^{(0)} + \Delta x_1)(x_2^{(0)} + \Delta x_2) + 5.5(x_1^{(0)} + \Delta x_1) - 2(x_2^{(0)} + \Delta x_2)] \\ & + \lambda_1[(x_1^{(0)} + \Delta x_1 + 0.4)^2 - (x_2^{(0)} + \Delta x_2) + 0.5 + (x_3^{(0)} + \Delta x_3)^2] \\ & + \lambda_2[(x_1^{(0)} + \Delta x_1) + (x_2^{(0)} + \Delta x_2) - 0.8 + (x_4^{(0)} + \Delta x_4)^2], \end{aligned} \tag{7.130}$$

where x_3^2 and x_4^2 are slack variables associated with constraints.

Taking partial derivatives of the above Lagrangian and setting them equal to zero, we get

$$\left[\begin{array}{cccc|cc} 12 & 4 & 0 & 0 & 2(x_1^{(0)}+0.4) & -1 \\ 4 & 6 & 0 & 0 & -1 & 1 \\ 0 & 0 & 0 & 0 & 2x_3^{(0)} & 0 \\ 0 & 0 & 0 & 0 & 0 & 2x_1^{(0)} \\ \hline 2(x_1^{(0)}+0.4) & -1 & 2x_3^{(0)} & 0 & 0 & 0 \\ -1 & 1 & 0 & 2x_4^{(0)} & 0 & 0 \end{array}\right] \left[\begin{array}{c} \Delta x_1 \\ \Delta x_2 \\ \Delta x_3 \\ \Delta x_4 \\ \Delta \lambda_1 \\ \Delta \lambda_1 \end{array}\right]$$

$$-\left[\begin{array}{l} 12x_1^{(0)} - 4x_2^{(0)} - 5.5 - 2\lambda_1^{(0)}(x_1^{(0)} + 0.4) + \lambda_2^{(0)} \\ -4x_1^{(0)} - 6x_2^{(0)} + 2 + \lambda_1^{(0)} - \lambda_2^{(0)} \\ -2x_3^{(0)}\lambda_1^{(0)} \\ -2x_4^{(0)}\lambda_2^{(0)} \\ g_1(\mathbf{x}^{(0)}) \\ g_2(\mathbf{x}^{(0)}) \end{array}\right] = 0, \quad (7.131)$$

which can be written as

$$[\mathbf{A}]\Delta\mathbf{z} - \nabla L(\mathbf{x}^{(0)}) = 0, \quad (7.132)$$

where the definitions of matrices **A**, $\nabla L(^0\mathbf{x})$, and $\Delta\mathbf{z}$ are obvious.

Equations in (7.131) represent a set of linear equations. [9] It may first appear that the system can be solved easily for corrections $\Delta\mathbf{x}$. However, that is not the case because of the transversality conditions indicated by (7.115) and (7.116). Because of these transversality conditions, either x_3 or x_4 or both of them can be zero. Hence the matrix on the left-hand side of (7.131) will be singular because the column corresponding to either x_3 or x_4 will be zero. Consequently, these equations cannot be solved in a straightforward manner without considering the transversality conditions.

Solution of Resulting Linear Equation System. The necessary conditions for the minima of (7.124) are that the partial derivatives of the Lagrangian (7.130) are zero. Substitute $Y_1 = x_3^2$ and $Y_2 = x_4^2$ in (7.130) and use the linearized constraint equations. The following necessary conditions for a minimum can be obtained by

[9] Compare the matrix on the left-hand side of the above equation to that of the Jacobian in (7.123).

partial differentiation (algebra omitted):

$$\frac{\partial L}{\partial \Delta x_1} = L(\mathbf{x}^{(0)}) + 12\Delta x_1 + 4\Delta x_2 + 2(\Delta x_1 + 0.4)\lambda_1 - \lambda_2 = 0, \qquad (7.133)$$

$$\frac{\partial L}{\partial \Delta x_2} = L(\mathbf{x}^{(0)}) + 4\Delta x_1 + 6\Delta x_2 - \lambda_1 + \lambda_2 = 0, \qquad (7.134)$$

where $L(\mathbf{x}^{(0)})$ represents the value of the Lagrangian function evaluated at the starting or previous iteration values of the vector $\mathbf{x}$,
subject to

$$g_1(\mathbf{x}^{(0)}) + 2(x_1 + 0.4)\Delta x_1 - \Delta x_2 + Y_1 = 0, \qquad (7.135)$$

$$g_2(\mathbf{x}^{(0)}) - \Delta x_1 + \Delta x_2 + Y_2 - 0.8 = 0, \qquad (7.136)$$

$$Y_1\lambda_1 = 0, \qquad (7.137)$$

$$Y_2\lambda_2 = 0. \qquad (7.138)$$

Equations (7.133) to (7.138) are a set of linear equations whose solution results in the minimum. The method of their solution is to include artificial variables α_1 and α_2 in (7.133) and (7.134) as

$$\frac{\partial L}{\partial \Delta x_1} = L(\mathbf{x}^{(0)}) + 12\Delta x_1 + 4\Delta x_2 + 2(\Delta x_1 + 0.4)\lambda_1 - \lambda_2 + \alpha_1 = 0, \qquad (7.139)$$

$$\frac{\partial L}{\partial \Delta x_2} = L(\mathbf{x}^{(0)}) + 4\Delta x_1 + 6\Delta x_2 - \lambda_1 + \lambda_2 + \alpha_2 = 0. \qquad (7.140)$$

Then, a solution is obtained to the linear system using the following LP procedure:

Choose $\Delta x_1, \Delta x_2, Y_1, Y_2, \lambda_1, \lambda_2, \alpha_1, \alpha_2$ in order to
Minimize $(\alpha_1 + \alpha_2)$
subject to (7.135), (7.136), (7.137), (7.138), (7.139), and (7.140).

Of these, constraints (7.137) and (7.138), despite being nonlinear, can be easily handled by the simplex method. These imply that either Y_1 or λ_1 should be included in the basis and not both. Similarly, either Y_2 or λ_2 should be in the basis. Such a programming accommodation can be easily built around the simplex method (see Section 3.7). We demonstrate this method by using an Excel spreadsheet to solve the linear program.

Example: The nonlinear constrained problem to
Minimize

$$f(\mathbf{x}) = 6x_1^2 + 3x_2^2 + 4x_1x_2 + 5.5x_1 - 2x_2, \qquad (7.141)$$

subject to

$$g_1(\mathbf{x}) = (x_1 + 0.4)^2 - x_2 + 0.5 \leq 0, \qquad (7.142)$$

and

$$g_2(\mathbf{x}) = -x1 + x_2 \leq 0.8 \qquad (7.143)$$

is linearized and transformed to the following LP:

Minimize $(\alpha_1 + \alpha_2)$ subject to the relations

$$\frac{\partial L}{\partial \Delta x_1} = \nabla L(\mathbf{x}^{(0)}) + 12\Delta x_1 + 4\Delta x_2 + 2(\Delta x_1 + 0.4)\lambda_1 - \lambda_2 + \alpha_1 = 0, \quad (7.144)$$

$$\frac{\partial L}{\partial \Delta x_2} = \nabla L(\mathbf{x}^{(0)}) + 4\Delta x_1 + 6\Delta x_2 - \lambda_1 + \lambda_2 + \alpha_2 = 0, \quad (7.145)$$

$$g_1(\mathbf{x}^{(0)}) + 2(x_1 + 0.4)\Delta x_1 - \Delta x_2 + Y_1 = 0, \quad (7.146)$$

$$g_2(\mathbf{x}^{(0)}) - \Delta x_1 + \Delta x_2 + Y_2 - 0.8 = 0, \quad (7.147)$$

$$Y_1\lambda_1 = 0, \quad (7.148)$$

$$Y_2\lambda_2 = 0. \quad (7.149)$$

As explained above, when one writes an algorithm from scratch, the last two equations in the above system can be easily incorporated into the simplex algorithm. However, here the Excel Solver algorithm cannot be altered. Therefore we use the nonlinear capability of the Solver to accommodate these two relations.

LP for QP. Xls

In the spreadsheet "LP for QP.Xls" depicted in Figure 7.10, an initial value is assigned for $x_1^{(0)}, x_2^{(0)}$, in cells A2, B2. The components of (7.144), (7.145), (7.146), and (7.147) for this initial guess (term with $^{(0)}$ superscript) are computed in cells A8:A11. The guess for the correction vector Δx_1, Δx_2 and for $Y_1, Y_2, \lambda_1, \lambda_2, \alpha_1$, and α_2 are entered in cells A5:H5.[10] The change components in equations (7.144) to (7.147) (terms with Δ) are in cells D8:D11. Consequently, the left-hand side of these equations is given by the sum of cells A8:A11 and D8:D11, which is shown in cells E8:E11. Cells H9 and H10 indicate the transversality conditions that is required to be zero.

Row 14 shows the updated values for $\mathbf{x}$ and shows the resulting values for other variables. The Solver is asked to solve the following LP. The goal or the objective is to minimize cell H2, which is equal to $\alpha_1 + \alpha_2$ = cell G5 + H5. Choice variables are $\Delta x_1, \Delta x_2, Y_1, Y_2, \lambda_1, \lambda_2, \alpha_1$, and α_2 in cells A5:H5. Constraints are that the values of $Y_1, Y_2, \lambda_1, \lambda_2, \alpha_1$, and α_2 are positive and that the values of cells E8:E11 are equal to zero to correspond with (7.144),(7.145),(7.146), and (7.147).

[10] The user can enter any values in this cell. We suggest the entry of 1 in all—that is, in cells A2, B2, and A5:H5. The ensuing discussion assumes that the user has done so.

	A	B	C	D	E	F	G	H
1	X_1	X_2	y1	y2	λ1	λ2		Minimize
2	-0.44231	0.35769	0.17795	0.00000	0.00000	0.81154		0.00000
3								
4	Δx_1	Δx_2	y1	y2	λ1	λ2	α_1	α_2
5	0.00000	0.00000	0.17795	0.00000	0.00000	0.81154	0.00000	0.00000
6								
7				Delta				
8	0.81154	$L(x_1^0)$		-0.81154	0.00000	$\partial L/\partial \Delta X_1$		Transversality
9	-0.81154	$L(x_2^0)$		0.81154	0.00000	$\partial L/\partial \Delta X_2$	$Y_1{*}\lambda_1$	0.00000
10	-0.17795	$g_1(x^0)$		0.17795	0.00000		$Y_2{*}\lambda_2$	0.00000
11	0.00000	$g_2(x^0)$		0.00000	0.00000			
12								
13	$x_1+\Delta x_1$	$x_2+\Delta x_2$	y1	y2	λ1	λ2	α_1	α_2
14	-0.44231	0.35769	0.17795	0.00000	0.00000	0.81154	0.00000	0.00000
15								
16	Final Solution							
17	X_1	X_2	y1	y2	λ1	λ2		
18	-0.44231	0.35769	0.35661	0.00000	-0.00034	1.62258		

Figure 7.10. LP solution to a constrained quadratic problem.

The first pass with the Solver execution (with the above starting values) results in an updated value in row 14 as follows:

$$-0.44231 \quad 0.35769 \quad 1.21808 \quad 0.00000 \quad 0.00000 \quad 0.81154 \quad 0.00000 \quad 0.00000$$

Observe that the values for α_1 and α_2 are zero. The values of the first two elements in cells A14 and B14 are now taken to be the updates $\mathbf{x}^{(1)}$ of $\mathbf{x}^{(0)}$. Hence their values are pasted in cells A2 and B2.

The next execution of the Solver with this updated $\mathbf{x}$ gives the following values in row 14:

$$-0.44231 \quad 0.35769 \quad 0.17795 \quad 0.00000 \quad 0.00000 \quad 0.81154 \quad 0.00000 \quad 0.00000$$

This is the final solution. However, since we have used $Y_1 = z_1^2$ and $Y_2 = z_2^2$, the values of Y and the Lagrange multiplier associated with the binding constraint are half the values. It is expected of the reader to ponder why this is the case.

7.11 FEASIBLE DIRECTIONS

Algorithms under the category of directional search methods start with an initial guess that is within a feasible region. That means that the initial guess does not violate any constraints. However, it is highly unlikely that the guess is the optimum point. The procedure then is to successively move to better and better locations in the feasible region. In order to move within the feasible region, certain directions are feasible and some others are infeasible.

A direction at any point **x** of the solution space is *feasible* when a small move in that direction does not violate any constraints. Furthermore, if a movement in a feasible direction, however small, decreases the value of the function (or increases for maximization problems), then such a direction in an algorithmic procedure is called a *usable* direction. Thus, at any point **x**, when a small step is taken along the feasible direction, the new point will still be within the feasible region. By taking a small step along such successive directions, one attempts to successively reduce the value of the objective function by movements along usable feasible directions **s**. The principle behind the method of feasible directions is to exploit this property.

Clearly, if one is already at the optimal point, there should be no feasible direction that will minimize f further. Consider a situation when one is at a point on the boundary of two constraints. This means that we have two binding constraints. At that point, multiplying the relation in (7.33) (arising from Kuhn–Tucker conditions) by the feasible direction vector **s**, we get

$$-\mathbf{s}^T \nabla f(\mathbf{x}) = \lambda_1 \mathbf{s}^T \nabla g_1(\mathbf{x}) + \lambda_2 \mathbf{s}^T \nabla g_2(\mathbf{x}). \tag{7.150}$$

By definition, the gradient vector of a constraint g indicates the direction in which its value increases the most. A vector **s** is a feasible direction if, by taking a small step from present $\mathbf{x}_t$, we do not go out of the feasible region that is bound by constraints of the problem. The product of **s** and ∇g indicates the increase of g along **s**. Since the constraint g is less than or equal to zero (with k moved to the left-hand side) as indicated in (7.17), the feasible direction **s** should satisfy the relation

$$\mathbf{s}^T \nabla g_1 < 0, \tag{7.151}$$

because by moving in the direction of **s**, ∇g_1 should not be violated (i.e., become > 0). Under some circumstances (Rao, 1996), one can accept equality in the above expression.

This can be pictured in two-dimensional world as a direction **s** which makes an angle greater than $\frac{\pi}{2}$ and less than π with the gradient vector ∇g, because only then the product of the two can be <0. Another way of saying this is that the feasible direction makes an obtuse angle with the normal of the constraint relation (or the gradient of the constraint). If there are two binding constraints, then a feasible direction has to make an obtuse angle with the gradients of both binding constraints.

The object of an algorithm exploiting feasible directions to solve nonlinear programs is to find search directions that are feasible. The computational efficiency of an algorithmic procedure is in selecting the best successive feasible directions for fast convergence to the minima. One makes an initial guess for **x** or arrives at a point

$\mathbf{x}^{(t)}$ in iteration t. The intent now is to find an improved $\mathbf{x}$ that minimizes f even further, if possible. Hence, it is necessary to know the direction of movement from $\mathbf{x}^{(t)}$ to $\mathbf{x}^{(t+1)}$. Certain directions are feasible in that they reduce f, and some others are infeasible as a movement in such directions cannot reduce f. Some directions might even increase f. We shall examine the properties of feasible directions via an example.

7.11.1 Example

In the problem Maximize

$$f(\mathbf{x}) = -x_1^2 - x_2^2 + x_1x_2 + 7x_1 + 4x_2,$$

subject to

$$\begin{aligned} g_1(\mathbf{x}) &= 2x_1 + 3x_2 \le 24, \\ g_2(\mathbf{x}) &= -5x_1 + 12x_2 \le 24, \\ g_3(\mathbf{x}) &= x_2 \le 4 \\ x_1 \ge 0, &\qquad x_2 \ge 0, \end{aligned}$$

find a suitable usable feasible direction at the following points:

$$\mathbf{x}_1 = [1, 1]^T, \qquad \mathbf{x}_2 = [6, 4]^T.$$

Solution: In the given problem, we have

$$\nabla f = \begin{pmatrix} -2x_1 + x_2 + 7 \\ -2x_2 + x_1 + 4 \end{pmatrix},$$

$$\nabla g_1 = \begin{pmatrix} 2 \\ 3 \end{pmatrix}, \quad \nabla g_2 = \begin{pmatrix} -5 \\ 12 \end{pmatrix}, \quad \nabla g_3 = \begin{pmatrix} 0 \\ 1 \end{pmatrix}.$$

Since constraints are all linear, ∇g_i contain constant terms and do not contain x_1 or x_2.

See Figure 7.11, which represents this problem.

1. Consider $\mathbf{x}_1 = [1, 1]^T$. At $\mathbf{x}_1$ we have $\nabla f = [6, 3]^T$, $g_1 = 5$, $g_2 = 7$, and $g_3 = 1$. Consequently, all constraints are satisfied. Hence $\mathbf{x}_1 = [1, 1]^T$ is a point within the feasible region. Thus $\nabla f = [6, 3]^T$ is a usable feasible search direction $\mathbf{s}$ in order to search for increasing values for $f(\mathbf{x})$.
2. Consider $\mathbf{x}_2 = [6, 4]^T$. At $\mathbf{x}_2$ we get $\nabla f = [-1, 2]$ and $g_1 = 24$, $g_2 = 18$, and $g_3 = 4$, indicating that the first and third constraint are binding. This is the point of intersection of two constraints as shown in Figure 7.11.

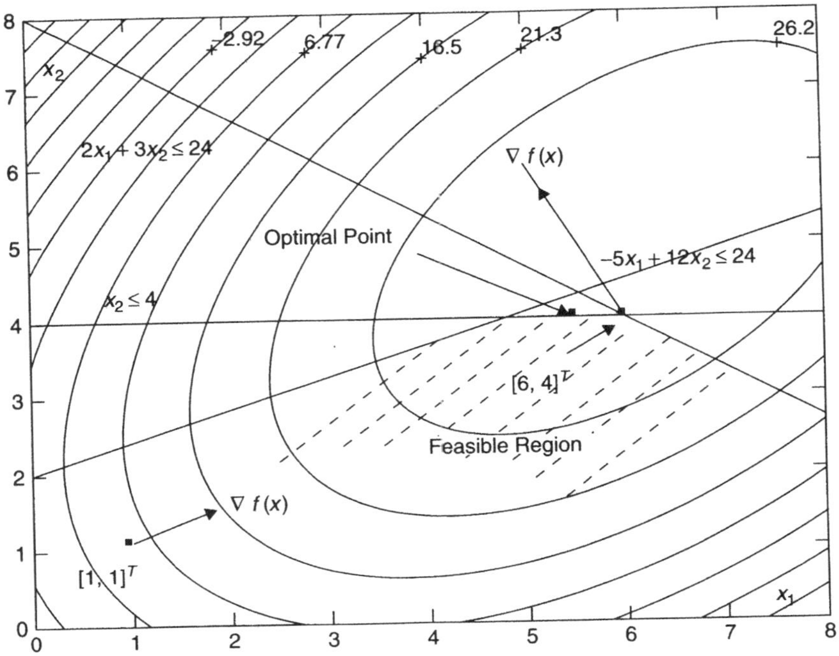

Figure 7.11. Feasible region and directions of example problem.

At $\mathbf{x}_2$, a useful feasible directionsshould have the following two properties:

$$\mathbf{s}^T \nabla g_j(\mathbf{x}_2) \le 0, \qquad j = 1, 3, \tag{7.152}$$

and since this is a maximization problem, we have

$$\mathbf{s}^T \nabla f(x_2) > 0. \tag{7.153}$$

In two conditions pertaining to constraints of (7.152), we permit equality since constraints are linear, and by keeping $\mathbf{s}$ along the constraint line, a movement on the constraint line will result in $\nabla g = 0$ while the value of f can be increasing. This can be seen in Figure 7.11, in which one of the directions can be along the line $x_2 = 4$ from $\mathbf{x}_2 = [6, 4]$ to the left. Another circumstance under which we can permit equality for (7.152) is when the constraint is nonlinear but outward curving or concave [see Rao (1996) for proof].

Testing for these conditions, if we keep $\mathbf{s} = [-1, 2]^T$, the gradient of the function, we obtain $\mathbf{s}^T \nabla g_1 = 4$, $\mathbf{s}^T \nabla g_2 = 29$, and $\mathbf{s}^T \nabla g_3 = 2$. Thus the first condition (7.152) is violated and we cannot accept $|\nabla f(\mathbf{x})|_{\mathbf{x}=\mathbf{x}_2}$ as the direction to proceed in order to increase the value of the function. This is evident from Figure 7.11.

We now embark on the search for a suitable direction. In this simple example, from Figure 7.11 we know that proceeding left along the line $x_2 = 4$ is a usable feasible direction that satisfies all the conditions. However, to search for a suitable direction we need a mathematical procedure.

A "brute-force" method is that of generating random vectors and testing them for above conditions. The vector that satisfies the two conditions of (7.152) and (7.153) can be accepted as the direction to proceed. Of course, how far we have to proceed in that direction is another question. Obviously, such a procedure is neither elegant nor computationally efficient. An alternative is the classical *direction finding problem* that is used in some algorithms, notably in the procedure of Zoutendik's method. We attempt neither a detailed explanation of such algorithms nor the process of finding a search direction **s** that is the most efficient in terms of reaching the goal. Nonetheless, the rudiments of the direction finding problem is discussed in the following.

7.11.2 Direction Finding Problem

First of all, we need a direction **s** such that $\mathbf{s}^T \nabla f(\mathbf{x})$ is as much greater than zero[11] as possible. Hence, we need a non zero parameter α that needs to be maximized such that

$$s_1 \frac{\partial f}{\partial x_1} + s_2 \frac{\partial f}{\partial x_2} - \alpha \geq 0, \tag{7.154}$$

where s_1 and s_2 are components of **s**. Clearly, larger the value of α larger is the value of $\mathbf{s}^T \nabla f(\mathbf{x})$.

The condition that $\mathbf{s}^T \nabla g \leq 0$ can be satisfied by adding any positive number on the left-hand side of this relation. The general way of expressing this is that we want the left-hand side to be less than or equal to $-\mu\alpha$, where μ is a scalar, making the constraint as $\mathbf{s}^T \nabla g_i \leq -\mu\alpha$. If we choose $\mu = 0$, we obtain zero on the right-hand side. Alternatively, we can choose μ to be any positive scalar. Thus the conditions regarding constraints turn out to be

$$s_1 \frac{\partial g_1}{\partial x_1} + s_2 \frac{\partial g_1}{\partial x_2} + \mu\alpha \leq 0, \tag{7.155}$$

$$s_1 \frac{\partial g_2}{\partial x_1} + s_2 \frac{\partial g_2}{\partial x_2} + \mu\alpha \leq 0, \tag{7.156}$$

and

$$s_1 \frac{\partial g_3}{\partial x_1} + s_2 \frac{\partial g_3}{\partial x_2} + \mu\alpha \leq 0. \tag{7.157}$$

In order to avoid large numbers for the components of **s**, we put bounds on s_1 and s_2 as

$$s_1, s_2 \geq -1 \tag{7.158}$$

[11]For maximization, we need to search for increasing $\mathbf{s}^T \nabla f(\mathbf{x})$. For minimization problems, we search for decreasing values $\mathbf{s}^T \nabla f(\mathbf{x})$.

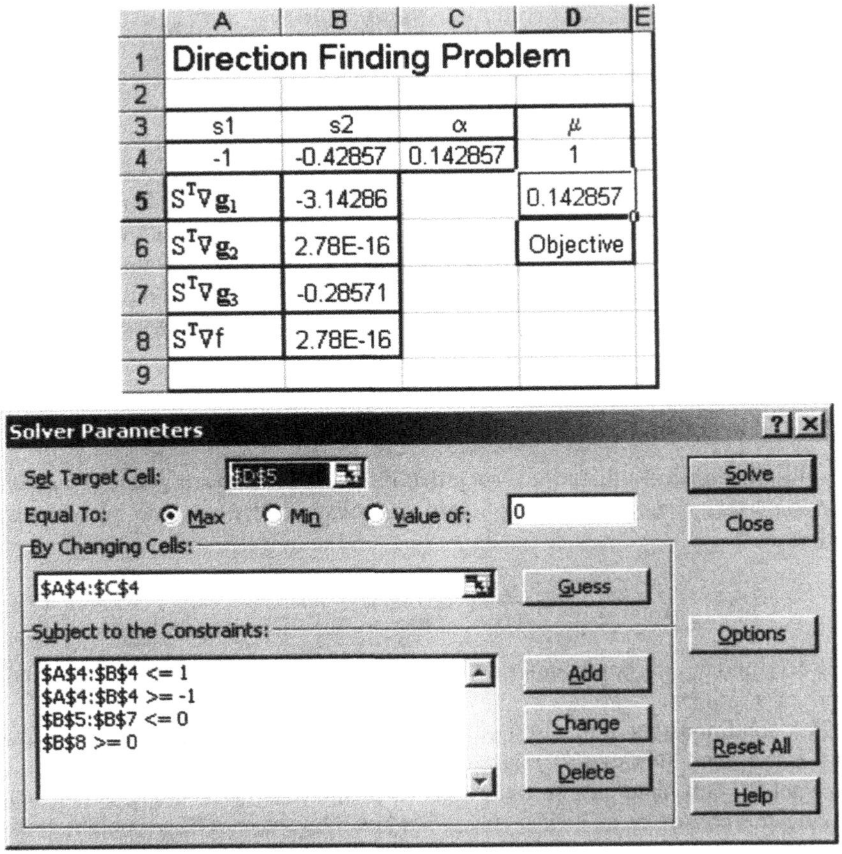

Figure 7.12. Direction finding problem.

and

$$s_1, s_2 \leq 1. \tag{7.159}$$

The above is a linear optimization problem to find s_1 and s_2 in order to minimize $-\alpha$ (or maximize α) subject to (7.155), (7.156), (7.157), (7.158), and (7.159).

Direct-ion.Xls

The spreadsheet "Direction.Xls" illustrates the solution. The Linear Program contained in the spreadsheet is also shown in Figure 7.12. We first assume $\mu =$ 1 (value in cell D4) with an initial guess of 1 for s_1, s_2, and α. The result of optimization gives $\mathbf{s} = [-1, -.42857]^T$. Confirm by an examination of Figure 7.11 that this makes sense.

In the spreadsheet, cells B5:B7 contain $\mathbf{s}^T\nabla g$, and cell B8 contains $\mathbf{s}^T\nabla f$. The objective function is in cell D5. The Solver screen in Figure 7.12 shows all constraints.

Finally, to render $\mu = 0$, we enter corresponding number in cell D4. Then the Solver solution gives $\mathbf{s}^T = [-1, -0.4166]$.

The above illustrates the process of finding directions. This principle is incorporated into the Zoutendijk's method. The details of this method can be found in Rao (1996) and Bazaraa et al. (1979).

7.12 GRADIENT PROJECTION METHODS – GENERALIZED REDUCED GRADIENT METHOD

The Generalized Reduced Gradient Method (GRG) enjoyed popularity before the introduction of algorithms such as the Sequential Quadratic Program (SQP) method. Originally, GRG was applied to problems with linear constraints and was extended later to include nonlinear constraints. We shall discuss and illustrate the underlying principles of the method rather than its algorithmic details.

The principal idea of the methods is to "stay on" the set of nonlinear constraints in each iteration and search in the feasible region to further reduce the objective function. In effect, the procedure searches for a usable feasible direction in each iteration while staying on the constraint surface.

The effect of the requirement to stay on constraints (to satisfy constraints) is to reduce the dimensionality of the problems because some choice variables are constrained by the relationship in the constraint equations as explained under the method of constrained variations in Section 7.4.

For example, consider a single constraint

$$f(x_1, x_2) = 5x_1x_2 \exp(x_2 - x_1) = 0 \tag{7.160}$$

in some two-variable minimization problem in the $x_1 - x_2$ plane.

Then, (7.160) defines a relationship between x_1 and x_2. Therefore, one variable can be used to solve for another. Consequently, the minimization problem has only one degree of freedom in which given the value of one variable, the value of the other can be determined using a univariate procedure. Theoretically, in an m variable problem, one variable can be reduced from the set for each equality constraint. Such a transformation projects the objective function π into the reduced subspace of dimension $m - p$, where p represents the number of equalities. The differences among several algorithms that employ this technique arise from the procedure by which constraints are exactly satisfied in each iteration (binding) while searching for better points in the subspace.

7.12.1 Procedure of GRG Method

In the following, we intertwine the theoretical development with an example to help comprehension.

Consider the problem if selecting $x_i,\ i = 1, 2, \ldots, m$ in order to minimize

$$\pi = f(\mathbf{x}), \tag{7.161}$$

subject to n inequalities

$$g_j(\mathbf{x}) \leq 0, \qquad j = 1, 2, \ldots, n, \tag{7.162}$$

and p equalities

$$h_k(\mathbf{x}) = 0, \qquad k = 1, 2, \ldots, p, \tag{7.163}$$

and

$$\mathbf{x}_i^{(l)} \leq \mathbf{x}_i \leq \mathbf{x}_i^{(u)}, \qquad i = 1, 2, \ldots, m, \tag{7.164}$$

where $\mathbf{x}^{(l)}$ and $\mathbf{x}^{(u)}$ are the lower bound and the upper bounds on choice variables.

GRG Method—Example. The following example is used throughout the explanation of the GRG method.

Select x_1, x_2, and x_3 ($m = 3$) to minimize

$$f(\mathbf{x}) = f(x_1, x_2, x_3) = 4x_1^2 + 3x_2^2 + 5x_3^2 - 6x_1x_2 + x_1x_3 + 15, \tag{7.165}$$

subject to two inequalities ($n = 2$) and one equality ($p = 1$) as follows:

$$g_1(\mathbf{x}) = 2x_1 + 3x_2 + x_3 \leq 8, \tag{7.166}$$

$$g_2(\mathbf{x}) = 4x_1 + 5x_2 - x_3 \leq 9, \tag{7.167}$$

$$h(\mathbf{x}) = 2x_1 + 4x_3 = 8. \tag{7.168}$$

Here, only linear constraints have been included in order to elucidate the principle of the method. Later, we examine the changes required to include nonlinearity. In this problem, $m = 3$, $n = 2$, and $p = 1$. The upper and lower bounds for $\mathbf{x}$ are taken to be unlimited in this example. Any finite bound will only influence the choice of γ, the optimum search length in the algorithmic procedure to be described subsequently.

Procedure of GRG Method—Continued. By adding nonnegative slack variables, the inequalities (7.162) are converted to equalities, and the problem can be stated as follows:

Minimize a function of m variables, x_i, $i = 1, \ldots, m$

$$\pi = f(\mathbf{x}), \tag{7.169}$$

subject to n inequalities transformed into equalities

$$g_j(\mathbf{x}) + \mathbf{x}_{m+j} = 0, \qquad j = 1, 2, \ldots, n, \tag{7.170}$$

and p equalities

$$h_k(\mathbf{x}) = 0, \qquad k = 1, 2, \ldots, p, \tag{7.171}$$

and the bounds

$$\mathbf{x}_i^{(l)} \leq \mathbf{x}_i \leq \mathbf{x}_i^{(u)}, \qquad i = 1, 2, \ldots, m, \tag{7.172}$$

$$\mathbf{x}_{m+j} \geq 0, \qquad j = 1, 2, \ldots, n. \tag{7.173}$$

Choice variables now are the previous x_i, $i = 1, \ldots, m$ plus the newly introduced slack variables $x_{(m+j)}$ of (7.170) corresponding to j inequalities of (7.162).

The idea behind the GRG method is the elimination of variables using the equality constraints, one variable for each equality constraint. In theory, since now there are $n + p$ equality constraints represented by (7.170) and (7.171), $n + p$ variables can be eliminated from the set of $m + n$ variables. Variables are generally classified into two arbitrarily chosen sets of design and dependent variables as

$$\mathbf{x} = \begin{bmatrix} \mathbf{y} \\ \mathbf{z} \end{bmatrix}, \tag{7.174}$$

where

$$\mathbf{y} = \begin{bmatrix} y_1 \\ y_2 \\ \vdots \\ y_{m-p} \end{bmatrix}, \tag{7.175}$$

and

$$\mathbf{z} = \begin{bmatrix} z_1 \\ z_2 \\ \vdots \\ z_{n+p} \end{bmatrix}. \tag{7.176}$$

In the above, $\mathbf{y}$ are the design variables that are completely independent, and variables $\mathbf{z}$ are dependent on the design variables which are used to satisfy constraints of (7.170) and (7.171).

Now, the first variations of the constraint and the objective function are given by

$$df(\mathbf{x}) = \sum_{i=1}^{m-p} \frac{\partial f}{\partial y_i} dy_i + \sum_{i=1}^{n+p} \frac{\partial f}{\partial z_i} dz_i = \nabla_{\mathbf{y}}^T f(\mathbf{x})\, d\mathbf{y} + \nabla_{\mathbf{z}}^T f(\mathbf{x})\, d\mathbf{z}, \tag{7.177}$$

where

$$\nabla_{\mathbf{y}} f(\mathbf{x}) = \begin{bmatrix} \dfrac{\partial f(\mathbf{x})}{\partial y_1} \\ \dfrac{\partial f(\mathbf{x})}{\partial y_2} \\ \vdots \\ \dfrac{\partial f(\mathbf{x})}{\partial y_{m-p}} \end{bmatrix}, \tag{7.178}$$

$$\nabla_{\mathbf{z}} f(\mathbf{x}) = \begin{bmatrix} \dfrac{\partial f(\mathbf{x})}{\partial z_1} \\ \dfrac{\partial f(\mathbf{x})}{\partial z_2} \\ \vdots \\ \dfrac{\partial f(\mathbf{x})}{\partial z_{n+p}} \end{bmatrix}, \tag{7.179}$$

and

$$dg_r(\mathbf{x}) = \sum_{i=1}^{m-p} \frac{\partial g_r}{\partial y_i} dy_i + \sum_{i=1}^{n+p} \frac{\partial g_r}{\partial z_i} dz_i, \qquad r = 1, 2, \ldots, n+p. \tag{7.180}$$

In (7.180), n inequalities of (7.170) and p equalities of (7.171) are combined and represented as g. Equation (7.180) can be written as

$$dg = \mathbf{C}\, d\mathbf{y} + \mathbf{D}\, d\mathbf{z}, \tag{7.181}$$

where

$$\mathbf{C} = \begin{bmatrix} \dfrac{\partial g_1(\mathbf{x})}{\partial y_1} & \cdots & \dfrac{\partial g_1(\mathbf{x})}{\partial y_{m-p}} \\ \vdots & & \vdots \\ \dfrac{\partial g_{n+p}(\mathbf{x})}{\partial y_1} & \cdots & \dfrac{\partial g_{n+p}(\mathbf{x})}{\partial y_{m-p}} \end{bmatrix}, \tag{7.182}$$

$$\mathbf{D} = \begin{bmatrix} \dfrac{\partial g_1(\mathbf{x})}{\partial z_1} & \cdots & \dfrac{\partial g_1(\mathbf{x})}{\partial z_{n+p}} \\ \vdots & & \vdots \\ \dfrac{\partial g_{n+p}(\mathbf{x})}{\partial z_1} & \cdots & \dfrac{\partial g_{n+p}(\mathbf{x})}{\partial z_{n+p}} \end{bmatrix}, \tag{7.183}$$

$$d\mathbf{z} = \begin{bmatrix} dz_1 \\ dz_2 \\ \vdots \\ dz_{n+p} \end{bmatrix}, \tag{7.184}$$

and

$$d\mathbf{y} = \begin{bmatrix} dy_1 \\ dy_2 \\ \vdots \\ dy_{m-p} \end{bmatrix}. \tag{7.185}$$

If we start the iterations with an initial feasible guess $\mathbf{x}^{(0)}$ such that all constraints are satisfied, any change to the vector $d\mathbf{x} = [d\mathbf{y}, d\mathbf{z}]^T$ must satisfy (7.181) to keep $dg(\mathbf{x}) = 0$. Hence, from (7.181) we get

$$d\mathbf{z} = -\mathbf{D}^{-1}\mathbf{C}\,d\mathbf{y}. \tag{7.186}$$

Substitution for $d\mathbf{z}$ from (7.186) into (7.177) gives the change in the objective function due to changes $d\mathbf{x}$ as

$$df(\mathbf{x}) = \left[\nabla_{\mathbf{y}}^T f(\mathbf{x}) - \nabla_{\mathbf{z}}^T f(\mathbf{x})\mathbf{D}^{-1}\mathbf{C}\right] d\mathbf{y}. \tag{7.187}$$

The above can be written as

$$\frac{df(\mathbf{x})}{d\mathbf{y}} = \mathbf{G}_R, \tag{7.188}$$

where

$$\mathbf{G}_R = \left[\nabla_{\mathbf{y}} f(\mathbf{x}) - [\mathbf{D}^{-1}\mathbf{C}]^T \nabla_{\mathbf{z}} f(\mathbf{x})\right] \tag{7.189}$$

is called the *generalized reduced gradient*. The reduced gradient can be described in geometrical terms as the projection of the original m-dimensional gradient onto $(m-p)$-dimensional feasible region of independent design variables.

The procedure for solution is to start with an initial vector $\mathbf{x}$ such that all constraints in (7.170) and (7.171) are satisfied. Then the search direction $\mathbf{s}$ for the independent variables x_1 and x_2 designated as $\mathbf{y}$ is $-G_R$, the negative of the reduced gradient. The change $d\mathbf{z}$ of the dependent variables is related to $d\mathbf{y}$ according to (7.186). Hence in order to determine $d\mathbf{z}$, we have to first decide on the length $\gamma\mathbf{s}$ to be traversed in the direction of $\mathbf{s}$ such that $d\mathbf{y} = \gamma\mathbf{s}$. Then $d\mathbf{z}$ can be computed from (7.186).

The optimal length to be traversed in the search direction $\mathbf{s}$ can be determined by the univariate procedures, by Newton's method, or by other procedures discussed in the earlier sections of this book. Irrespective of the value of γ, the changes to constraints will always be zero because of (7.186). However, if large steps along $\mathbf{s}$ are taken, dependent variables that are required to be positive may be driven negative.

Several articles have been written on the subject of determining the optimal length γ to be traversed along $\mathbf{s}$. In fact, numerous alternatives are proposed some of which are suited to particular classes of problems. It in not within our scope to discuss such algorithmic details. The reader may refer to Rao (1996), Bazaraa et al. (1979), and Pierre (1969) for details.

In our example, a method of determining γ will be shown. While such a method works for this example, it may not always yield fast convergence for all problems.

GRG Method: Example of (7.165)—Continued. By the addition of slack variables, our continuing example transforms to: Choose x_1, x_2, x_3, x_4 and x_5 ($m = 5$) to minimize

$$f(\mathbf{x}) = f(x_1, x_2, x_3) = 4x_1^2 + 3x_2^2 + 5x_3^2 - 6x_1x_2 + x_1x_3 + 15, \tag{7.190}$$

subject to equalities

$$g_1(\mathbf{x}) = 2x_1 + 3x_2 + x_3 + x_4 - 8 = 0, \tag{7.191}$$

$$g_2(\mathbf{x}) = 4x_1 + 5x_2 - x_3 + x_5 - 9 = 0, \tag{7.192}$$

$$h(\mathbf{x}) = 2x_1 + 4x_3 - 8 = 0. \tag{7.193}$$

For this problem, we have $m = 3$, $n = 2$, and $p = 1$. Hence $n + p = 3$ variables can be eliminated from $m + n = 5$ variables. Thus, from among x_1 to x_5, one may choose any two-vector $\mathbf{y}$. In this case, $[x_1, x_2]^T$ is chosen arbitrarily as the vector $\mathbf{y}$, resulting in $\mathbf{z} = [x_3, x_4, x_5]^T$.

Hence we obtain

$$\nabla_y f(\mathbf{x}) = \begin{bmatrix} 8x_1 - 6x_2 + x_3 \\ 6x_2 - 6x_1 \end{bmatrix} \tag{7.194}$$

and

$$\nabla_z f(\mathbf{x}) = \begin{bmatrix} 10x_2 + x_1 \\ 0 \\ 0 \end{bmatrix}, \tag{7.195}$$

$$\mathbf{D} = \begin{bmatrix} 1 & 1 & 0 \\ -1 & 0 & 1 \\ 4 & 0 & 0 \end{bmatrix}, \tag{7.196}$$

$$\mathbf{C} = \begin{bmatrix} 2 & 3 \\ 4 & 5 \\ 2 & 0 \end{bmatrix}. \tag{7.197}$$

GRG.Xls Refer to the Excel spreadsheet "GRG.Xls," which is shown in Figure 7.13. Cells J3:J4 and J5:J7 contain values of $\nabla_y f(\mathbf{x})$ and $\nabla_z f(\mathbf{x})$, respectively. Furthermore, from (7.181) we get for dg

$$dg = \begin{bmatrix} 2 & 3 \\ 4 & 5 \\ 2 & 0 \end{bmatrix} \begin{bmatrix} dx_1 \\ dx_2 \end{bmatrix} + \begin{bmatrix} 1 & 1 & 0 \\ -1 & 0 & 1 \\ 4 & 0 & 0 \end{bmatrix} \begin{bmatrix} dx_3 \\ dx_4 \\ dx_5 \end{bmatrix}, \tag{7.198}$$

where the identification of $\mathbf{C}$ and $\mathbf{D}$ as the two matrices on the right-hand side is obvious.[12]

[12] Note the similarity of this equation in which choice or design variables are written in terms of the other variables with that of (3.32).

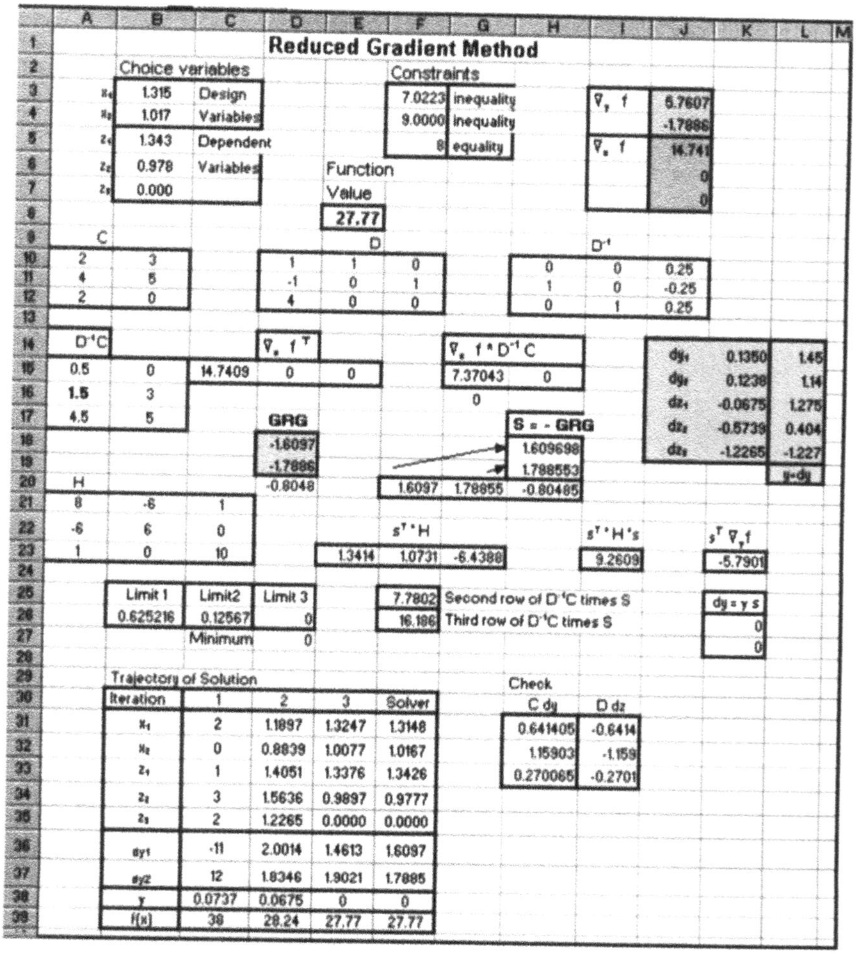

Figure 7.13. Example using the generalized reduced gradient method.

In this example, we have arbitrarily chosen x_1 and x_2 to represent $\mathbf{y}$ and have chosen x_3, x_4, and x_5 to represent $\mathbf{z}$. The former are the independent variables we intend to search, while the latter are related to these according to (7.186). The independent and dependent variables are in cells B3:B4, and B5:B7. Constraints expressed in terms of these variables corresponding to (7.191), (7.192), and (7.193) are in cells F3:F5. Matrices $\mathbf{C}$, $\mathbf{D}$, $\mathbf{D}^{-1}$, and $\mathbf{D}^{-1}\mathbf{C}$ are in cells A10:B12, D10:F12, H10:J12, and A15:B17, respectively. The GRG and the search direction (which is the negative of GRG) are in cells D18:D19 and H18:H19.

Any values for the independent variables dx_1 and dx_2 entered in cells K14 and K15 is transformed to dz_1, dz_2, and dz_3 in cells K16:K18 by a multiplication with $\mathbf{D}^{-1}\mathbf{C}$ per (7.186).

The optimum length of traverse, γ, along $\mathbf{s}$ is determined from following considerations.

The first consideration uses (6.17) of Section 6.7.1 to find the optimum length in the x_1, x_2, x_3 space. This is because there are no bounds on the values of these three variables. To this purpose, the Hessian matrix is entered in cells A21:C23, the denominator of (6.17) is computed in cell I23 and the numerator in cell K23. Hence the first value for γ is obtained in cell B26.

If we used this value of γ to compute $d\mathbf{y} = \gamma\mathbf{s}$, it is possible that $\mathbf{y} + d\mathbf{y}$ will make slack variables x_4 and x_5 (corresponding to z_2 and z_3) negative.[13] In order to avoid that, we have additional considerations as follows.

We want the new values

$$z_2^{\text{update}} = x_4^{\text{update}} = z_2^{\text{current}} + \gamma dz_2 \geq 0$$

and

$$z_3^{\text{update}} = x_5^{\text{update}} = z_3^{\text{current}} + \gamma dz_3 \geq 0.$$

Since $d\mathbf{z}$ is related to $d\mathbf{y}$ according to (7.186), which in turn is related to $\mathbf{s}$ by the relation $d\mathbf{y} = \gamma\mathbf{s}$, the above requirement translates to (using the values of the appropriate rows of the $\mathbf{D}^{-1}\mathbf{C}$ matrix)

$$z_2 - \gamma(1.5s_1 + 3s_2) \geq 0, \tag{7.199}$$

$$z_3 - \gamma(4.5s_1 + 5s_2) \geq 0, \tag{7.200}$$

or

$$\gamma \leq \frac{z_2}{1.5s_1 + 3s_2} \tag{7.201}$$

and

$$\gamma \leq \frac{z_3}{4.5s_1 + 5s_2}. \tag{7.202}$$

The minimum of γ from the above three considerations is taken as the appropriate value to proceed in the search direction $\mathbf{s}$.

Cells C26 and D26 compute the above value for γ and the minimum of the three values is entered in cell D27.

In order to run the program in the spreadsheet, we first select a starting value that satisfies all constraints. The spreadsheet shows starting values as 2, 0, 1, 3, and 2 (see cells C31:C35). Enter these values in cells B3:B7. With these values, the three constraints in cells F3:F6 are satisfied. Appropriate γ to proceed is 0.0737 for which the associated values for $d\mathbf{y}$ are in cell K26:K27. Copy this (values only

[13] Confirm that this indeed is the case for this example by entering such values for dy_1 and dy_2 in the spreadsheet.

by using paste special option) onto cells K14 and K15, and the updated values for the design and dependent variables will be given in cells L14:L18. To proceed to the next iteration, paste the updated values for choice variables from L14:L18 on to cells B3:B7. Continue this process to obtain the final solution shown in cells E31:E35.

During this procedure, observe that constraints are satisfied in every iteration even though the function value may not be at its minimum. This is because of (7.181). This can be confirmed by observing cells H31:H33 in each iteration. That is why it was said earlier that the idea of GRG method is to "stay on" constraints in each iteration.

The values obtained for variables at each iteration are shown at the bottom of Figure 7.13. The solution obtained in the second GRG iteration is in cells D31:D37. The solution of iteration 3 is in cells E31:E37. After the third iteration γ was zero indicating that the movement in the search direction is zero. Hence, the process is deemed to have converged to its minimum.

Also embedded into the spreadsheet is the Solver solution. Constraints for the Solver are formulated as in (7.166), (7.167), and (7.168). Click on the Solver option under tools and run it to obtain the optimal solution shown in cells F31:F35. This solution is very close to that obtained at iteration 3 using the GRG method. Further observe in the sensitivity sheet that only two constraints corresponding to (7.167), and (7.168) are binding. Lagrange multipliers associated with these binding constraints are -0.3577 and 3.5957, respectively. The pertinence of this observation will be evident in Section 7.15.

Remark: The choice of appropriate γ, it has been said earlier, is the subject of many technical publications. In that sense, it is possible to conceive of other ways of determining γ. A trivial procedure is to use a very small value for γ in each iteration. Of course, such a procedure will not be computationally efficient. Rao (1996) and Gill et al. (1981) suggest some preferred procedures.

The suggested procedure for this problem has one shortcoming in that if either z_2 or z_3 is assumed to be zero for a start, γ will be equal to zero. Under such circumstances, it may behoove to assume a very small value for γ and perturb the solution such that the resulting changes dz_1 and dz_2 do not render either z_1^{update} or z_2^{update} negative. Alternatively, it is better not to assume either $z_1^{(0)}$ or $z_2^{(0)}$ to be zero.

Affect of Nonlinear Constraints: Equation (7.186) is valid for linear constraints. However, it is a linearized approximation to nonlinear constraints obtained from (7.180) and (7.181). Consequently, if constraints are nonlinear, we may find that its value may not be exactly equal to zero after moving a distance $\gamma\mathbf{s}$ in the search direction. This means that $dg \neq 0$. The idea of GRG is to stay on constraints at all iterations. Therefore, when $\mathbf{y}$ is chosen to be $\mathbf{y} = \gamma\mathbf{s}$, in order that

$$g_i(\mathbf{x}) + dg_i(\mathbf{x}) = 0, \qquad i = 1, 2, \ldots, n + p \tag{7.203}$$

we need

$$g(\mathbf{x}) + dg(\mathbf{x}) = \mathbf{0}. \tag{7.204}$$

Substituting dg from (7.181) in the above equation, we get

$$d\mathbf{z} = \mathbf{D}^{-1}(-g(\mathbf{x}) - \mathbf{C}d\mathbf{y}). \tag{7.205}$$

Hence, when constraints are nonlinear, for a given $d\mathbf{y} = \gamma\mathbf{s}$, instead of using (7.186) to obtain $d\mathbf{z}$ as an update for $\mathbf{z}$, we use (7.205) to obtain new values for $\mathbf{z}$ as

$$\mathbf{z}^{update} = \mathbf{z}^{current} + d\mathbf{z}. \tag{7.206}$$

Example with Nonlinear Constraints: The following problem is similar to (7.190) to (7.193), but with the introduction of nonlinearity in the last constraint as well.

Choose x_1, x_2, x_3, x_4, and x_5 ($m = 5$) to minimize

$$f(\mathbf{x}) = f(x_1, x_2, x_3) = 4x_1^2 + 3x_2^2 + 5x_3^2 - 6x_1x_2 + x_1x_3 - 3x_1 - 2x_2 + 15, \tag{7.207}$$

subject to equalities

$$g_1(\mathbf{x}) = 2x_1 + 3x_2^3 + x_3 + x_4 - 5 = 0, \tag{7.208}$$

$$g_2(\mathbf{x}) = 4x_1 + 5x_2 - x_3^3 + x_5 - 7 = 0, \tag{7.209}$$

$$h(\mathbf{x}) = 2x_1^2 + 4x_3 - 12 = 0. \tag{7.210}$$

The solution to this problem is shown in the spreadsheet "GRG-NL.Xls" a snapshot of which is shown in Figure 7.14.

GRG-NL.Xls

The design and dependent variables are in cells B3:B7. The function value per (7.207) is evaluated in cell E8. As in the previous example, the matrices **C**, **D**, $\mathbf{D}^{-1}\mathbf{C}$ are computed and appropriately labeled in the spreadsheet. A matter to note is that matrix **C** is no longer a constant matrix as in the previous example; $C(3, 1) = 4x_1$ because of (7.210). The reduced gradient and the search direction are obtained as in the previous example.

At the outset, we can easily obtain the solution to this problem by using Excel Solver (Click on tools, then Solver to obtain the solution). The function has a minimum value of 23.748, and the values for variables will be those in cells L27:L31.

The departure from the previous example is that we use (7.205) to compute $d\mathbf{z}$ instead of (7.186) in cells L10:L12. As before, $d\mathbf{y}$ is obtained as a suitable multiplier times **s**, that is, $d\mathbf{y} = \gamma\mathbf{s}$. In order to obtain fast convergence, efficient algorithms search for the best γ. Since our purpose in this example is that of explaining the mathematical principle of GRG method, γ has been selected in an ad hoc manner.

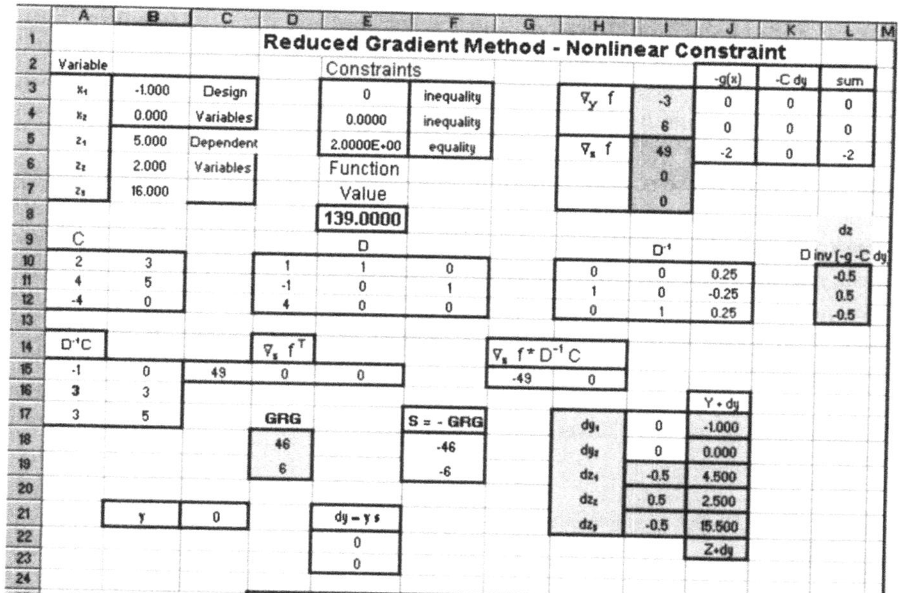

Trajectory of Solution

Variables	start	ITERATIONS 1	Local 1	2	Local 2	3	Local 3	4	Local 4	Solver
x_1	0.000	-1	-1	-3.025	-3.025	-2.659417	-2.6594	-3.134	1.32475	-3.005
x_2	0.000	0	0	-0.3	-0.3	-0.627	-0.627	-0.993	1.00773	-3.005
z_1	5.000	5	4.5	2.475	0.42469	1.530576	1.46375	0.203	1.33763	0.484
z_2	0.000	2	2.5	9.475	11.5253	10.66926	10.7361	14.043	0.9897	19.542
z_3	12.000	16	15.5	23.075	21.0247	22.30324	22.2364	24.701	0.0000	34.531
s1	-5	-46	-40.5	-45.793	18.2791	-17.649	-15.8	22.373	26	-2E-04
s2	0	-6	-6	-16.35	-16.35	-12.194	-12.194	-12.844	-12.84	-1E-05
γ		0.2		0.05		0.02		0.03	0	na
f(x)	140	139	115.75	69.5688	46.04	42.105	41.28	38.147	38.32	23.748

Figure 7.14. Spreadsheet of GRG method with nonlinear constraint.

The entries of the table at the bottom of the spreadsheet in Figure 7.14 show the trajectory of solution in each iteration. Also shown are the chosen values of γ chosen for each iteration. The starting point is to search for a set of values for variables such that constraints are satisfied. Recall that the GRG procedure remains on constraints in each iteration. The starting values indicated in cells C27:C31 satisfy this requirement giving $f(\mathbf{x}) = 140$, indicating the search directions $dy_1 s_1 = -5$, and $dy_2 s_2 = 0$.

Iteration 1: Enter the desired value of γ in cell C21. With $\gamma = 0.2$, the required changes to variables is computed in cells I17:I21, and the corresponding new values for variables are in cells J17:J21. Copy and paste these values (only values) in cells B3:B7. The function value decreases to 139.

Local Iteration: After using these updated values for variables, we observe that the equality constraint is not satisfied; cell E5 displays a value of 2.0, instead of desired zero, because the equality constraint is nonlinear. Such would not be the case if all constraints were linear. In order to stay on constraints, a local iteration is now performed.

Matrix **C**, because of nonlinear constraints, does not have constant entries; some of its entries contain x_1. Therefore the elements of this matrix change from iteration to iteration since x_1 changes in each iteration. Hence, we compute changes to $\mathbf{z}$ necessary to make cell E5 equal to zero. To that end, in order to hold x_1 and x_2 fixed, we make $d\mathbf{y} = 0$ (by entering zero in cell C21). Accounting for changes in the **C** matrix, vector $d\mathbf{z}$ is now obtained in cells I19:I21, giving us new values for choice variables $[z_1, z_2, z_3]^T$ in cells J19:J21.[14] The values obtained are shown in cells E27:E31 in the trajectory table. Paste these values in cells B3:B7 when all constraints will be exactly satisfied.[15] The value of the function now reduces to 115.75.

Iteration 2: Next we use a value of 0.05 for γ (enter this value in cell C21) to compute new values for variables in cells J17:J21 (values obtained are shown in cells F27:F31 under iteration 2). When these values are substituted in cells B3:B7, we get $f(\mathbf{x}) = 69.568$, but the constraint function value in cell E5 will not be zero.

Local Iteration: As in the previous iteration, a local iteration is performed. We make $d\mathbf{y} = 0$ by entering zero in cell C21 and obtain values of new variables in cells J17:J21 (values obtained are shown in cells G27:G31), which, when substituted in B3:B7, satisfies all constraints and gives a function value of 46.04.

As shown in the trajectory, we continue the above process for two more iteration with values of 0.02 and 0.03 for γ culminating in a final value of 38.32. We are still some distance from the solution obtained by the Solver. A continuation of this procedure will ultimately converge to the Solver solution.

Problem 4 in Section 7.18 asks the reader to modify the above example of GRG by making more than one constraint nonlinear, and to program the spreadsheet according to (7.205).

Further details on handling nonlinearities and the determination of γ can be found in Rao (1996), Gill et al. (1981), Pierre (1969), and Bazaraa et al. (1979).

7.13 SEQUENTIAL LINEAR PROGRAMMING METHOD

The Sequential Linear Programming (SLP) Method is one in which, as the name implies, the nonlinear programming problem is solved by a series of sequential linear programs. Each linear program is formed by the linear representation of

[14] The values of x_1 and x_2 remain unchanged as we have desired that $d\mathbf{y} = 0$.

[15] Since the nonlinearity is fairly simple (quadratic) in this example, we are able to satisfy constraints in one iteration. In problems where constraints are highly nonlinear, more than one local iteration may be required in order to obtain convergence.

the objective and constraints by their first-order Taylor expansion about $\mathbf{x}^{(t)}$, the current vector of choice variable in iteration t. The process is started from an initial guess for choice variables. The LP problem in each step is solved to obtain a new vector of choice variables $\mathbf{x}^{(t+1)}$. If $\mathbf{x}^{(t+1)}$ does not fulfill a specified tolerance of convergence criteria, the problem is relinearized about $\mathbf{x}^{(t+1)}$, and the procedure is repeated untill $\bar{\mathbf{x}}$ is obtained.

Procedure

1. Start with an initial guess $\mathbf{x}^{(t)}$ in iteration $t = 1$. The guessed $\mathbf{x}^{(t)}$ need not be feasible.
2. Linearize the objective function as

$$f(\mathbf{x}) \simeq f(\mathbf{x}^{(t)}) + \nabla f(\mathbf{x}^{(t)})^T(\mathbf{x} - \mathbf{x}^{(t)}) \tag{7.211}$$

and linearize constraints of (7.16) and (7.17) as

$$h_j(\mathbf{x}) \simeq h_j(\mathbf{x}^{(t)}) + \nabla h_j(\mathbf{x}^{(t)})^T(\mathbf{x} - \mathbf{x}^{(t)}) \ , \qquad j = 1, 2, \ldots, p, \tag{7.212}$$

and

$$g_i(\mathbf{x}) \simeq g_i(\mathbf{x}^{(t)}) + \nabla g_i(\mathbf{x}^{(t)})^T(\mathbf{x} - \mathbf{x}^{(t)}), \qquad i = 1, 2, \ldots, n. \tag{7.213}$$

3. Formulate and solve the following LP to obtain the vector $\mathbf{x}^{(t+1)}$:
Minimize

$$f(\mathbf{x}^{(t)}) + \nabla f(\mathbf{x}^{(t)})^T(\mathbf{x} - \mathbf{x}^{(t)}) \tag{7.214}$$

subject to

$$g_i(\mathbf{x}^{(t)}) + \nabla g_i(\mathbf{x}^{(t)})^T(\mathbf{x} - \mathbf{x}^{(t)}) \leq 0 \tag{7.215}$$

and

$$h_j(\mathbf{x}^{(t)}) + \nabla h_j(\mathbf{x}^{(t)})^T(\mathbf{x} - \mathbf{x}^{(t)}) = 0. \tag{7.216}$$

4. Find the values of

$$g_i(\mathbf{x}^{(t+1)}), \quad i = 1, 2, \ldots, n; \quad h_j(\mathbf{x}^{(t+1)}), \quad j = 1, 2, \ldots, p.$$

For a specified tolerance τ, if $g_i(\mathbf{x}^{(t+1)}) \leq \tau$ for $i = 1, 2, \ldots, n$, and $h_j(\mathbf{x}^{(t+1)}) \leq \tau$ for $j = 1, 2, \ldots, p$, stop the procedure and set $\bar{\mathbf{x}} = \mathbf{x}^{(t+1)}$. Else:

5. If $g_i(\mathbf{x}^{(t+1)}) > \tau$ for some i, or $h_j(\mathbf{x}^{(t+1)}) > \tau$ for some j, find the constraint that violates the constraint most; that is, find

$$g_k(\mathbf{x}^{(t+1)}) = \max_{i:\ i=1,\cdots,p} [g_i(\mathbf{x}^{(t+1)})],$$

and

$$h_k(\mathbf{x}^{(t+1)}) = \max_{j:\ j=1,\cdots,p} [h_j(\mathbf{x}^{(t+1)})].$$

Relinearize the maximum violating inequality or equality constraint. For example if the maximum violating constraint is an inequality, linearize it as

$$g_k(\mathbf{x}) \simeq g_k(\mathbf{x}^{(t+1)}) + \nabla g_k(\mathbf{x}^{(t+1)})^T(\mathbf{x} - \mathbf{x}^{(t+1)}) \leq 0, \tag{7.217}$$

and add this as the $(n+1)$th inequality constraint (or the $(p+1)$th equality if the maximum violating constraint is an equality) to the previous LP problem.
6. Set the iteration number as $t = t + 1$ with total number of inequality constraints equal to $n + 1$ and equality constraints equal to p. Go to step 3.

Example: The following example will make the above procedure clear.
Minimize the objective function:

$$f(\mathbf{x}) = 6x_1^2 + 3x_2^2 + 4x_1x_2 + 5.5x_1 - 2x_2, \tag{7.218}$$

subject to constraints

$$g_1(\mathbf{x}) = -x_1 + x_2 \leq 0.8, \tag{7.219}$$

$$g_2(\mathbf{x}) = 0.5 + (x_1 + 0.4)^2 - x_2 \leq 0. \tag{7.220}$$

SLP.Xls

The SLP solution procedure is set up on an Excel spreadsheet "SLP.Xls" shown in Figure 7.15. Also, in order to facilitate an understanding of the process, The contours of the objective function and constraints are plotted in Figure 7.16.

At any point $\mathbf{x}(t)$, we have

$$\nabla f(\mathbf{x}) = \begin{bmatrix} 12x_1 + 4x_2 + 5.5 \\ 6x_2 + 4x_1 - 2 \end{bmatrix}, \tag{7.221}$$

$$\nabla g_1(\mathbf{x}) = \begin{pmatrix} -1 \\ 1 \end{pmatrix}, \tag{7.222}$$

and

$$\nabla g_2(\mathbf{x}) = \begin{pmatrix} 2(x_1 + 0.4) \\ -1 \end{pmatrix}. \tag{7.223}$$

	A	B	C	D	E	F	G	H
1		**Sequential Linear Program Solution**						
2								
3	-1.95544			x_1	x_2		first min	
4			Choice	-0.29877	0.501226		**-2.64571**	
5							1.44E-14	
6	$g_1(x)$	1.44E-14		**∇f**	**∇g₁**	**∇g₂**	-0.08025	
7	$g_2(x)$	0.009021		3.919618	-1	0.202452		
8				-0.18774	1	-1		
9								
10								
11	$x^{(1)}$	-0.7	0.099996	$X-X^{(1)}$	0.401226	0.40123		
12								
13	$\nabla g_2(x^{(1)})$	-0.60001			**Sequential constraint values**			
14		-1			-0.15196			
15								
16	$x^{(2)}$	-0.39375	0.40625	x-x(2)	0.094976	0.094976		
17								
18	$\nabla g_2(x^{(2)})$	0.0125						
19		-1			3.72E-14			
20								
21	$x^{(3)}$	-0.29877	0.501226	x-x(3)	-7.7E-14	-9.8E-14		
22								
23	$\nabla g_2(x^{(3)})$	0.202452						
24		-1			0.009021			
25								
26	$\mathbf{x}^{(4)}$	**-0.28746**	**0.512537**					
27								

Figure 7.15. Excel spreadsheet of solution using the Sequential Linear Program.

Sequence 1: Enter a starting point of $\mathbf{x}^{(0)} = (0, 0)^T$ in choice variable cells D4 and E4. The constraint values (7.219), (7.220) derived from these values of $\mathbf{x}^{(0)}$ are in cells B6 and B7. $\nabla f(\mathbf{x})$ is in cells D7:D8, $\nabla g_1(\mathbf{x})$ in cells E7:E8, and $\nabla g_2(\mathbf{x})$ is computed in cells F7:F8. These correspond to (7.221), (7.222), and (7.223). The function value $f(\mathbf{x})$ is in cell A3, which turns out be zero for this initial guess.

Since $f(\mathbf{x}^{(0)}) = 0$, from (7.214) we observe that the objective is to minimize $\nabla f(\mathbf{x}^{(0)})^T(\mathbf{x} - \mathbf{x}^{(0)})$. From (7.221) and (7.214), the function to minimize becomes $5.5x_1 - 2x_2$ which is entered in cell G4.

At $\mathbf{x}^{(0)}$, we have $g_1(\mathbf{x}^{(0)}) = -0.8$, and $g_2(\mathbf{x}^{(0)}) = 0.66$. Using (7.212), (7.213) on constraints (7.129), (7.120), since the gradients are given by (7.222) and (7.223), at $\mathbf{x}^{(0)} = [0, 0]^T$ we get

$$g_1 = -0.8 - x_1 + x_2,$$

$$g_2 = 0.66 + 0.8x_1 - x_2.$$

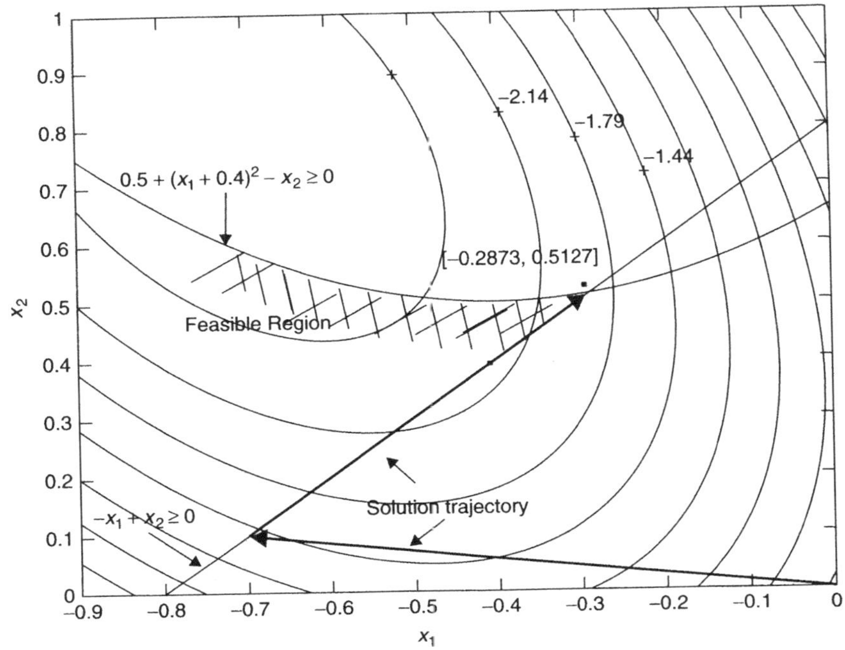

Figure 7.16. Contour plot of example solved by the SLP procedure.

These are entered in cells G5 and G6. The first sequence of solution is to minimize cell G4 subject to constraints that cells G5 and G6 are ≤ 0. Select "Assume Linear Model" in the options screen of the Solver. This obtains a solution of $\mathbf{x}^{(1)} = (-0.7, 0.1)$ with a value for $f(\mathbf{x}^{(1)}) = -1.36$ in cell A3. The value for second right-hand-side component of (7.211) in cell G4, which is our objective, is -4.05. Values of zero and 0.49 result (cells B6 and B7) for g_1 and g_2. The gradients of constraints per (7.222) and (7.223) for this value of $\mathbf{x}^{(1)}$ are indicated in cells E7:E8, and F7:F8.

In the spreadsheet "SLP.Xls," the values obtained in each iteration are shown as a table in columns J to O.

Sequence 2: The value of 0.49 for g_2 is the maximum violation of constraint. Therefore, from (7.217), we have

$$g_k(\mathbf{x}) = g_2 \simeq 0.49 + \nabla g_2^T(\mathbf{x}^{(t+1)})(\mathbf{x} - \mathbf{x}^{(t+1)}) \leq 0. \qquad (7.224)$$

The values for ∇g_2 given in cells F7 and F8 are copied onto cells B13:B14. The values of $(\mathbf{x}^{(1)} - \mathbf{x}^{(2)})$, with $\mathbf{x}^{(2)}$ yet to be determined, are in cells E11 and F11, and the value of (7.224) is computed in cell E14. This becomes the additional constraint. Hence, in this sequence, the Solver is asked to minimize cell G4 with the constraint that G5, G6, and E14 be ≤ 0. We obtain a new set of results with

	A	B	C	D	E
1	Solver Solution				
2	-1.91091		x1	x2	
3	Objective		-0.2873	0.512701	
4					
5	g(x)	0.8			
6	h(x)	3.87E-07			

Figure 7.17. Solution obtained directly using excel solver.

$\mathbf{x}^{(2)} = [-0.3937,\ 0.4062]^T$. Again, g_2 in cell B7 is the most violated constraint with a value of 0.0937.

Sequence 3: As in the previous case we obtain $\nabla g_2(\mathbf{x}^{(2)}) = [0.0125,\ -1]^T$ using which we evaluate (7.224) in cell E19. The Solver is now asked to minimize cell G4 subjects to cells G5, G6, E14, and E19 being ≤ 0. We now obtain a $\mathbf{x} = [-0.2987,\ 0.5012]^T$.

By continuing this process as indicated in the spreadsheet, we obtain a final value of $\mathbf{x} = [-0.28746,\ 0.51253]^T$, and $f(\mathbf{x}) = -1.9109$ in cell A3.

Figure 7.16 shows the path of the solution. After the first iteration, the solution travels along g_1 which happens to be a linear constraint.

Of course, this solution can be obtained in a straightforward manner using the nonlinear optimization capability of the Excel Solver routine. This is demonstrated in sheet 2 of SLP.Xls. In sheet 2, ensure that the nonlinear model is selected by clicking the appropriate options button of the Solver frame.

Figure 7.17 depicts sheet 2. The objective function in cell A3 is minimized subject to constraints in cells B5 and B6 [which correspond to (7.219) and (7.220)]. The resulting solution for $\bar{\mathbf{x}}$ is identical to the value of $\mathbf{x}^{(4)}$ obtained above by the SLP procedure.

7.14 SEQUENTIAL QUADRATIC PROGRAMMING (SQP)

It should be obvious from the above sections that a solution to a constrained nonlinear program can be obtained by solving a sequence of Quadratic Programs (SQP). The details of the algorithm procedure can be found in Rao (1996) and Bazaraa et al. (1979). Because of interest in and popularity of this method, we sketch some theoretical underpinnings behind this method.

Recall from (7.110) that a quadratic program can be written as

$$f(\mathbf{x}) = \mathbf{d}^T\mathbf{x} + \frac{1}{2}\mathbf{x}^T\mathbf{H}\mathbf{x}, \tag{7.225}$$

subject to the set of nonlinear constraints linearized as

$$h(\mathbf{x}) + [\nabla h]^T \Delta\mathbf{x} = \mathbf{0}. \tag{7.226}$$

Here, we have converted all the inequality constraints to equalities by the addition of slack variables. Hence all the n inequality constraints are linearized and represented along with p linearized equalities as $(n + p)$ constraints in (7.226).

The Kuhn–Tucker conditions for the above are given by [see (7.118)]

$$\nabla L = \mathbf{d} + H\mathbf{x} + \mathbf{G}\lambda = 0, \tag{7.227}$$

$$h(\mathbf{x}^{(j)}) + [\nabla h(\mathbf{x}^{(j)})]^T \Delta\mathbf{x} = 0, \tag{7.228}$$

where $\mathbf{G}$ has been defined earlier in (7.122), and $h(\mathbf{x}^{(j)})$ and $\nabla h(\mathbf{x}^{(j)})$ are constraints and their gradients evaluated at some initial guess for $\mathbf{x}$ at iteration j.

Now consider the problem of minimizing a quadratic function Q:

$$Q(\mathbf{x}) = \nabla f^T \Delta\mathbf{x} + \frac{1}{2}\Delta\mathbf{x}^T \nabla^2 L \Delta\mathbf{x}, \tag{7.229}$$

with the same set of constraints (7.226).

From the Lagrangian function we get the Kuhn–Tucker conditions [as in (7.227)]

$$\nabla L = \nabla f + [\nabla^2 L]\Delta\mathbf{x} + \mathbf{G}\boldsymbol{\lambda} = 0, \tag{7.230}$$

$$h_k(\mathbf{x}^{(j)}) + [\nabla h(\mathbf{x}_k^{(j)})]^T \Delta\mathbf{x} = 0, \quad k = 1, 2, \ldots, n + p. \tag{7.231}$$

Equations (7.230) and (7.231) can be identified to be the same as that of (7.95) in Section 7.9. But (7.95) in Section 7.9 is an algorithmic procedure for solving any nonlinear problem. Therefore, the process of finding successive iterates in a nonlinear programming problem represents a sequence of solving corresponding quadratic problems as in (7.229). The solution to quadratic problems, in turn, can be obtained from the solution to a corresponding linear program. The following examples reinforce this procedure.

Examples

1. Consider the example of Section 7.9.1:

Select x_1, x_2, and x_3 to minimize

$$f(\mathbf{x}) = f(x_1, x_2, x_3) = 4x_1^2 + 3x_2^2 + 5x_3^2 - 6x_1x_2 + x_1x_3 - 3x_1 - 2x_2 + 15, \tag{7.232}$$

subject to

$$g(\mathbf{x}) = x_1^2 + x_3^2 - 30 \leq 0, \tag{7.233}$$

$$h_2(\mathbf{x}) = 2x_1 + 3x_2^3 + x_3 - 8 = 0, \tag{7.234}$$

$$h_3(\mathbf{x}) = 4x_1 + 5x_2 - x_3^3 - 9 = 0. \tag{7.235}$$

By adding a nonnegative slack variable, x_4^2 to the inequality of (7.233) it is converted to an equality $h_1(\mathbf{x})$ as

$$h_1(\mathbf{x}) = x_1^2 + x_3^2 - 30 + x_4^2 = 0. \tag{7.236}$$

SQP.Xls

The spreadsheet "SLP.Xls" in Figure 7.18 outlines the solution procedure. We have

	A	B	C	D	E	F	G	H
1		**Sequential Quadratic Program**						
2								
3		x1	x2	x3	x4	λ_1	λ_2	λ_3
4	Corrections $\Delta\mathbf{x}$	-6.0114E-06	0.0000	-6.4E-05	-6.9E-05	0	0	0
5	Starting Values $\mathbf{x}$	0.610843706	1.311325	0.013547	5.443111	0	0	0
6								
7	$\nabla f(\mathbf{x})$			$\nabla^2 L$				
8	-5.967655773		8	-6	1	0		
9	2.20289061		-6	6	0	0		
10	0.746317964		1	0	10	0		
11	0		0	0	0	0		
12								
13				$\nabla h_1(\mathbf{x})$	$\nabla h_2(\mathbf{x})$	$\nabla h_3(\mathbf{x})$		
14	$h_1(\mathbf{x})$	0.00076646		1.221687	2	4		
15	$h_2(\mathbf{x})$	3.68361E-07		0	15.47617	5		
16	$h_3(\mathbf{x})$	-2.9521E-07		0.027095	1	-0.00055		
17				10.88622	0	0		
18	$h_1(x) + \nabla h_1(\mathbf{x})\,\Delta\mathbf{x}$	9.99987E-07						
19	$h_2(x) + \nabla h_2(\mathbf{x})\,\Delta\mathbf{x}$	1.04001E-11					4.09E-08	1.12E-09
20	$h_3(x) + \nabla h_3(\mathbf{x})\,\Delta\mathbf{x}$	2.52791E-14				$\nabla \mathbf{x}^T \nabla^2 L\, \Delta x$		2.1E-08
21						$\nabla f(\mathbf{x})\,\Delta\mathbf{x}$		-8.7E-07
22	0.999993989	1.000004861	0.999936	0.999931			sum	**-8.5E-07**
23								
24	0.610837694	1.311330335	0.013484	5.443041				
25								
26	Solver Solution							
27	x1	x2	x3	x4				
28	0.610837777	1.311330269	0.013485	5.443041		**12.39917**	Function Value	
29	$f(\mathbf{x})$	12.39916431						
30	$h_1(\mathbf{x})$	4.03366E-08						
31	$h_2(\mathbf{x})$	5.46863E-08						
32	$h_3(\mathbf{x})$	-4.3179E-10						
33								

Figure 7.18. Solution using the sequential quadratic programming procedure.

$$\nabla^2 L = \begin{bmatrix} 8+2\lambda_1 & -6 & 1 & 0 \\ -6 & 6+18\lambda_2 x_2 & 0 & 0 \\ 1 & 0 & 10+2\lambda_1+6\lambda_3 x_3 & 0 \\ 0 & 0 & 0 & 2\lambda_1 \end{bmatrix}, \tag{7.237}$$

$$\nabla f = \begin{bmatrix} 8x_1 - 6x_2 + x_3 - 3 \\ 6x_2 - 6x_1 - 2 \\ 10x_3 + x_1 \end{bmatrix} \tag{7.238}$$

and

$$\mathbf{G} = [\nabla h_1, \nabla h_2, \nabla h_3] = \begin{bmatrix} 2x_1 & 2 & 4 \\ 0 & 9x_2^2 & 5 \\ 2x_3 & 1 & -3x_3^2 \\ 2x_4 & 0 & 0 \end{bmatrix}. \tag{7.239}$$

In the spreadsheet, a starting value for $\mathbf{x}$ and $\boldsymbol{\lambda}$ is assumed. It is irrelevant what values for $\boldsymbol{\lambda}$ will be assumed. Start with an assumed value of unity for $\mathbf{x}$ and zeros for $\boldsymbol{\lambda}$ in cells B5:H5. The gradients of $f(\mathbf{x})$, gradients of three constraints $\nabla h_1, \nabla h_2, \nabla h_3$, and $\nabla^2 L$, are entered in cells A8:A11, D14:F17, and C8:F11, respectively. The value of the three linearized constraints according to (7.231) are in cells B18:B20. Two right-hand-side components of (7.229) are in cells H20 and H21, and their sum is in cell H22. The goal is to minimize H22.

If an algorithm were written from scratch, one could incorporate a successive linear programming procedure to solve the above quadratic program at hand. However, our purpose is to demonstrate that successive representations of a nonlinear program by the above quadratic program converge to the correct solution.

In order to demonstrate this indeed is the case, we use the capability of the Excel Solver. Open the Solver routine in the spreadsheet "SQP.Xls" which is programmed to choose corrections $\Delta\mathbf{x}$ and $\boldsymbol{\lambda}$ in cells B4:H4 in order to minimize cell H22. Since all are converted to equalities, constraints are that cells B18:B20 are equal to zero. The solution obtained in row 4 is added to a value of one (since that was our initial guess) to obtain values indicated in cells A22:D22. These are the updated values for $\mathbf{x}$ that are now pasted (values only) in cells B5:E5. Using the Solver, new corrections are obtained. The corrections are added to the values in cells B5:D5 to obtain values shown in cells A24:D24 that now becomes the updated $\mathbf{x}$. These values are again pasted in cells B5:E5.

Continuing this procedure, the results of Figure 7.18 are obtained. The solution obtained by using the Excel Solver directly to minimize the function value in cell B29 subject to constraints of the original problem in cells B30:B32 is shown in cells A28:D28. Compare this with cell F28 which is the value of the function obtained by the quadratic procedure. As expected, the Solver solution is identical to that obtained by solving sequential quadratic programs for corrections.

This example, therefore, proves that starting from an initial guess, a nonlinear optimization problem can be solved by computing corrections in a sequence of quadratic programming problems structured according to (7.230).

2. Consider the following problem: Minimize

$$u(\mathbf{x}) = 6x_1^2 + 3x_2^2 + 4x_1x_2 + 5.5x_1 - 2x_2 + 8x_1^2x_3x_2, \tag{7.240}$$

subject to

$$h_1(\mathbf{x}) = (x_1 + 0.4)^2 - x_2 + x_4^2 = 0, \tag{7.241}$$

$$h_2(\mathbf{x}) = -x1 + x_2 - 0.8 + x_5^2, \tag{7.242}$$

and

$$h_3(\mathbf{x}) = x_2 + x_3^2 = 4.0, \tag{7.243}$$

in which (7.241) and (7.242) are inequalities represented as equalities by the addition of nonnegative slack variables x_4^2 and x_5^2.

Confirm using the Excel Solver that the solution to this problem is $x_1 = -0.31602$, $x_2 = 0.298135$, and $x_3 = 1.924023$.

To demonstrate that SQP procedure obtains the same solution, assume a starting value of $\mathbf{x}^{(0)} = [x_1, x_2, x_3, x_4, x_5] = [1, 1, 1, 1, 1]$. Then, we get

$$\nabla(u^{(0)}) = \begin{bmatrix} 12x_1 + 4x_2 + 5.5 + 16x_1x_2x_3 \\ 6x_2 + 4x_1 - 2 + 8x_1^2x_3 \\ 8x_1^2x_2 \\ 0 \\ 0 \end{bmatrix}_{\mathbf{x}^{(0)}} = \begin{bmatrix} 37.5 \\ 16 \\ 8 \\ 0 \\ 0 \end{bmatrix}, \tag{7.244}$$

$$\mathbf{H}^{(0)} = \nabla^2 L^{(0)} = \begin{bmatrix} 12 + 16x_3x_2 & 4 + 16x_1x_3 & 16x_1x_2 & 0 & 0 \\ 4 + 16x_1x_3 & 6 & 16x_1^2 & 0 & 0 \\ 16x_1x_2 & 8x_1^2 & 0 & 0 & 0 \\ 0 & 0 & 0 & 2\lambda_1 & 0 \\ 0 & 0 & 0 & 0 & 2\lambda_2 \end{bmatrix}_{\mathbf{x}^{(0)}}$$

$$= \begin{bmatrix} 28 & 20 & 16 & 0 & 0 \\ 20 & 6 & 16 & 0 & 0 \\ 16 & 8 & 0 & 0 & 0 \\ 0 & 0 & 0 & 2 & 0 \\ 0 & 0 & 0 & 0 & 2 \end{bmatrix}, \tag{7.245}$$

and

$$\mathbf{G}_{\mathbf{x}^{(0)}} = \begin{bmatrix} (2x_1 + 0.4) & -1 & 0 \\ -1 & 1 & 0 \\ 0 & 0 & 2x_3 \\ 2x_4 & 0 & 0 \\ 0 & 2x_5 & 0 \end{bmatrix}_{\mathbf{x}^{(0)}} = \begin{bmatrix} 2.4 & -1 & 0 \\ -1 & 1 & 0 \\ 0 & 1 & 2 \\ 2 & 0 & 0 \\ 0 & 2 & 0 \end{bmatrix}. \tag{7.246}$$

The constraint values at this initially assumed $\mathbf{x}^{(0)}$ are $h_1(\mathbf{x}^{(0)}) = 1.96$, $h_1(\mathbf{x}^{(0)}) = 0.2$, and $h_1(\mathbf{x}^{(0)}) = -2$.

The first quadratic programs to be solved is, therefore, the following: Choose $\Delta\mathbf{x} = [\Delta x_1, \Delta x_2, \Delta x_3, \Delta x_4, \Delta x_5,]$ in order to minimize

$$u(\mathbf{x}) = \begin{bmatrix} 37.5 \\ 16 \\ 8 \\ 0 \\ 0 \end{bmatrix}^T \Delta\mathbf{x} + \frac{1}{2}\Delta\mathbf{x}^T \begin{bmatrix} 28 & 20 & 16 & 0 & 0 \\ 20 & 6 & 16 & 0 & 0 \\ 16 & 8 & 0 & 0 & 0 \\ 0 & 0 & 0 & 2 & 0 \\ 0 & 0 & 0 & 0 & 2 \end{bmatrix} \Delta\mathbf{x}, \tag{7.247}$$

subject to constraints linearized from (7.241) to (7.243) that correspond to (7.231) as

$$\begin{bmatrix} 1.96 \\ 0.6 \\ -2.0 \end{bmatrix} + \mathbf{G}^T \begin{bmatrix} \Delta x_1 \\ \Delta x_2 \\ \Delta x_3 \\ \Delta x_4 \\ \Delta x_5 \end{bmatrix} = \mathbf{0}. \tag{7.248}$$

As in the previous example, the solution to the above problem can be obtained via successive corrections to the initially guessed $\mathbf{x}$.

It is expected of the reader to confirm the solution to this problem using an Excel spreadsheet.

Remark: There are more efficient ways of writing an algorithm to solve sequential quadratic programs than indicated in the above examples. It is wasteful to compute the Hessian matrix in every iteration. Rao (1996), for example, shows that one can start out with any positive definite **H** matrix and, only toward the end of the solution, make it coincide with the actual values of the Hessian matrix of the problem. In the above examples, linearizing constraints does not pose any problems. However, for some highly nonlinear constraints, the linearized boundary may cut off the solution space for the minimum. There are ways of avoiding such outcomes.

This method is continually being researched for improvements. Indicated references and continuing publications are sources of such details.

7.15 SYNTHESIS OF LAGRANGE MULTIPLIERS FROM SOLUTION

In Section 7.5 it was shown that, at the optimal point, the negative gradient of the objective function can be expressed as a linear combination of the gradients of active constraints. Furthermore, exploiting this property, (7.64) has shown a method of computing the Lagrange multipliers from the binding constraints of an optimization problem. This is a very useful relationship because it offers a method to determine if the solution to a problem is the optimum if constraints of an optimization problem are known.[16]

Two examples given below reinforce the concept of this procedure.

Example 1: Consider the following problem of Section 7.9.1:
Minimize

$$f(x_1, x_2) = 0.25x_1^2 + x_2^2, \tag{7.249}$$

subject to

$$h_1(x_1, x_2) = 5 - x_1 - x_2 = 0 \tag{7.250}$$

and

$$g(x_1, x_2) = x_1 + 0.2x_2 - 3 \leq 0. \tag{7.251}$$

It was shown that the solution to this problem is (see Figure 7.7) $\overline{\mathbf{x}} = [2.5,\ 2.5]^T$. We now synthesize Lagrange multipliers.

From a substitution of the values for x_1 and x_2 in constraint equations h and g, we note that both constraints are binding. Therefore there are two nonzero Lagrange multipliers.

Matrix **G** and vector **F** are now given by

$$\mathbf{G}|_{\overline{\mathbf{x}}} = \begin{bmatrix} -1 & 1 \\ -1 & 0.2 \end{bmatrix}, \qquad \mathbf{F}|_{\overline{\mathbf{x}}} = \begin{bmatrix} 1.25 \\ 5 \end{bmatrix}. \tag{7.252}$$

Substituting these in (7.64), we obtain $\lambda = [-5.9375,\ -4.6875]^T$, which is identical to the Lagrange multipliers indicated in the sensitivity sheet of the spreadsheet Second Eg.Xls. Since the signs of all Lagrange multipliers are positive, we infer that $(x_1, x_2) = (2.5,\ 2.5)$ represents the minimum (see Section 7.8).

Example 2: Consider the minimization problem of (7.165):
Choose x_1, x_2, x_3 $(m = 3)$ to minimize

$$f(\mathbf{x}) = f(x_1, x_2, x_3) = 4x_1^2 + 3x_2^2 + 5x_3^2 - 6x_1x_2 + x_1x_3 + 15, \tag{7.253}$$

[16]In the deregulated electric power industry, if the dispatch and transmission loading along with constraining paths are known, one can synthesize the Lagrange multipliers to audit system dispatch.

subject to inequalities

$$g_1(\mathbf{x}) = 2x_1 + 3x_2 + x_3 + x_4 \leq 8, \tag{7.254}$$

$$g_2(\mathbf{x}) = 4x_1 + 5x_2 - x_3 + x_5 \leq 9, \tag{7.255}$$

and the equality

$$h(\mathbf{x}) = 2x_1 + 4x_3 = 8. \tag{7.256}$$

The solution to the problem was obtained as $\bar{\mathbf{x}} = [x_1, x_2, x_3]^T = [1.3146, 1.0167, 1.3246]^T$. A substitution of these values in the above three constraint equations results in values of 7.022, 9, and 8 for $g_1(\mathbf{x})$, $g_2(\mathbf{x})$ and $h(\mathbf{x})$. Therefore, only $g_2(\mathbf{x})$, and $h(\mathbf{x})$ of (7.255) and (7.256) are binding.

Consequently, we have

$$\mathbf{G}|_{\bar{\mathbf{x}}} = \begin{bmatrix} \dfrac{\partial g_2}{\partial x_1} & \dfrac{\partial h}{\partial x_1} \\ \dfrac{\partial g_2}{\partial x_2} & \dfrac{\partial h}{\partial x_2} \\ \dfrac{\partial g_2}{\partial x_3} & \dfrac{\partial h}{\partial x_3} \end{bmatrix} = \begin{bmatrix} 4 & 2 \\ 5 & 0 \\ -1 & 4 \end{bmatrix}, \qquad \mathbf{F}|_{\bar{\mathbf{x}}} = \begin{bmatrix} -(8x_1 - 6x_2 + x_3) \\ -(6x_2 - 6x_1) \\ -(10x_3 + x_1) \end{bmatrix}. \tag{7.257}$$

A substitution of values of $\mathbf{x}$ into (7.64) obtains $[\lambda_2, \lambda_3] = [-0.3577, 3.5957]$ which is identical to the values in the sensitivity sheet of spreadsheet "GRG.Xls." Evidently, $\lambda_1 = 0$ since $g_1(\mathbf{x})$ is nonbinding.

We note that the sign of λ_2 associated with the inequality is positive. However, as discussed earlier, the sign of λ associated with an equality can either be positive or negative depending on whether $h(\mathbf{x}) > 0$ or $h(\mathbf{x}) < 0$ is binding. Problem 1 of Section 7.18 asks the reader to synthesize Lagrange multipliers in a manner similar to the above examples.

7.16 RATE OF CONVERGENCE

In the preceding chapters and sections on nonlinear programming, we have alluded to several procedures and algorithms. Though the intent is not to educate the reader to be a numerical analyst in order to develop algorithms, it is helpful to know how to quantify and compare the speed of execution of algorithms.

Toward that end, the rate of convergence of a procedure is defined as r, where r is given by the expression

$$\frac{\| \mathbf{x}^{(i+1)} - \bar{\mathbf{x}} \|}{\| \mathbf{x}^{(i)} - \bar{\mathbf{x}} \|^r} \leq k, \qquad k \geq 0, \quad r = \text{ integer}, \ r \geq 1. \tag{7.258}$$

Here, $\mathbf{x}^{(i)}$ and $\mathbf{x}^{(i+1)}$ represent the values of the solution vector of choice variables at iteration i and $i + 1$, and $\overline{\mathbf{x}}$ is the optimum vector and $\| \mathbf{x} \|$ denotes the Euclidean norm of the m vector given by

$$\| \mathbf{x} \| = \sqrt{x_1^2 + x_2^2 + \cdots + x_m^2}. \tag{7.259}$$

A method or procedure is said to be *linearly convergent* (slow convergence) if $r = 1$ and $0 \leq k \leq 1$. If $r = 2$, the method is said to be *quadratically convergent* (fast). If

$$\lim_{i \to \infty} \frac{\| \mathbf{x}^{(i+1)} - \overline{\mathbf{x}} \|}{\| \mathbf{x}^{(i)} - \overline{\mathbf{x}} \|} \to 0, \tag{7.260}$$

the method is said to have *superlinear convergence*.

Most second-order methods have the property of quadratic convergence as opposed to univariate and gradient methods, which are linearly convergent. Methods employing searches in conjugate directions are quadratically convergent. It can also be proved that a quadratic function of m variables can be solved after m cycles by searching in m directions. Newton's method, being a second-order method, converges to the solution of an unconstrained problem in just one iteration as shown in Section 6.8.

7.17 SUMMARY

In Chapters 6 and 7 we have discussed some methods of solution to unconstrained and constrained optimization problems. The discussion has been confined only to methods that are considered major, or popular.

Our purpose in these chapters has been that of exposing the reader to the theoretical basis behind some algorithms that are available commercially. We have not discussed a host of other excellent methods such as penalty function methods. Nor have we discussed integer and mixed integer programs as well as convex programming. Such methods merit further consideration by those that intend to develop and advance algorithms. Despite these intended omissions, it is hoped that the methods discussed provide a foundation for understanding optimization techniques.

In illustrative examples, we have indulged in some rudimentary programming using the Excel spreadsheet. It is hoped that the examples foster an understanding of inner workings of some commercial software. In short, the hope is that the foregoing chapters have improved the awareness of professionals above and beyond that of being mere users of software.

7.18 EXERCISE PROBLEMS

1. Consider a simple dispatch of five generators $i = 1, \ldots, 5$, to minimize $C_i P_i$, where P_i is the output of generators and C_i is the cost of generation, subject to constraints:

$$P_i \leq R_i, \qquad i = 1, \ldots, 5, \tag{7.261}$$

where R_i is the capability limit of generators set at 2, 1, 3, 2, and 6, respectively, and

$$\sum_{i=1}^{5} C_i = D = 11, \tag{7.262}$$

where D represents the demand. The costs of generation from five generators are 10, 20, 30, 40, and 50. The results of optimization yielded for five choice variables (outputs) values of 2, 1, 3, 2, and 3, respectively, indicating that the first four constraints of (7.261) and that of (7.262) are binding.

(a) Compute the Lagrange multipliers from the procedure indicated in section 7.15.

(b) Solve the above problem using Excel Solver and compare the above-computed λ with the Lagrange multipliers and reduced gradients in the sensitivity sheet.

2. For the optimization problem:

Minimize

$$f(x_1, x_2, x_3) = x_1^3 + 4x_1^2 + x_3,$$

subject to six constraints,

$$\begin{aligned} x_1 &\geq 0, \\ x_2 &\geq 0, \\ x_3 &\geq 0, \\ x_1^2 + x_2^2 - x_2 x_3 &\leq 0, \\ -x_1^2 - x_2^2 - x_3^2 + 4 &\leq 0, \end{aligned}$$

and

$$x_2 \leq 8.$$

Compute the Lagrange multipliers for the following solutions and indicate if the solutions represent the optimum:

(a) $\mathbf{x} = [0,\ 1.414213,\ 1.414213]^T$ (note that for this solution the first, fourth, and the fifth constraint equations will be binding),

(b) $\mathbf{x} = [0,\ 1,\ 1.73205]^T$ (note that the first and the fifth constraints will be binding).

Using Excel Solver, obtain the solution to the problem and the Lagrange multipliers.

3. Consider the example of Section 7.13 modified to include two nonlinear constraints as

$$f(\mathbf{x}) = 6x_1^2 + 3x_2^2 + 4x_1x_2 + 5.5x_1 - 2x_2,$$

subject to constraints

$$g_1 = 1.5x_1^2 + x_2 \leq 1,$$

$$g_2(\mathbf{x}) = 0.5 + (x_1 + 0.4)^2 - x_2 \leq 0.$$

Solve the problem as successive linear programs. Examine the trajectory of successive solutions and compare the result of SLP with that obtained by using Excel Solver.

4. For the problem of (7.240), formulate linearized constraints. Then solve a successive quadratic program with linearized constraints using Excel Solver. Confirm that the solution so obtained is the same as the earlier directly obtained solution using the Excel Solver.

5. Recall that for the problem in Section 6.8.1 without constraints, the starting point was critical in converging to a desired local minima in the proximity of the starting point.

Use the earlier problem of (6.27):

$$f(\mathbf{x}) = 6x_1^2 + 3x_2^2 + 3.5x_1^2x_2 + 4x_1x_2 + 5.5x_1 - 2x_2$$

and constraints (7.108) and (7.109). Solve the problem as a quadratic program. Use the procedure of the spreadsheet Quadnc.Xls and examine if the solution converges to the same value irrespective of different initial guesses for the solution.

6. Consider the following problem: Minimize

$$f(x_1, x_2) = x_1^2 + (x_2 - 7)^2,$$

subject to

$$x_1 + 4x_2 \leq 20,$$

$$-1 \leq x_i \leq 10, \qquad i = 1, 2.$$

Find a feasible direction at a point $\mathbf{x} = [4,\ 4]^T$ using the method of feasible directions suggested in Section 7.11.2. Readers who have access to MATLAB may find it advantageous to obtain a contour plot of the function and constraints to visualize the problem.

CHAPTER 8

SOLVED NONLINEAR OPTIMIZATION PROBLEMS

8.1 INTRODUCTORY REMARKS

In this chapter, we select a few problems that are typical of nonlinear programming and find a solution to them using Excel Solver. While most of the chosen examples are related to the electric power industry and its operation, nonlinear programming methods find application in several branches of engineering, science, and technology. Therefore, a couple of problems related to other industries are included.

An advanced reader may well attempt a solution to the problems of this chapter using other commercially available software such as MATLAB, IMSL, and GAMS. In addition to the solution, the sensitivity of the solution will play an important role in interpreting the results.

As demonstrated in previous chapters, the core of nonlinear programs is to minimize an objective function subject to a set of constraints. Recall that the Lagrange multiplier associated with each constraint plays an important role in that it signifies the sensitivity of the objective function to the change in the right hand side of that constraint. As the following examples demonstrate, the study of the sensitivity and answer sheets of a problem contributes to the determination of several decision parameters and market parameters in the deregulated electrical industry.

It is only the binding constraints that have nonzero Lagrange multipliers. Resulting from the conventions used in the Excel software, for minimization problems, a binding constraint of the less-than-or-equal-to type will have a negative Lagrange multiplier while the greater-than-or-equal-to type of binding constraint will have a positive Lagrange multiplier. Consider a constraint $0 \leq P_1 \leq 4$ that can be written

Optimization Principles, by Narayan S. Rau
ISBN 0-471-45130-4

as two separate constraints $P_1 \geq 0$ and $P_1 \leq 4$. If the solution requires that $P_1 = 0$, the associated Lagrange multiplier will be positive. However, if the solution requires $P_1 = 4$ to minimize an objective, the corresponding Lagrange multiplier will be negative. Of course, the Lagrange multiplier will have a value equal to zero if P_1 has a value between zero and four. This will be discussed in more detail later.

In connection with solved linear program problems in Chapter 4, for each problem its mathematical formulation was shown. In this chapter, we depart from this practice by omitting mathematical statement of the problem because of following reason. Showing the mathematical formulation would add to the mathematical rigor; and since most of the problems deal with the same electrical network, it would be repetitious to write out the same equations again and again. Nonetheless, the requisite mathematical explanation is included in the solutions, either as a description or as equations. However, it is a trivial matter for the reader to write a problem's formal mathematical formulation, if necessary.

8.1.1 Choice of Solver Options

The spreadsheets associated with solved problems have several options already selected. These can be changed if needed. To understand the choice of options, click on Tools in an Excel program and select Solver, and thence the Options button to obtain a screen that corresponds to Figure 8.1. We now discuss which are the appropriate choices to make in this screen.

Understandably, the choices to be made depend on the problem. Some problems, particularly those with local minima, may require very small numbers for precision, tolerance, and convergence. A general starting choice for problems is 0.001, to be changed later as experience dictates. In sample problems, the actual choice that was made will be apparent from a look at the options screen.

Unequivocally, for a nonlinear program the linear mode should not be checked. If it is known *a priori* that choice variables are all positive, check "assume nonnegative values." That way, entering several constraints that they are greater than zero can be avoided.

It is generally a good practice to choose automatic scaling option. Also it is normal practice to leave the "show iteration results" box unchecked. However, when one wants to proceed from iteration to iteration and watch the trajectory of solution, or certain trends, this box is checked.

One can choose either the tangent or quadratic approximation for estimates of the functions at a point. For highly nonlinear functions, the quadratic estimate may be a better choice. Similarly, for derivatives, either of the two choices will do with the central difference choice having the advantages discussed in Section 6.7.1.

As for choice for search directions, Newton's method approximates the objective function at a given point from a quadratic approximation while the conjugate choice chooses conjugate directions as explained in Section 6.5. The relative advantage or otherwise of one method over another will not be apparent to those that are not sophisticated users of this software, particularly when the results of each iteration are not tracked.

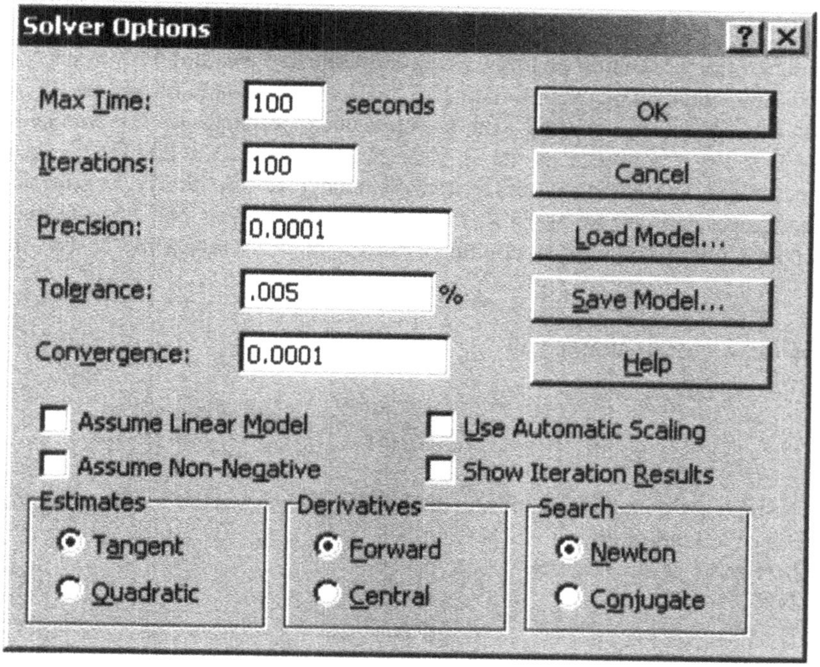

Figure 8.1. Screen to select options in Excel Solver.

Checks for Local Minima. Some problems may have several or local minima and may not have a unique global minimum (see Section 3.1). Checks based on convexity and concavity are applicable either to simple problems or to problems where the Hessian matrix can be checked for definiteness. Even then, the fact that the Hessian matrix is positive definite in the entire solution space, the problem will have a unique minimum. However, for problems with several extrema, the Hessian matrix is positive definite only in proximity of the local minimum. Thus one is not assured that there is not another extreme point at some other point elsewhere. For larger and complex problems, the recourse is to start with different sets of initial guesses. If the solution converges to the same value for all guesses, it confirms that a global minimum is attained. For problems that exhibit several local minima, one obtains several solutions and one with the lowest value is generally accepted as the solution. However, one may encounter difficulties with conflicting sensitivities, particularly if the local minima are in close proximity in the solution space.

8.2 EXAMPLES OF UNCONSTRAINED NONLINEAR OPTIMIZATION

In engineering practice, one encounters unconstrained optimization less frequently than constrained optimization problems. We solve the following problem as an illustration of the unconstrained optimization methods.

8.2.1 Transmission Voltage Optimization

Problem: It is decided to build a transmission line to transmit 300 MW of generation in a remote location to a load center. The losses in the three phase line are $3I^2R$, where I is the rms current and R is the resistance of each phase. The imputed cost of losses is \$0.15. The cost of the three-phase line towers is \$500,000 V, where V is the line to line voltage in kV. The cost of the conductor is inversely proportional to its resistance R given by $\$10^8/R$. Perform one iteration of Newton's method toward the optimum voltage and conductor resistance to minimize the cost of the project.

Solution: Line current is given by

$$I = \frac{300 \times 10^3}{\sqrt{3}V},$$

which can be written as k/V where the definition of the constant k is obvious, and V is the rms voltage in kilovolts.

The cost of losses is given by

$$0.45\frac{k^2R}{V^2}.$$

The total cost of the project is given by

$$\pi = 0.45\frac{k^2R}{V^2} + \frac{10^8}{R} + 500{,}000V.$$

The variables are R and V. We have

$$\nabla\pi = \begin{bmatrix} 0.45\frac{k^2}{V^2} - \frac{10^8}{R^2} \\ -0.9k^2\frac{R}{V^3} + 500{,}000 \end{bmatrix} \tag{8.1}$$

and

$$\mathbf{H} = \begin{bmatrix} 2\frac{10^8}{R^3} & -0.9\frac{k^2}{V^3} \\ -0.3\frac{k^2}{V^3} & 0.9k^2\frac{R}{V^4} \end{bmatrix}. \tag{8.2}$$

Tr-opt.Xls We assume a starting value of $R = 1$ and $V = 10$. Then we have $\mathbf{x}^{(0)} = [1, 10]^T$. Figure 8.2 shows the spreadsheet "Tr-opt.Xls" in which cells A4 and B4 indicate the starting values for R and V. The above expressions for ∇f and $\mathbf{H}$ are in cells B7:B8 and D7:E8. The objective function π is in cell G8.

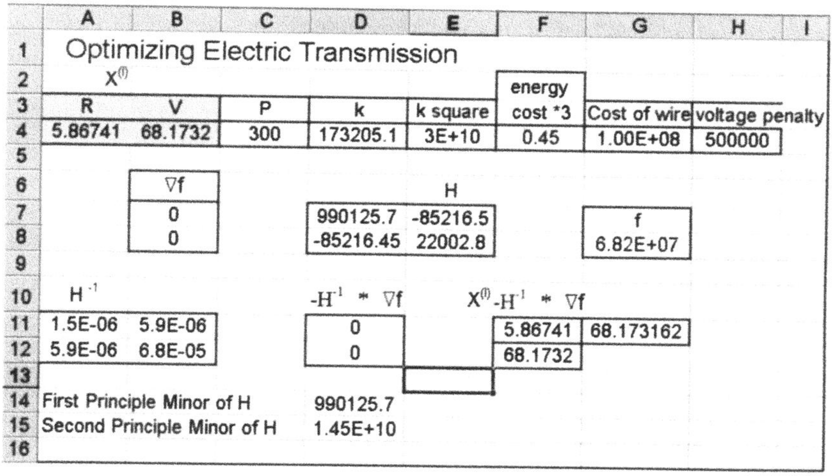

	A	B	C	D	E	F	G	H	I
1	Optimizing Electric Transmission								
2	X(0)					energy			
3	R	V	P	k	k square	cost *3	Cost of wire	voltage penalty	
4	5.86741	68.1732	300	173205.1	3E+10	0.45	1.00E+08	500000	
5									
6		∇f			H				
7		0		990125.7	-85216.5		f		
8		0		-85216.45	22002.8		6.82E+07		
9									
10	H -1			-H-1 * ∇f		X(0)-H-1 * ∇f			
11	1.5E-06	5.9E-06		0		5.86741	68.173162		
12	5.9E-06	6.8E-05		0		68.1732			
13									
14	First Principle Minor of H			990125.7					
15	Second Principle Minor of H			1.45E+10					
16									

Figure 8.2. Spreadsheet of Newton's method for minimizing transmission cost.

To facilitate an understanding of the solution space, Figure 8.3 indicates the contours of cost.

Before embarking on the process of finding the minimum, we confirm from the first and second principal minors of **H** in cells D14 and D15 that **H** is positive definite. Then we use the relation of the Newton's method

$$\mathbf{x}^{(i+1)} = \mathbf{x}^{(i)} - \mathbf{H}^{-1}\nabla f(\mathbf{x})$$

and paste values of cells F11 and G11 over the old values of R and V in cells A4 and B4. These values are 1.4848 and 14.88, respectively.

It is left as an exercise for the reader to confirm that repeated calculations for new values of R and V result in $R = 5.867\ \Omega$ and $V = 68.173$ kV. Note that the converged value corresponds to the minimum point of the contour in Figure 8.3.

An exercise problem related to this is give in Section 8.11.

8.3 EXAMPLES OF CONSTRAINED NONLINEAR OPTIMIZATION

8.3.1 Coal Mining Strategy

Consider the mining strategy in a hypothetical mine. The intent of this example is to illustrate how problems that involve choices can be formulated as optimization problems. Admittedly, this is a very simple example. In reality, expectations for profits and sales are built into a mathematical model in a probabilistic sense. Such complex models can be found in many disciplines, which contain a theme of optimization at their core. However, it is important to stress that the following example should not be viewed as being applicable only to mining or production

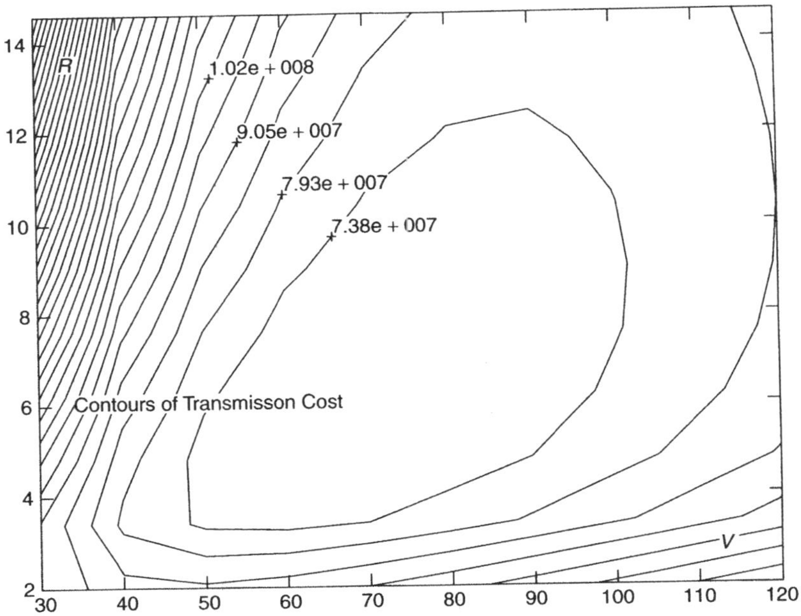

Figure 8.3. Contours of cost of transmission line.

related enterprises. The problem of fuel procurement and inventory in the electric power generation industry has mathematics similar to that of this example. Of course, such models of fuel procurement in the industry are more complex than this example and consider many other choices and constraints.

Problem Formulation: A coal mine has signed a contract to supply 50 tons of coal each month for the next six months. It has a maximum mining capability of 80 tons a month. Mined excess coal that is not shipped can be stored at a cost of \$ $0.5 + 0.1Q$, where Q is the quantity stored. Cost of mining the coal is given by $\$2.2 + 18Q + 0.2Q^2$. Find the optimum strategy for the quantity to be mined during the next six month's period in order to minimize cost. Assume that the market price for the stored coal at the end of six months is expected to be \$40.00 per ton.

Coalmine.Xls *Solution:* The spreadsheet "Coalmine.Xls" shows the setup of the problem. This is shown in Figure 8.4 as well. Choice variables for the six periods are the quantities to be mined which are in cells B5:G5. The supply contract being 50 tons per period, the cumulative contracted shipment is indicated in cells B8:G8. The cumulative quantity actually mined is in cells B6:G6. Consequently, the quantity stored is the difference between the cumulative quantity mined and shipped, which is shown in cells B7:G7. To download Excel files related to this book, see Wiley's ftp site at ftp://ftp.wiley.com/public/sci_tech_med/electric_power/.

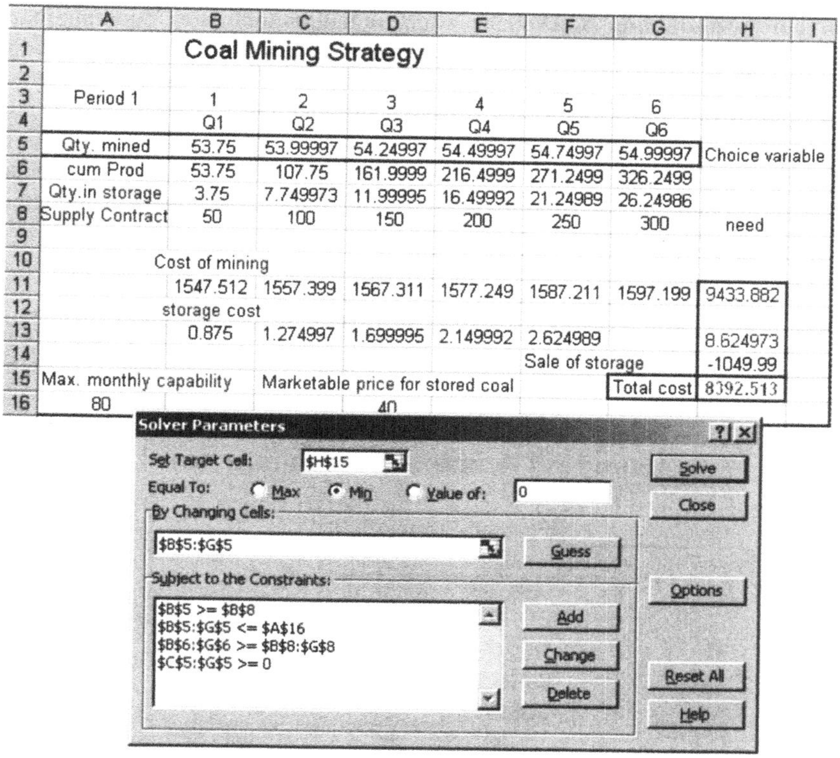

Figure 8.4. Solution to coal mining problem.

The cost of mining and storing for each period are computed in cells B11:G11 and B13:F13, respectively. The total cost to be minimized is the cost of mining, plus cost of storage, minus the cost recovered by selling stored coal. This is in cell H15.

Constraints: It is assumed that there is no coal in storage at the beginning. Then we require the quantity mined in the first period, cell B5, to be greater than or equal to the supply contract in cell B8. In a similar vein, the cumulative quantity mined B6:G6 in each period should be greater than or equal to the cumulative quantity of contracted sale shown in cells B8:G8. Finally, we require that the quantity mined in each period be greater than or equal to zero and that it cannot exceed the maximum mining capability entered in cell A16. The final results are shown in the spreadsheet.

8.3.2 Optimizing Apartment Development

Our next example is that of construction industry. It is one of optimizing the design of an apartment building. This example is based on an exercise problem in Rao (1996).

Problem Formulation: A developer wants to build a multi-story building on a lot of 500 ft × 500 ft with a living floor area of 800,000 sft. The city bylaws require that the apartments provide green space and parking lots of at least 15% of the total floor area of the living space. Furthermore, the building cannot exceed a height of 85 ft. It is determined that the height of each story is required to be 12 ft. The building cost is estimated to be $\$(500,000h + 2000f + 500g)$, where h is the total height of the building, f is the floor area of the living space, and g is the area of green space and parking lot.

Apartment .Xls

Determine the minimum cost design of the building.

Solution: Open the spreadsheet "Apartment.Xls." Figure 8.5 shows the spreadsheet. Choice variables are the number of stories n, the length l, and breadth b of

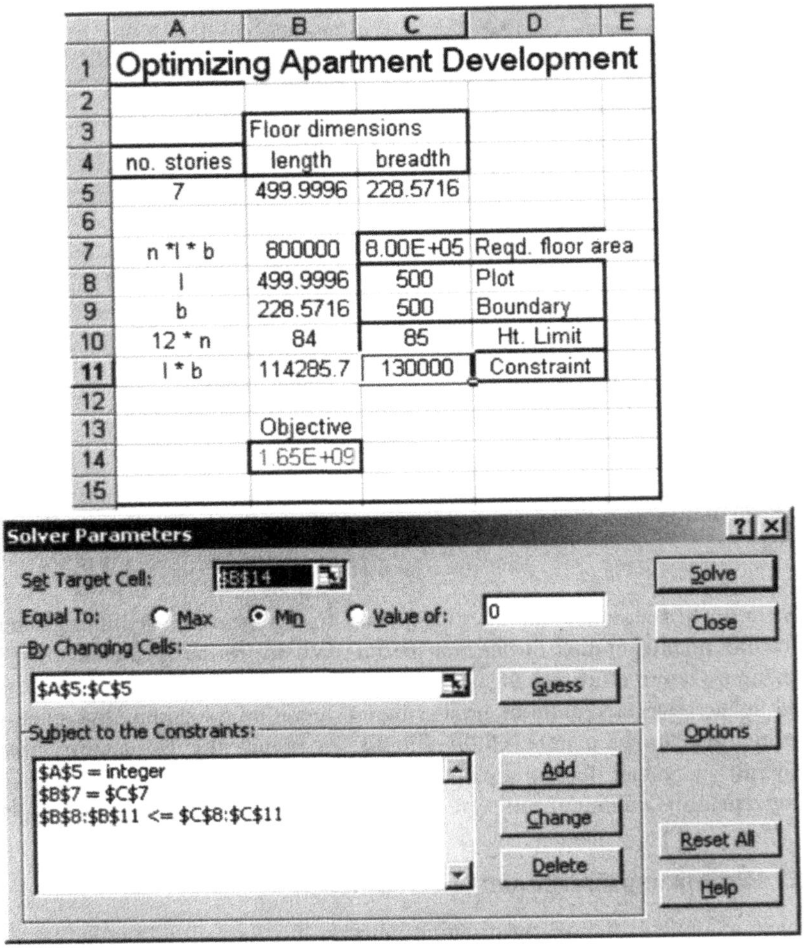

Figure 8.5. Minimum cost design of Apartment.

each floor as shown in cells A5:C5. All choice variables are positive; in particular, the number of stories n is an integer.[1] The total floor area required is equal to nlb and must be equal to 800,000 in an equality constraint. Since the dimension of the plot is fixed, l and b of the apartment are required to be less than the plot dimensions. The height of the building, $12n$, should be less than or equal to 85 feet.

The final constraint is that of the green space requirement. Since green space should at least be 15% of the floor area, we require $lb \leq 500^2 - (800{,}000 \times 0.15)$.

Figure 8.5 shows the Solver sheet with these constraints embedded and the final result requiring an apartment of seven stories.

8.3.3 Power System Dispatch

Purpose. In the control room of a power system, the operators are assisted by several software running concurrently. Each set of software performs a dedicated task. One set, called the Energy Management System (EMS), is dedicated to determine how much energy should be produced from each generator under the command of the operator. The mathematics of this procedure is called economic dispatch in power engineering parlance.

The object of the first set of problems under system dispatch is to expose the reader to a mathematical procedure involving nonlinear programming. Subsequently, problems that address sensitivities in order to focus on LMP and reactive power dispatch are presented.

Problem: Find the optimal output p_i, $i = 1, 5$, to minimize the cost of dispatch from five generators to supply a demand of 9.8 units with the following incremental cost functions $C(p_i)$, $i = 1, 5$, where

$$\begin{aligned} C(p_1) &= 12 + 5p_1 + 1.125p_1^2, \\ C(p_2) &= 5 + 2p_2 + 0.816p_2^2, \\ C(p_3) &= 10 + p_3 + 3.57P_3^2, \\ C(p_4) &= 35, \\ C(p_5) &= 32, \end{aligned} \tag{8.3}$$

subject to

$$0 \leq p_1 \leq 4, \tag{8.4}$$

$$3 \leq p_2 \leq 3.5, \tag{8.5}$$

$$0 \leq p_3 \leq 2.8, \tag{8.6}$$

[1] Integer and mixed integer programming methods have not been discussed in this book. We refer the reader to Hu (1969) and Hillier and Liberman (1974) for details. However, in the Solver routine, we select the corresponding choice variable to be an integer as can be seen in the spreadsheet. In this example, failure to do so will result in a fraction for the number of stories, which can be rounded up or down.

$$0 \leq p_4 \leq 2, \tag{8.7}$$

$$0 \leq p_5 \leq 3. \tag{8.8}$$

8.3.4 Solution Using Kuhn–Tucker Conditions

The condition that the total load be equal to 9.8 is an equality constraint because

$$p_1 + p_2 + p_3 + p_4 + p_5 = 9.8,$$

or

$$\sum_{i=1}^{5} p_i = 9.8. \tag{8.9}$$

Next we alter all the other constraints to equalities as follows.

The first constraint $0 \leq p_1 \leq 4$ can be written as two constraints $-p_1 \leq 0$ and $p_1 \leq 4$, which in turn can be converted to two equalities

$$0 - p_1 + y_1^2 = 0 \tag{8.10}$$

and

$$p_1 - 4 + x_1^2 = 0, \tag{8.11}$$

where y_1 and x_1 are artificially introduced variables.

Converting all inequalities in this manner, one gets

$$p_i^{\min} - p_i + y_i^2 = 0, \qquad i = 1, 5 \tag{8.12}$$

and

$$p_i - p_i^{\max} + x_i^2 = 0, \qquad i = 1, 5, \tag{8.13}$$

where $p_i^{\min}$ and $p_i^{\max}$ are the minimum and the maximum output of the ith generator.

The incremental cost function represents the cost of next unit of energy from that generator at a given output (or loading) level p_i of the ith generator. The incremental cost function is a nonlinear (convex as well) function of p_i. Therefore the cost of energy production from a generator at a loading level p_i is given by $\int_0^{p_i} C(p_i)\,dp_i$. The objective function to be minimized is

$$\pi = \sum_{i-1}^{5} \int_0^{p_i} f(p_i)\,dp_i. \tag{8.14}$$

Performing the integration of incremental cost functions of (8.3), we obtain the objective function as

$$\pi = \left(12p_1 + \frac{5}{2}p_1^2 + \frac{1.125}{3}p_1^3\right) + \left(5p_2 + p_2^2 + \frac{0.816}{3}p_2^3\right) + \left(10p_3 + \frac{1}{2}p_3^2 + \frac{3.57}{3}p_3^3\right) + 35p_4 + 32p_5, \tag{8.15}$$

and the Lagrangian function as

$$L = \left(12p_1 + \frac{5}{2}p_1^2 + \frac{1.125}{3}p_1^3\right) + \left(5p_2 + p_2^2 + \frac{0.816}{3}p_2^3\right) + \left(10p_3 + \frac{1}{2}p_3^2 + \frac{3.57}{3}p_3^3\right) + 35p_4 + 32p_5 + \lambda\left(9.8 - \sum_{i=1}^{5} p_i\right) + \sum_{i=1}^{5} \beta_i^{\min}(p_i^{\min} - p_i + y_i^2) + \sum_{i=1}^{5} \beta_i^{\max}(p_i - p_i^{\max} + x_i^2), \tag{8.16}$$

where λ, $\beta_i^{\max}$ and $\beta_i^{\min}$ are the Lagrange multipliers associated with the constraint of total demand, minimum and maximum output from generators.

The above can be written as

$$L = \left(12p_1 + \frac{5}{2}p_1^2 + \frac{1.125}{3}p_1^3\right) + \left(5p_2 + p_2^2 + \frac{0.816}{3}p_2^3\right) + \left(10p_3 + \frac{1}{2}p_3^2 + \frac{3.57}{3}p_3^3\right) + 35p_4 + 32p_5 + \lambda h_1(\mathbf{z}) + \beta_1^{\min}h_2(\mathbf{z}) + \beta_2^{\min}h_3(\mathbf{z}) + \beta_3^{\min}h_4(\mathbf{z}) + \beta_4^{\min}h_5(\mathbf{z}) + \beta_5^{\min}h_6(\mathbf{x}) + \beta_1^{\max}h_7(\mathbf{z}) + \beta_2^{\max}h_8(\mathbf{z}) + \beta_3^{\max}h_9(\mathbf{z}) + \beta_4^{\max}h_{10}(\mathbf{z}) + \beta_5^{\max}h_{11}(\mathbf{z}), \tag{8.17}$$

where $h_i(\mathbf{z})$ represents the 11 foregoing constraints [equality of (8.9) and 10 constraints resulting from (8.12) and (8.13)], and $\mathbf{z}$ represents the vector of choice variables.

The Kuhn–Tucker necessary conditions requires that the gradient of L be zero. Hence, we want

$$\nabla L = \begin{bmatrix} \partial L/\partial p_1 \\ \partial L/\partial p_2 \\ \partial L/\partial p_3 \\ \partial L/\partial p_4 \\ \partial L/\partial p_5 \\ \partial L/\partial y_1 \\ \partial L/\partial y_2 \\ \partial L/\partial y_3 \\ \partial L/\partial y_4 \\ \partial L/\partial y_5 \\ \partial L/\partial x_1 \\ \partial L/\partial x_2 \\ \partial L/\partial x_3 \\ \partial L/\partial x_4 \\ \partial L/\partial x_5 \\ \partial L/\partial \lambda \\ \partial L/\partial \beta_1^{\min} \\ \partial L/\partial \beta_2^{\min} \\ \partial L/\partial \beta_3^{\min} \\ \partial L/\partial \beta_4^{\min} \\ \partial L/\partial \beta_5^{\min} \\ \partial L/\partial \beta_1^{\max} \\ \partial L/\partial \beta_2^{\max} \\ \partial L/\partial \beta_3^{\max} \\ \partial L/\partial \beta_4^{\max} \\ \partial L/\partial \beta_5^{\max} \end{bmatrix} = \begin{bmatrix} 12 + 5p_1 + 1.125p_1^2 - \lambda - \beta_1^{\min} + \beta_1^{\max} \\ 5 + 2p_2 + 0.816p_2^2 - \lambda - \beta_2^{\min} + \beta_2^{\max} \\ 10 + p_3 + 3.57p_3^2 - \lambda - \beta_3^{\min} + \beta_3^{\max} \\ 35 - \lambda - \beta_4^{\min} + \beta_4^{\max} \\ 32 - \lambda - \beta_5^{\min} + \beta_5^{\max} \\ 2y_1\beta_1^{\min} \\ 2y_2\beta_2^{\min} \\ 2y_3\beta_3^{\min} \\ 2y_4\beta_4^{\min} \\ 2y_5\beta_5^{\min} \\ 2x_1\beta_1^{\max} \\ 2x_2\beta_2^{\max} \\ 2x_3\beta_3^{\max} \\ 2x_4\beta_4^{\max} \\ 2x_5\beta_5^{\max} \\ 9.8 - p_1 - p_2 - p_3 - p_4 - p_5 \\ 0 - p_1 + y_1^2 \\ 3 - p_2 + y_2^2 \\ 0 - p_3 + y_3^2 \\ 0 - p_4 + y_4^2 \\ 0 - p_5 + x_5^2 \\ p_1 - 4 + x_1^2 \\ p_2 - 3.5 + x_2^2 \\ p_3 - 2.8 + x_3^2 \\ p_4 - 2 + x_4^2 \\ p_5 - 3 + x_5^2 \end{bmatrix} = 0. \tag{8.18}$$

In the above equation, note that the last 10 rows are nothing but the constraint equations of (8.12) and (8.13).

In order to satisfy the above equation, the vector to be solved for is

$$\mathbf{z} = [p_1, p_2, p_3, p_4, p_5, y_1, y_2, y_3, y_4, y_5, x_1, x_2, x_3, x_4, x_5,$$
$$\lambda, \beta_1^{\min}, \beta_2^{\min}, \beta_3^{\min}, \beta_4^{\min}, \beta_5^{\min}, \beta_1^{\max}, \beta_2^{\max}, \beta_3^{\max}, \beta_4^{\max}, \beta_5^{\max}]^T. \tag{8.19}$$

Dispatch KT equations .Xls

To conform to equation (7.77), the above can be written as two blocks of vectors as

$$\mathbf{z} = [\mathbf{x}, \mathbf{\Lambda}]^T, \tag{8.20}$$

where **x** represents choice variables and Λ represents the associated Lagrange multipliers $[\lambda, \beta]^T$.

Equations in (8.18) represent a set of nonlinear equations. Its solution can be obtained by using the Jacobian matrix **J** as shown by (7.91).

From (7.94), we obtain **J** as

$$\mathbf{J} = \begin{bmatrix} [\nabla_{\mathbf{x}}^2 L] & [\mathbf{G}] \\ [\mathbf{G}]^T & [\mathbf{0}] \end{bmatrix}. \tag{8.21}$$

Hence from (8.18) we derive the matrix elements of **J** as follows:

$$\nabla_{\mathbf{x}}^2 L = \left[\begin{array}{ccccc|c|c} 5+2.25p_1 & 0 & 0 & 0 & 0 & & \\ 0 & 2+1.632p_2 & 0 & 0 & 0 & & \\ 0 & 0 & 1+7.14p_3 & 0 & 0 & 0 & 0 \\ 0 & 0 & 0 & 0 & 0 & & \\ 0 & 0 & 0 & 0 & 0 & & \\ \hline 0 & \cdots & & & 0 & Diag. & \\ . & . & . & . & . & (2\beta_i^{\min}) & 0 \\ 0 & \cdots & & & 0 & & \\ \hline 0 & \cdots & & & 0 & & Diag. \\ . & & & & . & 0 & (2\beta_i^{\max}) \\ 0 & \cdots & & & 0 & & \end{array}\right], \tag{8.22}$$

where and all the block matrices are of dimension 5×5 as indicated in cells A10:O24 in the attached spreadsheet "Dispatch KT Equations.Xls." **Dispatch KT Equations .Xls**

The matrix **G** is given by

$$\mathbf{G} = \begin{bmatrix} \frac{\partial h_1}{\partial p_1} & \frac{\partial h_2}{\partial p_1} & \cdot\cdot & \frac{\partial h_{11}}{\partial p_1} \\ \vdots & \vdots & \vdots & \vdots \\ \frac{\partial h_1}{\partial p_5} & \frac{\partial h_2}{\partial p_5} & \cdot\cdot & \frac{\partial h_{11}}{\partial p_5} \\ \frac{\partial h_1}{\partial y_1} & \frac{\partial h_2}{\partial y_1} & \cdot\cdot & \frac{\partial h_{11}}{\partial y_1} \\ \vdots & \vdots & \vdots & \vdots \\ \frac{\partial h_1}{\partial y_5} & \frac{\partial h_2}{\partial y_5} & \cdot\cdot & \frac{\partial h_{11}}{\partial y_5} \\ \frac{\partial h_1}{\partial x_1} & \frac{\partial h_2}{\partial x_1} & \cdot\cdot & \frac{\partial h_{11}}{\partial x_1} \\ \vdots & \vdots & \vdots & \vdots \\ \frac{\partial h_1}{\partial x_5} & \frac{\partial h_2}{\partial x_5} & \cdot\cdot & \frac{\partial h_{11}}{\partial x_5} \end{bmatrix}, \tag{8.23}$$

which gives

$$\mathbf{G} = \left[\begin{array}{|c|c|c|} \hline -1 & \text{Diag. }(-1) & \text{Diag. }(1) \\ \hline 0 & \text{Diag. }(2y_i) & 0 \\ \hline 0 & 0 & \text{Diag. }(2x_i) \\ \hline 0 & 0 & 0 \\ \hline \end{array} \right], \tag{8.24}$$

where all the block matrices are of dimension 5×5 except for the blocks (1,1), (2,1), and (3,1) which are of dimension 5×1.

The **G** matrix can be seen in cells P10:Z24 of the spreadsheet and the **J** matrix in full can be seen in cells A10:Z35 of the spreadsheet.

Solution: Examine the spreadsheet in which the Jacobian matrix and its inverse are computed. When matrices are large and sparse as this matrix is, it is efficient to exploit partitioned matrices as shown in Faddeeva (1959).[2]

The right-hand side of ∇L of (8.18) is entered in cells C69:C94 of the spreadsheet under the label ∇L.

Start the iterations by entering an initial guess of unity for all the elements of the **z** vector in cells A4:Z4. Do not enter zeros as that will render the Jacobian matrix singular. The correction to the **z** vector and the **z** vector for the next iteration are indicated in cells E69:E94 and N69:N94, respectively. Copy and paste the values of cells N69:N94 onto cells A4:Z4 to obtain updated values for **z** in N64:N94. Continue this process about four times to observe that ∇L goes to zero (first-order condition is satisfied) and there is no substantial change in the vector **z**. The final result for a total demand of 9.8 is shown in the spreadsheet. The cost of dispatch computed in cell U6 according to (8.15) is 180.99.

Discussion of Results. Figure 8.6 shows selected cells of the spreadsheet to portray the final results for **z**. Generator #4 has zero output because it is the most expensive machine.

The first Lagrange multiplier λ associated with the equality constraint is in cell P4. We note that generator #5 is the marginal generator and the output from #4 is zero. Therefore, the cost of supplying the next increment of energy is $32; this is called "system lambda." This means that all generators will be operating such that their incremental costs are the same with a value equal to system lambda.[3] Figure 8.7 also shows the incremental cost curves and the output ranges of the five generators. The outputs from the generators for a system lambda of $32 corresponding to cells A4:E4 are also indicated therein. Generators that have an incremental cost

[2]If a matrix is partitioned as $\mathbf{J} = \left[\begin{array}{c|c} \mathbf{A} & \mathbf{B} \\ \hline \mathbf{C} & \mathbf{D} \end{array}\right]$, its inverse is given by $\mathbf{J}^{-1} = \left[\begin{array}{c|c} \mathbf{K} & \mathbf{L} \\ \hline \mathbf{M} & \mathbf{N} \end{array}\right]$, where $\mathbf{K} = (\mathbf{A} - \mathbf{B}\mathbf{D}^{-1}\mathbf{C})^{-1}$, $\mathbf{M} = -\mathbf{D}^{-1}\mathbf{C}\mathbf{K}$, $\mathbf{N} = (\mathbf{D} - \mathbf{C}\mathbf{A}^{-1}\mathbf{B})^{-1}$, and $\mathbf{L} = -\mathbf{A}^{-1}\mathbf{B}\mathbf{N}$.

[3]This is true if there are no other constraints such as transmission limitations and minimum output requirements. These will be discussed later.

Dispatch of Five Generators Using KT Conditions

P_1	P_2	P_3	P_4	P_5	y_1	y_2	y_3	y_4	y_5	x_1	x_2	x_3	x_4	x_5
2.544	3.5	2.3463	0	1.4098	1.595	0.7071	1.5318	0	1.1873	1.2067	0	0.6736	1.4142	1.261

0	3	0	0	0	Min Output
4	3.5	2.8	2	3	Max. Output
32	21.996	32	35	32	Incremental cost at indicated output level

Demand 9.8 Actual dispatch 9.8

λ	β_1^{min}	β_2^{min}	β_3^{min}	β_4^{min}	β_5^{min}	β_1^{max}	β_2^{max}	β_3^{max}	β_4^{max}	β_5^{max}
32	0	0	0	3	0	0	10.004	0	0	0

Cost of Dispatch 181

Figure 8.6. Dispatch solution from Kuhn–Tucker conditions.

below system lambda will be at their full outputs. Therefore, from Figure 8.7 we see that the outputs of generators 1, 3, and 5 are such that their incremental costs are equal.

It is interesting to examine the Lagrange multipliers and their relation to the slack variables and generator output levels. Recall from (7.41), (7.42), and (7.43) that either (a) the constraint is fully satisfied ($g = 0$) when the associated Lagrange multiplier is nonzero or (b) if $g < 0$, the associated Lagrange multiplier is zero. Examine first the values of β_i^{min} in relation to the associated values of y_i. All generator outputs in cells A4:E4, except that of generator #4, are above their lower limits indicated in cells A6:E6. Consequently, y_1, y_2, y_3, and y_5 are nonzero and the square of their values is equal to the generator output in order to satisfy (8.10). However, since generator #4 is held at its lower limit of zero, $y_4 = 0$. The values of β_i^{min} associated with values of y_i portray the effect of relaxing (increasing the value) the lower limit associated with these constraints. Since generators 1, 2, 3, and 5, are above their lower limits, increasing the lower limit from zero to, say, 0.1 will not make any difference to the value of π. Hence the value of β^{min} associated with them is zero.

On the other hand, generator #4 is held at its lower limit. If the lower limit is increased from zero to a higher value, π will increase since this generator is more expensive than system lambda. The cost of energy from this generator is \$35; and with the system lambda being \$32, the increase in π would be the difference between these two, \$3.00. Hence, β_4^{min} has a value of 3.

Except for generator #2, which is at its upper limit, all generators are operating below their upper limit. Therefore, $x_2 = 0$ in order to satisfy (8.13). The value of β_2^{max} for this generator indicates the sensitivity of π to the relaxation of p_2^{max}. Relaxation here implies a reduction from the upper limit of 3.5 because of the negative sign associated with p_i^{max} in (8.13). This value is equal to \$10.00 being the difference between the system lambda and the incremental cost of generation at full output from generator #2.

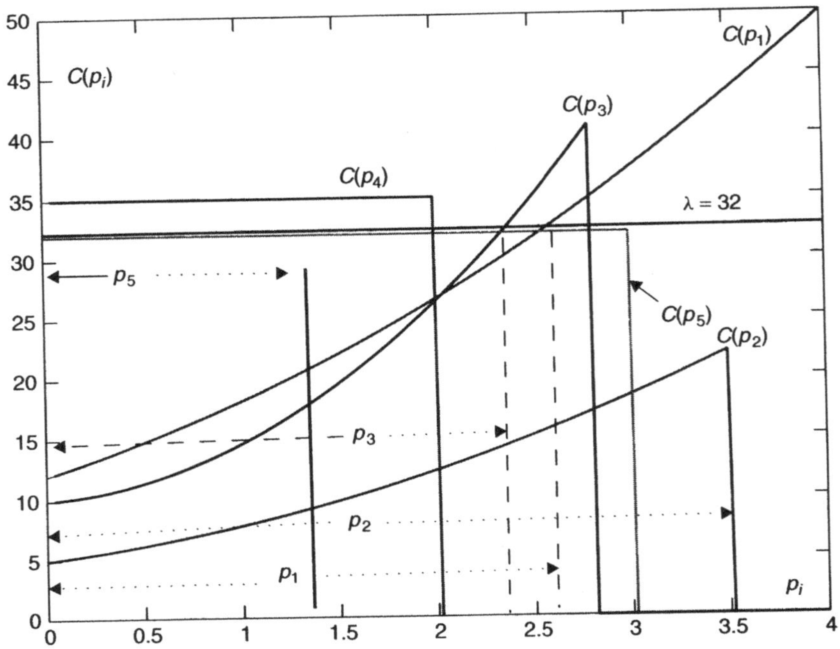

Figure 8.7. Portrayal of generator incremental costs and system lambda.

For generators #1, #2, #3, and #5, which are operating below their upper limits, the values of x are equal to the square root of the difference between the upper limit and the generator output. Correspondingly we note that the values of $\beta^{\max}$ for these generators are zero.

Generator #4 has zero output. Therefore, in order to satisfy (8.12), the associated y has a value of zero. Also, $\beta_3^{\min}$ has a value of three implying that if the lower limit of zero is relaxed (increased), the increase in cost will be equal to 3, which is the difference between the cost of generation from this generator and the marginal generator.

8.3.5 Solution Using Excel Solver

The solution to the above dispatch problem can be obtained easily by using Excel Solver. We first show a solution using a piecewise approximation for the incremental costs of generators followed by the solution using the nonlinear incremental cost characteristics in its entirety.

Piece-wise App-.Xls

Piecewise Approximation for Generation Cost. The attached spreadsheet "Piecewise App.Xls" shown in Figure 8.8 indicates the procedure. The incremental costs of generators are broken into segments of generator capacities in column D with an associated price obtained from incremental cost equations (8.3). For

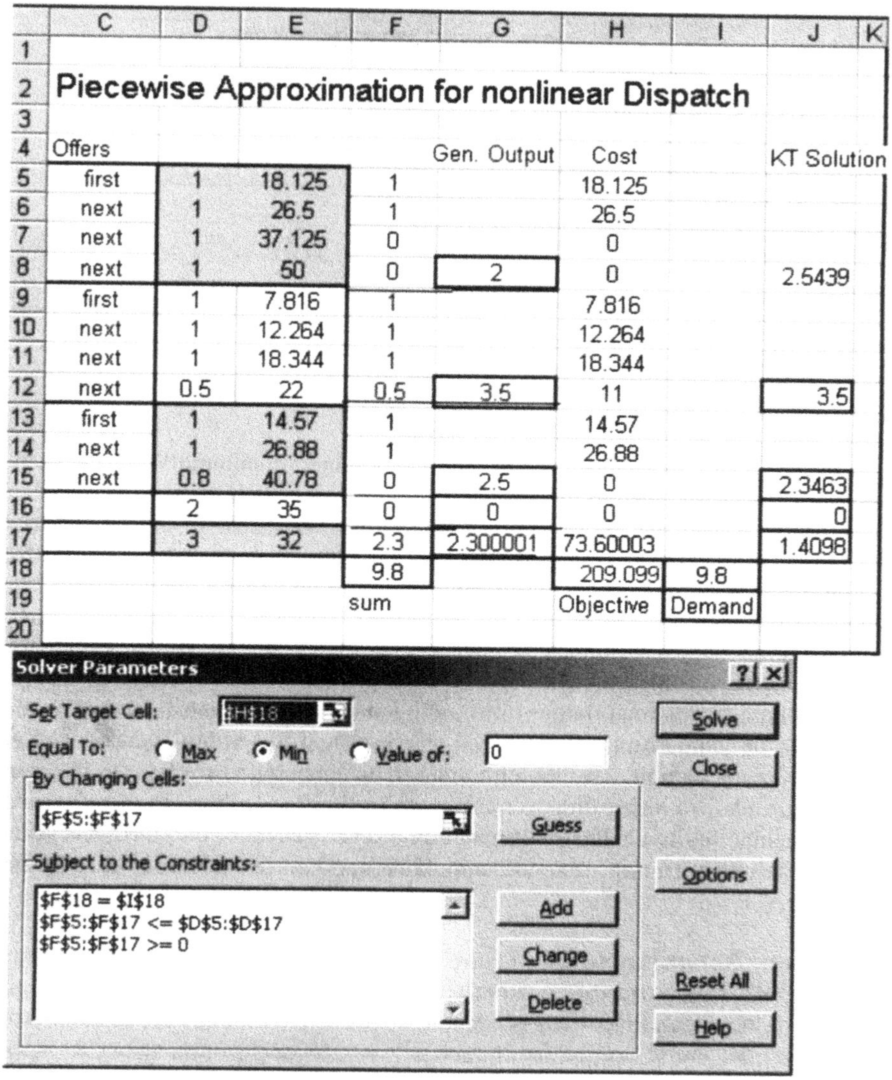

Figure 8.8. Dispatch solution using piecewise approximation to generation cost.

example, the offer to supply energy from generator #1 is as follows: the first block of 1 MW at \$18.125, the second block of 1 MW at \$26.5, the third at \$37.125, and the last block of 1 MW at \$50.00. These offer prices are derived from the cost characteristic $C(p_1)$ of (8.3) at the cumulative capacity offered.

The formulation of the problem is then simple. Choice variables in column F indicate the amount of offer accepted by the dispatcher. The cost of dispatch is the

	A	B	C	D	E	F	G	H	I	J	K	L	M
1					Dispatch without Transmission Model								
2	nodes	1	2	3	4	5	6	7	8				
3			Generators' Cost of offers - $/MWh										
4	$	32	0	21.996	0	32	35	0	32				
5	P	4	0	3.5	0	2.8	2	0	3	MW Capability			
6		52.8793	0	41.412	0	41.5871	0	0	45.1124	180.99087			
7				Demand at nodes									
8	P	1	1.5	1.7	0.5	0	3	2.1	0	Specified Load, P			
9	Q	-0.2	0.1	0.25	-0.1	0	0.1	0.22	0	Specified load, Q			
10													
11	Generation at nodes			Choice Variables									
12		P1	P2	P3	P4	P5	P6	P7	P8				
13		2.54391	0	3.5	0	2.34632	0	0	1.40976				
14													
15													
16	Result Of Dispatch Solution				Σ Generation			Σ Demand		Objective Function		180.9909	
17					9.8			9.8					
18													

Figure 8.9. Power dispatch considering nonlinear incremental cost of generation.

product of quantity accepted and its cost as computed in column H. The objective is to minimize the total cost of dispatch computed in cell H18.

Constraints are that the quantity accepted must be equal to the total demand in cell I18, the quantity accepted from each offer must be less than or equal to the quantity offered, and the quantity accepted has to be nonnegative. These formulations can be seen in the Solver sheet of Figure 8.8. Furthermore, observe that as in the case of examples of chapter 4, we have selected the option of using a linear model for this problem.

In column G, the total output from each generator is computed. Also shown in column J for comparison is the solution obtained by using Kuhn–Tucker equations in the previous section considering nonlinear incremental cost relations. Compare the two results. An interesting exercise to the reader is to reduce the block size for each machine (to divide the total capability of each generator into finer pieces) and observe that the result of piecewise approximation becomes closer to the nonlinear solution.

NLdis-patch.Xls

Nonlinear Generation Cost. Figure 8.9 shows sheet 1 of the spreadsheet "NLdispatch.Xls" which uses the nonlinear cost relationship of (8.3). There are some details on the spreadsheet that we ignore at present. We note that the generators #1, #2, #3, #4, and #5 are located in nodes 1, 3, 5, 6, and 8 of the network, respectively. For now, we ignore the network and the locations of demand in the network. However, we consider that the total demand is equal to 9.8 as in cell H17.

In row 4, the cost relations of generators according to (8.3) are embedded. Hence, the cost of dispatch per (8.15) is entered in cell L16. This forms the objective function. Choice variables are the generator outputs in row 13 with the constraint that they must be nonnegative and less than the capability entered in row 5. The equality constraint is that the total of selected outputs must be equal to the demand in cell H17.

The Solver solution using nonlinear option is identical to that obtained from Kuhn–Tucker conditions in Figure 8.6. The incremental cost of each generator is shown in row 4 and the cost of production from each generator is shown in row 6.

As an exercise, we suggest that the reader examine the sensitivity sheet associated with this spreadsheet. In particular, it is instructive to compare the Lagrange multipliers from this spreadsheet with those obtained in Figure 8.6.

Introduction of Transmission Network. In sheet 2 of the spreadsheet (shown in Figure 8.10), the transmission network of Figure 4.9 was used. The network is also inserted into the spreadsheet for ready reference. The spreadsheet contains all the network parameters such as the real and imaginary parts of the admittance matrix in the right-hand end.

Because of the network, it is now necessary that the dispatch of generation does not violate conditions such as voltage levels, or flow limits in branches or interfaces. Therefore, it is necessary to compute network conditions and flows.

First we address only the voltage limits. It is assumed that all the branches have ample capacity to permit power flow. Finite flow capabilities that enforce binding flow limits are included in a later example.

Toward that end, consider rows 25 to 50, columns A to H. In these cells we compute branch flows for given voltage and angle at nodes per (B.4). The voltage and angles at all nodes, cells B12:I12 and B14:I14, now become choice variables. Branch real power flows are computed in cells A25:A50; reactive power flows are in cells F25:F50. The total injection of real power at each node into the network given by (B.7) is computed in cells D27, D31, D35, D38, D41, D44, D47, and D50. The injection of reactive power is in corresponding cells in column H. Note that the injection into the network can be positive or negative. The latter indicates withdrawal from the network to serve demand.

As in the previous examples, the cost of generation and the generator capabilities are in rows 4 and 5. In addition, the modal location of demands is indicated in rows 8 and 9.

We now discuss the restrictions that the network imposes on the problem. As before, we desire the machine outputs to be nonzero and less than or equal to their capability. Additionally we require that the voltages at nodes are within some upper and lower specified bounds of cells K20 and L20. Cells B19:I19 show the external real power injection or withdrawal (load demand) at nodes. This is given by the sum of flows in all branches emanating from a node and the local demand at the node given in rows 8 and 9. Similarly, cells B21:I21 indicate the reactive power component.

Since there are no generators located at nodes 2, 4, and 7, we require the external injection of active and reactive power at those nodes to be equal to zero. The injection at other nodes has to be less than the machine capabilities.[4] These constraints can be seen in the Solver screen shown in Figure 8.10.

In the previous case without transmission, we had stipulated that the sum of power dispatched should be equal to the total demand. Now with intervening transmission, we cannot use such a constraint because of losses. The loss in the network is not known prior to obtaining the solution to all branch flows.

[4]The machine capability, in reality, is the volt–ampere output and not just the power output. The volt–ampere output is a Pythagorean addition of real and reactive power output. The constraint involving the volt–ampere output is introduced in a latter example.

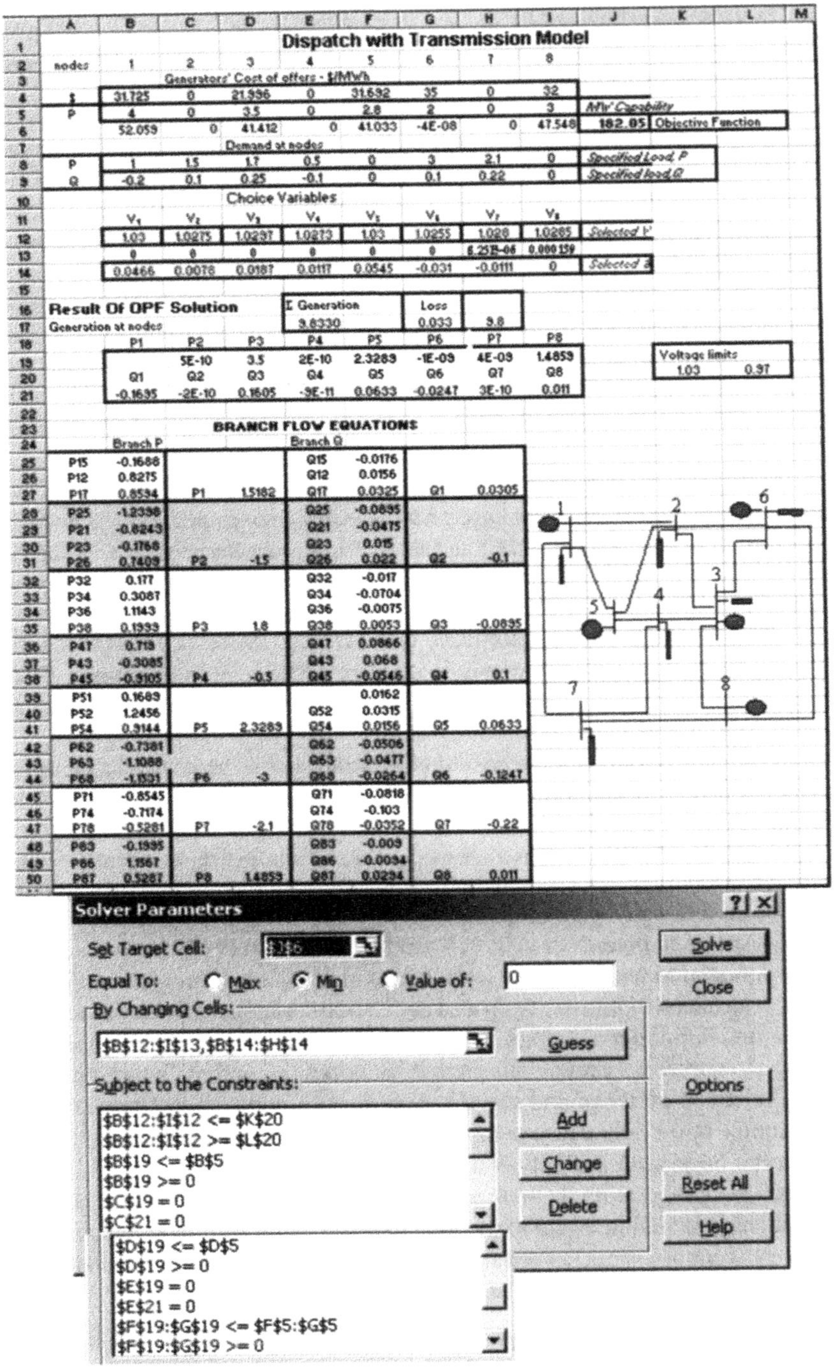

Figure 8.10. Optimal power dispatch considering transmission network.

From the branch flow solution, the total network loss is computed in cell G17 as the difference between the total power dispatched and the total demand. Figure 8.10 shows the results of the spreadsheet. The reader is encouraged to study the sensitivity sheet associated with this spreadsheet. A discussion of sensitivity with a view to compute locational marginal prices is deferred to a later example.

8.4 OPTIMAL POWER FLOW—FORMULATION AND INTERPRETATION

The optimal power flow formulation can take many forms, from the simplest to the most complex. Many professionals have spent a lifetime researching and developing better methods to solve such problems. Obviously, therefore, it is inconceivable to make one conversant in all issues of optimal power flow through a mere selection of representative examples. However, our purpose is to introduce the reader to simpler forms of such power flow solutions and develop the concept of locational prices in a network. Pricing energy on the basis of the location of its withdrawal or injection in a network [originally proposed in Caramanis et al. (1982)] appears to have a universal following in the evolving deregulated electric power industry.

AC OPF-1.Xls

Purpose. In the previous section, we have discussed economic dispatch. We have already considered the effect of the transmission network. The intent now is to extend the consideration of transmission, particularly in regard to sensitivities of the results. Aided by the sensitivity sheets, one can determine the cost of supplying the next unit of energy at each node. This is called the locational marginal price (LMP). The following problems are a study of effects of transmission restrictions on LMP. Later problems address the issue of reactive power production and consumption on LMP.

Problem: The spreadsheet "AC OPF-1.XLS" portrays an elementary form of OPF problem. We consider a dispatch situation where, as before, the cost of active power dispatch is minimized; that is, we assume the cost of reactive power supply for dispatch purposes (not in market settlement) is zero. Furthermore, to start with, we do not assume any line flow constraints.

The network inserted in Figure 8.11 is identical to that of Figure 4.9. The network parameters are shown in the extreme right-hand side of the spreadsheet in columns P:R. The admittance parameters are the same as those used for the ac load flow in Section B.3.2 with an X to R ratio of 10:1.

The ac power flow solution of in B.3.2 addressed the solution of a set of nonlinear flow equations to obtain specified nodal power injections. The injections at all nodes except the slack node were specified; the solution obtained voltages and angles at all nodes. The resulting injection at the slack bus was equal to total demand minus the sum of generation at other nodes, plus losses. In the OPF calculation, for

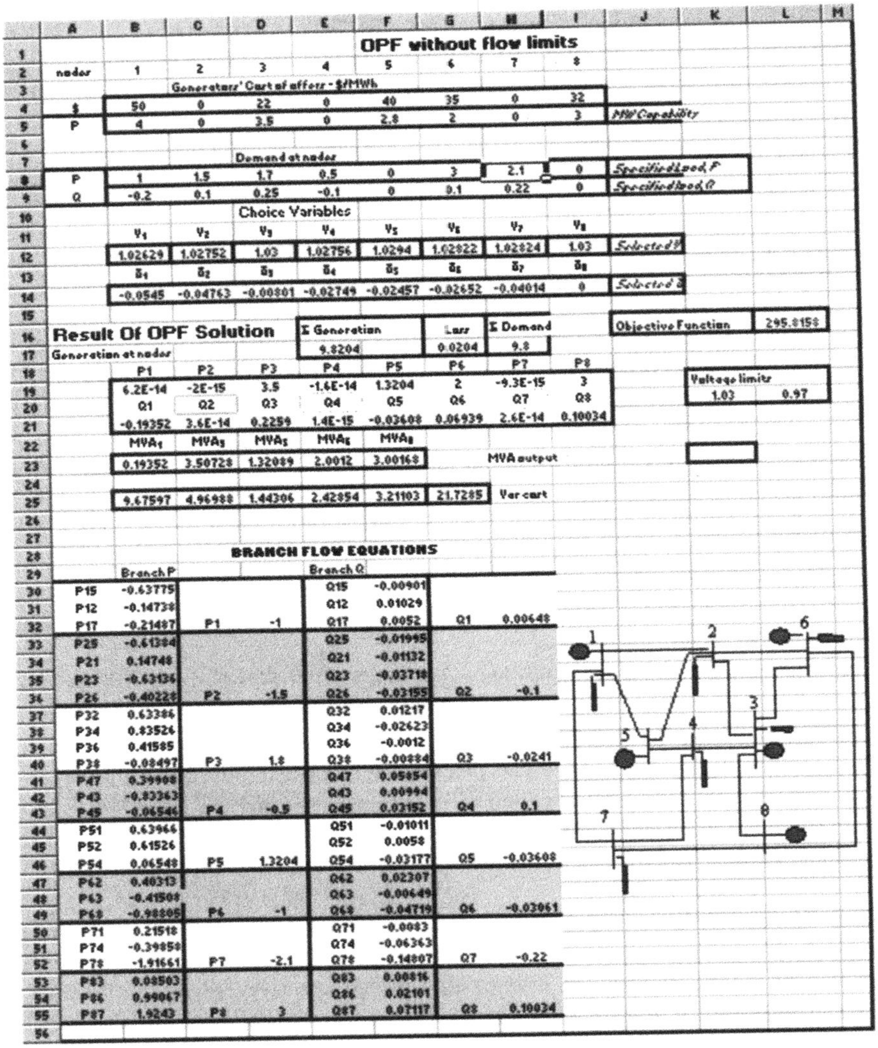

	A	B	C	D	E	F	G	H	I	J	K	L
1						OPF without flow limits						
2	nodes	1	2	3	4	5	6	7	8			
3			Generators' Cost of offers - $/MWh									
4	$	50	0	22	0	40	35	0	32			
5	P	4	0	3.5	0	2.8	2	0	3	MW Capability		
6												
7				Demand at nodes								
8	P	1	1.5	1.7	0.5	0	3	2.1	0	Specified Load P		
9	Q	-0.2	0.1	0.25	-0.1	0	0.1	0.22	0	Specified load Q		
10				Choice Variables								
11		V_1	V_2	V_3	V_4	V_5	V_6	V_7	V_8			
12		1.02629	1.02752	1.03	1.02756	1.0294	1.02822	1.02824	1.03	Selected V		
13		δ_1	δ_2	δ_3	δ_4	δ_5	δ_6	δ_7	δ_8			
14		-0.0545	-0.04763	-0.00801	-0.02749	-0.02457	-0.02652	-0.04014	0	Selected δ		
15												
16	Result Of OPF Solution				Σ Generation		Loss	Σ Demand		Objective Function		295.8158
17	Generation at nodes				9.8204		0.0204	9.8				
18		P1	P2	P3	P4	P5	P6	P7	P8		Voltage limits	
19		6.2E-14	-2E-15	3.5	-1.6E-14	1.3204	2	-9.3E-15	3		1.03	0.97
20		Q1	Q2	Q3	Q4	Q5	Q6	Q7	Q8			
21		-0.19352	3.6E-14	0.2259	1.4E-15	-0.03608	0.06939	2.6E-14	0.10034			
22		MVA_1	MVA_3	MVA_5	MVA_6	MVA_8						
23		0.19352	3.50728	1.32089	2.0012	3.00168		MVA output				
24												
25		9.67597	4.96988	1.44306	2.42854	3.21103	21.7285	Var cost				
26												
27												
28				BRANCH FLOW EQUATIONS								
29		Branch P			Branch Q							
30	P15	-0.63775			Q15	-0.00901						
31	P12	-0.14738			Q12	0.01029						
32	P17	-0.21487	P1	-1	Q17	0.0052	Q1	0.00648				
33	P25	-0.61384			Q25	-0.01995						
34	P21	0.14748			Q21	-0.01132						
35	P23	-0.63136			Q23	-0.03718						
36	P26	-0.40228	P2	-1.5	Q26	-0.03155	Q2	-0.1				
37	P32	0.63386			Q32	0.01217						
38	P34	0.83526			Q34	-0.02623						
39	P36	0.41585			Q36	-0.0012						
40	P38	-0.08497	P3	1.8	Q38	-0.00884	Q3	-0.0241				
41	P47	0.39908			Q47	0.05854						
42	P43	-0.83363			Q43	0.00994						
43	P45	-0.06546	P4	-0.5	Q45	0.03152	Q4	0.1				
44	P51	0.63966			Q51	-0.01011						
45	P52	0.61526			Q52	0.0058						
46	P54	0.06548	P5	1.3204	Q54	-0.03177	Q5	-0.03608				
47	P62	0.40313			Q62	0.02307						
48	P63	-0.41508			Q63	-0.00649						
49	P68	-0.98805	P6	-1	Q68	-0.04719	Q6	-0.03061				
50	P71	0.21518			Q71	-0.0083						
51	P74	-0.39858			Q74	-0.06363						
52	P78	-1.91661	P7	-2.1	Q78	-0.14807	Q7	-0.22				
53	P83	0.08503			Q83	0.00816						
54	P86	0.99067			Q86	0.02101						
55	P87	1.9243	P8	3	Q87	0.07117	Q8	0.10034				
56												

Figure 8.11. Example of AC OPF and its solution.

defined demand at nodes, our intent is to find injections of generation at all nodes in order to minimize total cost.[5] There is no such thing as a slack bus except for a reference bus to measure angles.

[5]Although the minimization of dispatch cost is the most common objective, OPF solutions may address several other concerns such as minimization of losses, minimization of reactive power requirements, and a host of other objectives.

8.4.1 Problem Formulation for Solver

The purpose of this example is to study the sensitivity to determine locational marginal prices and other characteristics. In the example of the previous section, we had assumed a nonlinear convex relationship between cost of energy and machine output. In this example, in order to make the ensuing analysis clearer, we have assumed that the cost of dispatching energy from the generators is constant at indicated values in row 4 at all loading levels of the generators. The objective to be minimized is the total cost of power dispatch, which is computed in cell L16 as a product of cost and the output of the generator. Choice variables are voltages and angles, with δ_8 arbitrarily chosen as the reference (zero).

As in the previous example, the constraints are as follows:

- The machine power output should be less than or equal to its capability.
- The injections from generators should be greater than or equal to zero, or the minimum generation level specified.
- The voltages at nodes should be within the upper and lower limits entered in cells K20, L20,
- The angles should be within the first or the second quadrant ($\pi/2$ to $-\pi/2$) to avoid large values for them.
- Net injection into the nodes 2, 4, 7, since there are no generators there, should be zero (cells C19:20, E19:20, and H19:20).

8.4.2 Results and Sensitivity Sheets

The results obtained from the Solver are shown in Figure 8.11. The cost of dispatch in cell L16 is \$295.8048. From these results, the system losses are computed in cell G17. The MVA outputs of the generators are computed in cells B23:F23. In several deregulated markets for electricity, suppliers of reactive power are given a proxy payment based on "lost opportunity cost." This is tantamount to a payment at the same or some fraction of the real power rate. Cell G25 shows the cost of supplying reactive power if its suppliers were compensated at the same rate as real power. It is assumed that both leading and lagging reactive power is compensated at the same rate. We will revisit this discussion later.

Refer to the sensitivity sheets in the spreadsheet AC OPF-1.Xls. The limits sheet indicates limits that were enforced. The answer sheet shows all constraints of the problem and which of them were binding. Associated with each constraint, the Lagrange multiplier is shown. Recall that the Lagrange multiplier is zero unless the associated constraint is binding.

In the sensitivity sheet, reduced gradients associated with variables are shown. These correspond to the reduced costs in linear programs. The reduced gradient indicates the rate of change of the objective function for a unit increase in the bounds (upper or lower) of the variable.

Figure 8.4.2 shows the answer report. The first table shows the final values of choice variables and the assumed values for them at the start of the iterations.

Target Cell (Min)

Cell	Name	Original Value	Final Value
$K$16	Objective Function	192.400000	295.815844

Adjustable Cells

Cell	Name	Original Value	Final Value
$B$15	d1	0	-0.05450113
$C$15	d2	0	-0.04763071
$D$15	d3	0	-0.00801223
$E$15	d4	0	-0.02748628
$F$15	d5	0	-0.02457022
$G$15	d6	0	-0.02652081
$H$15	d7	0	-0.04013536
$B$13	V1	1	1.026291199
$C$13	V2	1	1.027523747
$D$13	V3	1	1.03
$E$13	V4	1	1.027556884
$F$13	V5	1	1.029394697
$G$13	V6	1	1.028217444
$H$13	V7	1	1.028237498
$I$13	V8	1	1.03

Constraints

Cell	Name	Cell Value	Formula	Status	Slack
$B$19	P1	1.19238E-13	$B$19>=0	Binding	0
$F$19	P5	1.320396093	$F$19>=0	Not Binding	1.3204
$G$19	P6	2	$G$19>=0	Not Binding	2
$I$19	P8	3	$I$19>=0	Not Binding	3
$D$19	P3	3.5	$D$19>=0	Not Binding	3.5
$B$19	P1	1.19238E-13	$B$19<=$B$6	Not Binding	4
$C$19	P2	9.46576E-13	$C$19<=$C$6	Binding	0
$D$19	P3	3.5	$D$19<=$D$6	Binding	0
$E$19	P4	-8.57092E-14	$E$19<=$E$6	Binding	0
$F$19	P5	1.320396093	$F$19<=$F$6	Not Binding	1.4796
$G$19	P6	2	$G$19<=$G$6	Binding	0
$H$19	P7	5.6799E-13	$H$19<=$H$6	Binding	0
$I$19	P8	3	$I$19<=$I$6	Binding	0
$C$19	P2	9.46576E-13	$C$19=0	Not Binding	0
$C$21	Q2	2.13718E-15	$C$21=0	Not Binding	0
$E$19	P4	-8.57092E-14	$E$19=0	Not Binding	0
$E$21	Q4	8.52096E-15	$E$21=0	Not Binding	0
$H$19	P7	5.6799E-13	$H$19=0	Not Binding	0
$H$21	Q7	-2.41474E-15	$H$21=0	Not Binding	0
$B$15	d1	-0.054501131	$B$15<=0.6	Not Binding	0.6545
$C$15	d2	-0.047630707	$C$15<=0.6	Not Binding	0.64763
$D$15	d3	-0.008012233	$D$15<=0.6	Not Binding	0.60801
$E$15	d4	-0.027486278	$E$15<=0.6	Not Binding	0.62749

$F$15	d5	-0.024570218	$F$15<=0.6	Not Binding	0.62457
$G$15	d6	-0.026520814	$G$15<=0.6	Not Binding	0.62652
$H$15	d7	-0.040135364	$H$15<=0.6	Not Binding	0.64014
$B$13	V1	1.026291199	$B$13>=$F$17	Not Binding	0.05629
$C$13	V2	1.027523747	$C$13>=$F$17	Not Binding	0.05752
$D$13	V3	1.03	$D$13>=$F$17	Not Binding	0.06
$E$13	V4	1.027556884	$E$13>=$F$17	Not Binding	0.05756
$F$13	V5	1.029394697	$F$13>=$F$17	Not Binding	0.05939
$G$13	V6	1.028217444	$G$13>=$F$17	Not Binding	0.05822
$H$13	V7	1.028237498	$H$13>=$F$17	Not Binding	0.05824
$I$13	V8	1.03	$I$13>=$F$17	Not Binding	0.06
$B$15	d1	-0.054501131	$B$15>=-0.6	Not Binding	0.5455
$C$15	d2	-0.047630707	$C$15>=-0.6	Not Binding	0.55237
$D$15	d3	-0.008012233	$D$15>=-0.6	Not Binding	0.59199
$E$15	d4	-0.027486278	$E$15>=-0.6	Not Binding	0.57251
$F$15	d5	-0.024570218	$F$15>=-0.6	Not Binding	0.57543
$G$15	d6	-0.026520814	$G$15>=-0.6	Not Binding	0.57348
$H$15	d7	-0.040135364	$H$15>=-0.6	Not Binding	0.55986
$B$13	V1	1.026291199	$B$13<=$E$17	Not Binding	0.00371
$C$13	V2	1.027523747	$C$13<=$E$17	Not Binding	0.00248
$D$13	V3	1.03	$D$13<=$E$17	Binding	0
$E$13	V4	1.027556884	$E$13<=$E$17	Not Binding	0.00244
$F$13	V5	1.029394697	$F$13<=$E$17	Not Binding	0.00061
$G$13	V6	1.028217444	$G$13<=$E$17	Not Binding	0.00178
$H$13	V7	1.028237498	$H$13<=$E$17	Not Binding	0.00176
$I$13	V8	1.03	$I$13<=$E$17	Binding	0

Figure 8.12. Answer sheet from Excel Solver for the example problem.

Adjustable Cells

Cell	Name	Final Value	Reduced Gradient
$B$15	d1	-0.054501131	0
$C$15	d2	-0.047630707	0
$D$15	d3	-0.008012233	0
$E$15	d4	-0.027486278	0
$F$15	d5	-0.024570218	0
$G$15	d6	-0.026520814	0
$H$15	d7	-0.040135364	0
$B$13	V1	1.026291199	0
$C$13	V2	1.027523747	0
$D$13	V3	1.03	-0.420016807
$E$13	V4	1.027556884	0
$F$13	V5	1.029394697	0
$G$13	V6	1.028217444	0
$H$13	V7	1.028237498	0
$I$13	V8	1.03	-1.166871657

Constraints

Cell	Name	Final Value	Lagrange Multiplier
$B$19	P1	1.19E-13	9.759868
$F$19	P5	1.320396	0
$G$19	P6	2	0
$I$19	P8	3	0
$D$19	P3	3.5	0
$B$19	P1	1.19E-13	0
$C$19	P2	9.47E-13	-40.1844
$D$19	P3	3.5	-17.8677
$E$19	P4	-8.6E-14	-40.0241
$F$19	P5	1.320396	0
$G$19	P6	2	-5.01569
$H$19	P7	5.68E-13	-40.1238
$I$19	P8	3	-7.80403
$C$19	P2	9.47E-13	0
$C$21	Q2	2.14E-15	-0.00495
$E$19	P4	-8.6E-14	0
$E$21	Q4	8.52E-15	0.007149
$H$19	P7	5.68E-13	0
$H$21	Q7	-2.4E-15	-0.00895

Figure 8.13. Sensitivity sheet for the example problem.

The columns of the second table of the answer sheet under the heading "Formula" show all the constraint equations that were specified for the problem. Additionally, which constraints were nonbinding and which were binding is also indicated. The sensitivity report 1 shown in Figure 8.13 is pasted onto sheet 1 for ready reference. Compare the answer sheet (Figure 8.4.2) and the sensitivity sheet (Figure 8.13) to confirm that the nonbinding constraints have a zero Lagrange multiplier while the binding constraints have a nonzero multiplier.

There is one matter about equality constraints that might cause some confusion. The problem required that the net injection be zero at nodes that do not have a generator. For instance, power and var injection in node 7 represented by cells H19 and H21 $= 0$. The answer sheet in Figure 8.4.2 shows this to be a nonbinding constraint while in fact this is a binding requirement.

The explanation for this apparent anomaly is that the equality constraint is in fact two constraints; $P_7 = 0$ can be written as $P_7 \leq 0$ and $P_7 \geq 0$. In the solution, only one of these two will be binding. In order to understand this, re-run the spreadsheet by modifying the equality constraints for H19 and H21 by incorporating them as two inequalities. Although the same final result for the objective will be obtained, the answer sheet and the sensitivity sheet will be somewhat different. These can be seen as answer report 2 and sensitivity report 2. They are shown in Figures 8.14 and 8.15. Figure 8.14 shows only the portion of Answer sheet 2 that is relevant to our discussion.

Note in the answer sheet of Figure 8.14 that both $H19 < 0$ and $H19 > 0$ are shown to be binding indicating that equality was enforced. In the associated sensitivity sheet of Figure 8.15, a Lagrange multiplier of -40.1238 is shown for

Target Cell (Min)

Cell	Name	Original Value	Final Value
$L$16	Objective Function	295.8158	295.8158

Adjustable Cells

Cell	Name	Original Value	Final Value
$B$14	d1	-0.054501131	-0.054501131
$C$14	d2	-0.047630707	-0.047630707
$D$14	d3	-0.008012233	-0.008012233
$E$14	d4	-0.027486278	-0.027486278
$F$14	d5	-0.024570218	-0.024570218
$G$14	d6	-0.026520814	-0.026520814
$H$14	d7	-0.040135364	-0.040135364
$B$12	V1	1.026291199	1.026291199
$C$12	V2	1.027523747	1.027523747
$D$12	V3	1.03	1.03
$E$12	V4	1.027556884	1.027556884
$F$12	V5	1.029394697	1.029394697
$G$12	V6	1.028217444	1.028217444
$H$12	V7	1.028237498	1.028237498
$I$12	V8	1.03	1.03

Constraints

Cell	Name	Cell Value	Formula	Status	Slack
$C$21	Q2	2.14E-15	$C$21=0	Not Bindin	0
$E$19	P4	-8.6E-14	$E$19=0	Not Bindin	0
$E$21	Q4	8.52E-15	$E$21=0	Not Bindin	0
$H$19	P7	5.68E-13	$H$19<=0	Binding	0
$H$19	P7	5.68E-13	$H$19>=0	Binding	0
$H$21	Q7	-2.4E-15	$H$21>=0	Binding	0
$H$21	Q7	-2.4E-15	$H$21<=0	Binding	0
$B$14	d1	-0.0545	$B$14<=0	Not Bindin	0.654501
$C$14	d2	-0.04763	$C$14<=0	Not Bindin	0.647631
$D$14	d3	-0.00801	$D$14<=0	Not Bindin	0.608012
$E$14	d4	-0.02749	$E$14<=0	Not Bindin	0.627486
$F$14	d5	-0.02457	$F$14<=0	Not Bindin	0.62457
$G$14	d6	-0.02652	$G$14<=0	Not Bindin	0.626521
$H$14	d7	-0.04014	$H$14<=0	Not Bindin	0.640135
$B$12	V1	1.026291	$B$12>=$	Not Bindin	0.056291
$C$12	V2	1.027524	$C$12>=$	Not Bindin	0.057524
$D$12	V3	1.03	$D$12>=$	Not Bindin	0.06

Figure 8.14. Answer sheet with expanded equality constraint.

Adjustable Cells

Cell	Name	Final Value	Reduced Gradient
$B$14	d1	-0.054501131	0
$C$14	d2	-0.047630707	0
$D$14	d3	-0.008012233	0
$E$14	d4	-0.027486278	0
$F$14	d5	-0.024570218	0
$G$14	d6	-0.026520814	0
$H$14	d7	-0.040135364	0
$B$12	V1	1.026291199	0
$C$12	V2	1.027523747	0
$D$12	V3	1.03	-0.420013884
$E$12	V4	1.027556884	0
$F$12	V5	1.029394697	0
$G$12	V6	1.028217444	0
$H$12	V7	1.028237498	0
$I$12	V8	1.03	-1.16663428

Constraints

Cell	Name	Final Value	Lagrange Multiplier
$B$19	P1	1.19E-13	9.759882
$F$19	P5	1.320396	0
$G$19	P6	2	0
$I$19	P8	3	0
$D$19	P3	3.5	0
$B$19	P1	1.19E-13	0
$C$19	P2	9.47E-13	0
$D$19	P3	3.5	-17.8677
$E$19	P4	-8.6E-14	0
$F$19	P5	1.320396	0
$G$19	P6	2	-5.01567
$H$19	P7	5.68E-13	0
$I$19	P8	3	-7.80404
$C$19	P2	9.47E-13	-40.1844
$C$21	Q2	2.14E-15	-0.00495
$E$19	P4	-8.6E-14	-40.0241
$E$21	Q4	8.52E-15	0.007149
$H$19	P7	5.68E-13	-40.1238
$H$19	P7	5.68E-13	0
$H$21	Q7	-2.4E-15	0
$H$21	Q7	-2.4E-15	-0.00895

Figure 8.15. Sensitivity sheet with expanded equality constraint.

the former $H19 < 0$, while a Lagrange multiplier of zero for the latter $H19 > 0$ indicates that it is not binding.[6]

The interpretation of this is the same as in Section 3.8.2, which is worth repeating. This means that if the right-hand side of $H19 < 0$ is relaxed (increased, for example, from zero to 0.1), the objective function will decrease (because of the negative sign of the Lagrange multiplier) at a rate of 40.1238 per unit increase. The reader is expected to confirm this by changing this limit to, say, 0.1 and by calculating the rate of increase of the objective function.

The lower limit of generation plays a role in the deregulated electrical industry. Generators must be kept running at some minimum output level arising from several reasons. For instance, the output of a generator, particularly reactive vars, even though the generator's energy bid may not warrant running it, may be required to maintain voltage levels at certain nodes of the network. Another example is when a generator may not be able to reduce its output as required by the system operator because of insufficient response capability, minimum run time requirement, or minimum fuel burn requirement. The Lagrange multiplier associated with such constraints is used in the settlement procedure for charging or paying market participants a condign amount according to prevailing market rules.

[6]In Figure 8.13, despite the answer sheet of Figure 8.4.2 indicating that the above constraint is nonbinding, the Lagrange multiplier associated with this equality constraint is −40.1238, the same as in Figure 8.15 obtained by splitting the equality to two inequalities.

A cautionary note is that, despite the negative sign of the Lagrange multiplier, as the demand at node 7 increases, the objective function increases. This is because loads (demands) have a negative sign (cell D52), and hence the constraint for cell H19 is that the flow into the network plus the demand at that node (negative in sign) is equal to zero. Consequently, relaxing this constraint implies that the load is decreased. Therefore, the cost of supplying the next unit of energy at this node is +40.1238, which is the negative of the Lagrange multiplier. The reader can confirm this by changing the constraint to $H19 \leq 0.1$ and by observing the active power in cell D52, which reduces by 0.1 to 2.0.

While on the matter of Lagrange multipliers, we see that the reactive power consumption also changes the objective function, even though the dispatch to minimize cost is based only on the cost of real power offers. For example, associated with the constraint for the external reactive power injection into node 7 ($Q_7 = 0$), we get for cell H21 a Lagrange multiplier of −0.00894 (see Figure 8.13). This means that incremental cost of reactive (positive) var consumption is −\$0.00894. While this may be a small amount in this problem, it can be much larger under some circumstances.

In subsequent examples in this chapter, we will not dwell on aspects concerning answer sheets anymore, but will proceed directly to a discussion of results.

8.4.3 Locational Marginal Prices

It has been shown in Section 4.7 that the locational price defines the marginal cost of supplying the next unit of energy at a particular location (node). It was also shown that the LMP is obtained by subtracting the Lagrange multiplier from the bid price for energy. In our examples, since there are only offer prices for the supply of energy and there are no bid prices associated with demand, the LMP at nodes where there is only demand without any generation simply becomes equal to the negative of the Lagrange multiplier.

The LMP computed using the above logic is shown in Figure 8.16. This figure is also attached to the lower portion of the AC OPF-1.Xls spreadsheet. In Figure 4.10 of Section 4.7, we have computed the LMP at the nodes of the same network using a dc load flow. The LMP at all nodes was the same as that of the marginal generator, namely, \$40.00. However, in Figure 8.16, the effect of losses and the use of ac network flows have created differences in LMP at nodes.

Note that voltage constraints also may involve additional cost in dispatching the system. Because of the requirement that the voltage of nodes be within limits 0.97 to 1.03, a finite reduced gradient associated with V_3 and V_8 can be seen in the sensitivity sheets of Figure 8.13 or 8.15. Consequently, relaxing the upper limit of voltage limits at these nodes will also result in a change in value of the objective function. The reader is encouraged to change these limits by an appropriate amount and observe resulting changes to the objective function.

Node	Bid	λ	LMP
1	50	9.75987	40.2401
		0	
		0	
		0	
		0	
		0	
2	0	-40.1844	40.1844
3	22	-17.8677	39.8677
4	0	-40.0241	40.0241
5	40	0	40
6	35	-5.01569	40.0157
7	0	-40.1238	40.1238
8	32	-7.80403	39.804

Figure 8.16. Calculation of locational prices from Lagrange multipliers.

8.4.4 Finite Branch Flow Limits and Other Constraints

In previous examples we had assumed that all the branches had ample transfer capability and, therefore, did not pose any restrictions on generator dispatch. Now we consider the fact that flows in some branches (or sum of flows in some branches representing an interface) cannot exceed certain values. Such limits may arise from thermal ratings of lines and terminal equipment, voltage collapse considerations, or stability limits. In any event, such considerations are generally translated to a certain MW or MVA flow in appropriate branches.

AC OPF Trans Limit.Xls

Open the spreadsheet "AC OPF Trans Limit.Xls." Branch flow MVAs are computed in cells I30:I55. If the flow in a particular branch, or the sum of flows in some branches (called interface flows), should not exceed a certain value, this information is now incorporated as a constraint.

This example is the same as the earlier ones except that we require the flow in branch 7–8 not to exceed 1.6 in either direction. The flow limit in either direction for all branches is entered in cells K30:L55. In this example, large limits are set for all branches except for branch 7–8 (which is restricted to 1.6 pu). The constraint is entered into the Solver as the computed MVA flow in cell I52 or I55 should be greater than the entry in cell L52 or less than the entry in cell K52.

Start the computation with an initial guess of one for voltages and zero for the other choice variables. Obtain the solution by minimizing the energy dispatch cost in cell L16. From the sensitivity report note that the Lagrange multiplier for cell I55 associated with the flow constraint is 16.82. This means that if the constraint were to be relaxed, the resulting savings would be $16.82. The objective function has increased to $301.19 from $295.81 of the unconstrained case in Figure 8.11.

In terms of LMP, the transmission restriction has created a wider spread in their values; they vary from a high of $45.13 to a low of $22.00. This, computed as in the previous example, is shown in Figure 8.17.

Node	Bid	λ	LMP
1	50	8.893	41.107
		0	
		0	
		0	
		0	
		0	
2	0	-39.0101	39.0101
3	22	0	22
4	0	-40.8879	40.8879
5	40	0	40
6	35	-0.51174	35.51174
7	0	-45.1308	45.1308
8	32	0	32

Figure 8.17. Locational marginal prices arising from transmission limit.

There is another constraint in the spreadsheet—that of machine reactive power limits. The requirement is that their reactive power outputs must be within some limit. Such a limit is generally expressed as a relation between the real and reactive power outputs of generators. Here, we use a simple relationship between P and Q output of generators—that the reactive power output must be within a range. In cells P60:64 of the spreadsheet "AC OPF Trans Limit.Xls," reactive power outputs of machines are computed. Cells O60:O64 and Q60:Q64 indicate the lower and upper bounds incorporated into constraints. Since large numbers are entered for these upper and lower bounds, these constraints are not binding. Hence they do not affect the solution. The reader may wish to reduce these limits and examine consequences of doing so.

8.5 OPTIMAL POWER FLOW — PHASE ANGLE REGULATORS

Phase angle regulators (PAR) are installed in electrical networks to control flows of active and reactive power in selected branches or interfaces. A phase angle regulator introduces a difference between the phase angles of the input and output voltages. In its simplest form, it can be represented just by the difference in angles obtainable from such a device.[7]

In the simulation of the spreadsheet "AC OPF PAR.Xls," we assume that a PAR is introduced in branch 7–8 of the previous example to control the flow in that branch. The effect of a PAR on the expression for injected power in (B.7) is

AC OPF PAR.Xls

$$
\begin{aligned}
S_i = |Y_{ii}||V_i^2|\angle\theta_{ii} - |Y_{ij}||V_i||V_j|\angle(\theta_{ij} + \delta_j - \delta_i) \\
-Y_{ik}||V_i||V_k|\angle(\theta_{ik} + \delta_k - (\delta_i - \delta_P)) \\
-|Y_{il}||V_i||V_l|\angle(\theta_{il} + \delta_l - \delta_i), \qquad (8.25)
\end{aligned}
$$

[7] A more detailed representation would involve the inclusion of the internal impedances of such devices, and voltage differences between the input and the output described by a suitable equivalent circuit.

where δ_P is the angle introduced by a PAR located in the branch i–k (see Figure B.2).

Without the installation of a PAR, choice variables in the previous examples were the power outputs from generators.[8] However, now we have δ_P as an additional choice variable in cell L14. In the spreadsheet "AC OPF PAR.Xls," cells B52, B55, F52, and F55 now incorporate necessary changes to include this new choice variable according to (8.25). Evidently, δ_P is the angle by which node 8 is advanced in relation to node 7. Additionally, we include constraint equations in the Solver routine requiring the range of control for the PAR is ∓ 10 degrees. All other limits and constraints remain the same as in the previous example.

The result of the spreadsheet indicates that node 8 is made to lag node 7 by an angle of −0.0287 radians, or 1.644 degrees. Also, we note that the more expensive generation is now reduced, because the PAR is used to limit the flow on branch 7–8 to 1.6 pu. In comparison with a production cost of $301.19 without PAR (spreadsheet "AC OPF Trans Limit.Xls"), the production cost has decreased to $295.85.

8.6 OPTIMAL POWER FLOW—ACTIVE AND REACTIVE POWER OPTIMIZATION

In all previous problems, we implicitly assumed that the cost of reactive power for dispatch purposes was zero. Furthermore, we had assumed in the constraint equations that the power output was to be less than the MW ratings of generators. In other words, no distinction was made between the MW rating and the MVA rating of generators.

In this problem we will remove these restrictions. Although we can model a separate market for reactive power with its own offer prices (Rau, 2000), we will assume that the offer price for reactive power, both leading and lagging, is the same as that of energy. This assumption does not taint any of the observations that we are about to make in regard to the optimal distribution of reactive power.

AC OPF Vars.Xls

To this end, open the spreadsheet "AC OPF Vars.Xls". The relevant part of this spreadsheet is shown in Figure 8.18 for ready reference. The branch flow equations and the network diagram are omitted in this figure because these are identical to the previous example. However, they can be found in the spreadsheet.

There are some differences in the optimization. In row 5 we define machine output capability in MVA, instead of MW. Furthermore, for choice variables, in addition to node voltages and angles (cells B12:I12 and B14:H14), we include the real and reactive power injections from generators at nodes 1, 3, 5, 6, and 8. These are in cells B16:F18.

In cell E26, the objective function of dispatch cost assumes that the price of positive and negative vars from a machine is the same as the offer price for energy.

[8] Choice variables were voltages and angles at nodes that specified the necessary power output from generators.

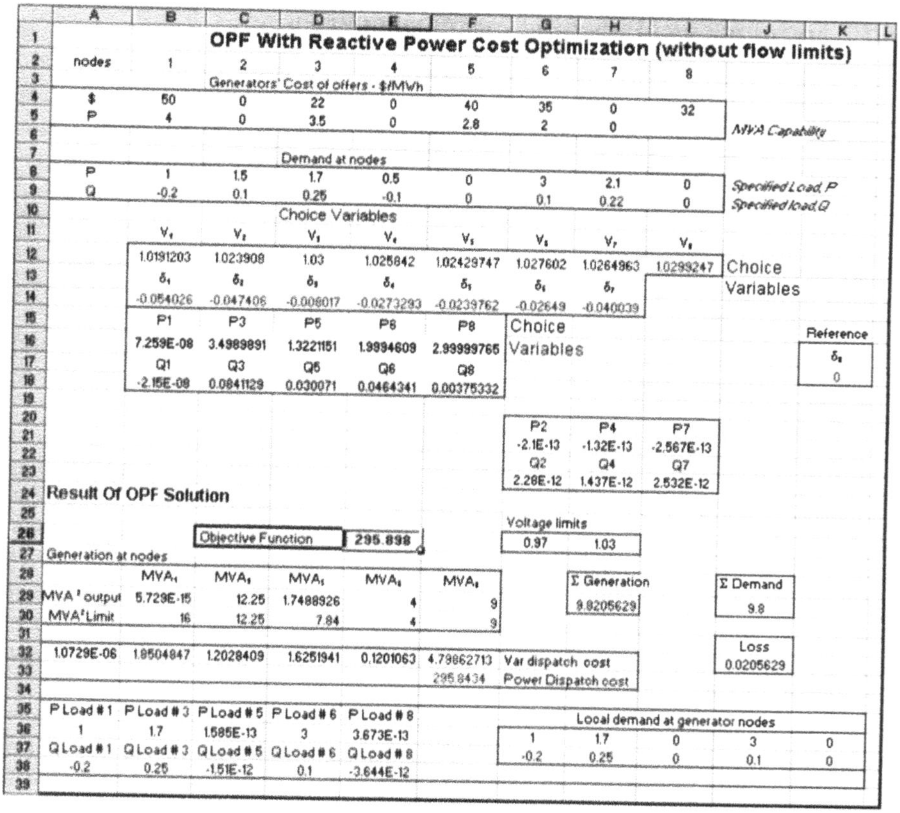

Figure 8.18. Optimal dispatch considering the cost of reactive power.

Consequently, the dispatch cost is equal to volt–ampere output of the machine (Pythagorean addition of real and reactive power) times the offer cost of energy. This procedure of minimizing cell E26, therefore, draws vars from sources to minimize the sum of active and reactive power costs rather than dispatching reactive power from sources indiscriminately.

Constraints regarding voltage and angle limits and nonnegativity of generation are the same as in the previous example. The following changes are made to other constraints in order to obtain appropriate sensitivity values. The power generation at a node minus the net flow into (or from) a network must be equal to the local demand at a node. The difference between generation (choice variables) minus the net flow into the network is computed in cells A36:E36 and A38:E38. The constraint equation requires these cells to be equal to the local loads indicated in cells G36:K37. The square of MVA output of generators is calculated from choice variables in cells B29:F30. We constrain square of the MVA output to be less than or equal to the square of machine MVA capabilities in row 5.

8.6.1 Advantages of Reactive Power Optimization

Before a discussion of sensitivity, we discuss results and compare them to results if vars were indiscriminatingly procured. Toward that end, compute a solution of "AC OPF Vars.Xls" to minimize cell E26. Set the initial guess for all voltages at 1.0; for all angles, machine power, and reactive powers set the initial guess at zero.[9] The result reveals a dispatch cost of $295.898. In cells F32 and F33, the components of this cost have been broken down to the cost of reactive power dispatch and the cost of real power dispatch. These components, assuming that the real and imaginary component of power are priced at the offer bid for energy, are $4.798 and $295.843.

From the results we note that generator #5 is the marginal generator, generators #3, #6, and #8 are infra-marginal, and generator #1 with an output of zero is supra-marginal.

What would happen if we were to assume that the cost of reactive power is zero? In order to examine this, with the end result obtained above as the starting values, run the Solver to minimize the real power cost in cell F33 instead of cell E26. We then obtain an active power dispatch cost that is lower than before ($295.8285). However, cell F27 will indicate that the compensation for vars[10] has increased to $11.286. This increase assumes that reactive power suppliers are compensated at the same rate as energy in settlement of market accounts. The total compensation shown in cell E26 now is, therefore, $299.189. This increase in compensation for suppliers (assuming that the compensation is based on offers and not locational marginal prices) is because of indiscriminate supply of reactive power. It is important to observe that the reactive power outputs of the machines change depending on which objective is minimized. That is, the reactive power distribution changes if the dispatch procedure optimizes var distribution.

Is this solution unique? No, because when one optimizes just the power dispatch cost, the algorithm cannot track the compensation to var suppliers. To confirm this, run the Solver with different initial guesses for all choice variables and minimize active power dispatch cost. Then minimize the cost of active and reactive power alternately. Despite slight differences in the answers, the conclusion is that optimizing reactive power results in a different distribution of its supply and reduces cost further. However, the nature of the problem is such that a robust estimate for the cost of reactive power cannot be obtained from the sensitivity reports as we will discuss next.

[9] This problem has several minima, but, fortunately, they all lie on a more or less shallow surface with little difference in the objective function (second or higher decimal places). Nonetheless, the voltage conditions, although within their limits, can be different. In order to study and understand the solution to this problem, try different initial guesses for choice variables and observe the results.

[10] This assumes that the suppliers of reactive power are compensated at the energy rate for the vars supplied. However, this results in an overpayment since the sum of real power and reactive power is greater that the volt amperes supplied. A way of avoiding this is to pay suppliers for the volt amperes supplied at the energy offer rate. But, that is another subject of discussion since none of the deregulated electric energy markets have an explicit payment for reactive power at the time of writing this book.

8.6.2 Locational Prices of Active and Reactive Power

The locational marginal prices can be computed from the sensitivity sheet which shows the Lagrange multipliers associated with constraints. Toward this, run the program in the spreadsheet "AC OPF Vars.Xls" with an initial guess of unity for all voltages and zero for other choice variables. Along with the solution, obtain sensitivity report 1 in the FTP site offered by Wiley. This report and the associated constraints are shown in Figure 8.19.

First consider active loads in A36 to E36 shown in the last five rows of Figure 8.19. The constraints, as can be seen in the last column and last row, are of the form $\$E\$36 = \$K\$36$. Since the Lagrange multiplier indicates the rate of change in the objective function with the rate of change of the right-hand side, the rate of increase of cost in cell E26 for the load increase at node 8 (cell K36) is 40.24124. Therefore, the locational marginal price of active power at these nodes is equal to the Lagrange multiplier in rows 17 to 21. From a similar reasoning, the Lagrange multipliers associated with rows 11 to 15 indicate the marginal cost of supplying reactive power at these nodes.

We now address nodes where there are no generators. These are nodes 2, 4, and 7. The constraint at these nodes, for example in row 7 of Figure 8.19, is of the form $G21 = 0$, implying that there is no injection from any generators. If there is an injection, the right-hand side would increase and would be > 0. Hence the associated Lagrange multiplier (-39.74) indicates that the objective function would decrease at that rate if there is generation injection (at zero cost) at this node. Correspondingly, therefore, if there were to be an incremental load at these

	Cell	Name	Final Value	Lagrange Multiplier	Formula
1	$B$29	MVA output MVA1	1.00342E-14	0	$B$29<=$B$30
2	$C$29	MVA output MVA3	12.25	-2.595806736	$C$29<=$C$30
3	$D$29	MVA output MVA5	1.747828028	0	$D$29<=$D$30
4	$E$29	MVA output MVA6	4.000000009	-1.244609214	$E$29<=$E$30
5	$F$29	MVA output MVA8	9	-1.37385737	$F$29<=$F$30
6	$I$23	Q7	-1.08346E-11	-9.133323827	$I$23=0
7	$G$21	P2	2.77287E-11	-39.75333934	$G$21=0
8	$G$23	Q2	-7.22334E-11	-9.14359514	$G$23=0
9	$I$21	P7	5.25535E-12	-39.83046231	$I$21=0
10	$H$23	Q4	-5.71937E-11	-9.138358126	$H$23=0
11	$A$38	Q Load #1	-0.2	9.189685822	$A$38=$G$37
12	$B$38	Q Load #3	0.25	9.099694954	$B$38=$H$37
13	$C$38	Q Load #5	4.78026E-10	9.164813788	$C$38=$I$37
14	$D$38	Q Load #6	0.1	9.127823552	$D$38=$J$37
15	$E$38	Q Load #8	-4.3836E-11	9.141622118	$E$38=$K$37
16	$H$21	P4	1.92255E-11	-39.96097819	$H$21=0
17	$A$36	P Load #!	1	39.67937045	$A$36=$G$36
18	$B$36	P Load #3	1.7	40.16226599	$B$36=$H$36
19	$C$36	P Load #5	7.65321E-12	39.99332809	$C$36=$I$36
20	$D$36	P Load #6	3	39.97063118	$D$36=$J$36
21	$E$36	P Load #8	5.26867E-12	40.24125799	$E$36=$K$36

Figure 8.19. Locational marginal prices for active and reactive power.

nodes, the marginal cost of supplying them would be the negative of the Lagrange multiplier. Thus we conclude that the negative of the Lagrange multiplier in rows 6 to 10 and row 16 gives the LMP at nodes 2, 4, and 7 for real and reactive power.

Finally, Lagrange multipliers in rows 1 to 4 show the effect of increasing the MVA rating of the generators on the dispatch cost. Increasing the MVA capability of generators at nodes 1 and 3 would have no effect on the cost; since generator #1 is supra-marginal, and generator #3 is the marginal unit.

Sensitivity of Lagrange Multipliers for Reactive Power. That this problem has local minima has been alluded to earlier. As before, obtain a solution to the problem first, starting with initial guesses of unity for voltages and zero for other choice variables, to minimize the cost of MVA dispatch in cell E26. Then with the resulting solution as the starting values, minimize cell F33, the cost of real power dispatch. Using the resulting solution as the starting value, again minimize cell E26 and obtain a sensitivity sheet (shown as sensitivity report 2 in the FTP site offered by IEEE). A comparison of sensitivity report 2 and sensitivity report 1 shows that the change in most Lagrange multipliers is small. However, the incremental cost of supplying reactive power has changed considerably; they are now in the range of $3.98 as opposed to the previous value of about $9.14.

We encourage the reader to switch from one objective function to the other–that is, from minimizing volt–ampere costs to that of minimizing the real energy cost. An examination of resulting sensitivity reports will reveal large variations in the cost of reactive power. Therefore, it is not surprising that the search for an appropriate method to price reactive power is a topic of current research.

8.6.3 Cost Associated with Transmission Congestion

In the deregulated electricity industry, there is concern about allocation of congestion costs. Cost of congestion is the increased cost of dispatch arising from transmission limitations. For example, in the problem of Section 8.4 shown in Figure 8.11, the objective function (cost of dispatch) was $295.8158. For this case when branch 7–8 flow was restricted to 1.6, the cost of dispatch becomes $301.1987. Consequently, the difference ($301.19 − $295.81 = 5.38) can be attributed to congestion. However, we have to ensure that any constraint other than those of specified flow limits is not binding. Because of the limit on branch flow, there may be demands on reactive power requirements from sources to maintain voltage levels. Such demands result in the robbing of machine MVA capability, which, in turn, could lead to increased cost. As a result, any such binding constraint could also increase the cost of dispatch. Therefore, one has to exercise caution in declaring what the cost of congestion really is.

It is also important to note that the cost of congestion cannot be obtained from Lagrange multipliers associated with the flow limit ($16.82 in the example of Section 8.4.4). This issue is demonstrated in Rau (2002). Additionally, it is difficult to derive congestion costs from dispatch cost if there is demand side bidding by loads. In such a scenario, consumption is related to price. Because of transmission limitations, the changes in price at delivery points results in consumption changes. Therefore, one cannot determine congestion cost by a mere comparison of dispatch costs with and without transmission limitation.

In Section 4.5, dispatch of a system with demand-side bids was considered using dc flow equations. The next example extends the procedure to the case of an ac network.

8.7 DISPATCH OF AN AC NETWORK IN THE PRESENCE OF DEMAND-SIDE BIDS

The spreadsheet "demandbid.Xls" shows an example in which the network of the earlier examples is used. Figure 8.20 shows relevant portion of the spreadsheet.

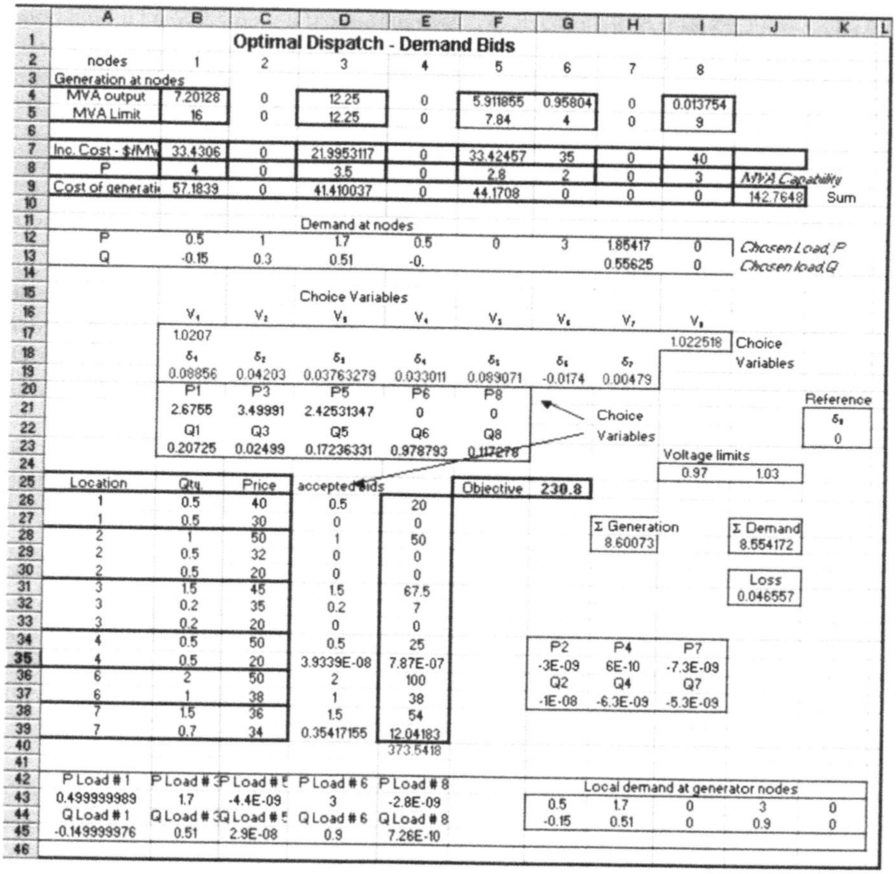

Figure 8.20. Optimal system dispatch—demand-side bids.

8.7.1 Problem Formulation

demandbid.xls In this problem, we could have modeled offer bids for supply of energy in blocks. Since that has been done in earlier sections of linear optimization, we now assume the offer bids to be in the form of a nonlinear function. The cost functions for generators #1, #2, and #3 are identical to the relations expressed by (8.3) in Section 8.3.3. Recall that the cost of energy output from generators #4 and #5 is \$35 and \$40, respectively, at all outputs. In Figure 8.20, the generator output limits are indicated in row 5, and the cost of energy dispatched from each generator is computed in row 9. The sum of the cost of dispatch is in cell J9.

Cells A26:C39 indicate the quantity price bids for consumption at different nodes of the network. As in the previous examples, the ac branch flow equations are in A50:H75.

8.7.2 Solution

As shown in Section 4.5, the objective is to maximize the area between the demand and supply curves subject to constraints of the network and machine limits. The area between the two curves is given by the area under the demand function minus the area under the supply function. Hence we need to maximize

$$\Sigma Q_{L_i} C_{L_i} - \Sigma_i^5 \int_{p_i=0}^{p_i} C(p_i)\,dp,$$

where Q_{L_i} is the quantity and C_{L_i} is the price associated with i bids for consumption in cells B26:C39, and cell J9 indicates the sum of cost of production at power output p_i from each generator corresponding to (8.15). The difference between demand and supply curves is computed in cell G25 which is the objective function. Choice variables are the network voltages and angles, real and reactive outputs from machines, and the quantities to be selected from the demand bids. Constraints are identical to the previous problems in terms of machine MVA and voltage limits, and that choice variables are nonnegative. In the interest of avoiding repetition, we leave it to the reader to examine the Solver screen and the cells for embedded formulas.

The results in Figure 8.20 show the selected bids. A total quantity of 8.554 (cell J28) is supplied with a resulting network loss of 0.04565571. Note that machine #3 is fully loaded (compare appropriate cells in rows 4 and 5) and machines #1 and #5 are operating at the same marginal cost of generation. The real power output from machine #6 is zero. However, it is called on to supply reactive power. This is the case for the generator #8 as well. The cost of supplying reactive power is zero.

We suggest that the reader study this spreadsheet carefully and perturb some cost values and bid values. Such an exercise will foster a deeper understanding of the underlying mechanism. Also, it is informative to obtain and study the sensitivity sheets. An exercise problem at the end of this chapter requires the enforcement of transmission constraints.

8.8 LOSS ALLOCATION

The identification of responsibility toward network losses is an important concern in the deregulated electric networks. Different methods are used in different markets to allocate responsibility for loss. The question is whether responsibility for loss should be based on incremental loss or the actual losses incurred. The calculation of incremental loss based on a quadratic approximation can be made by the traditional method using "B coefficients" (Wood and Wollenberg, 1984; Stevenson, 1962). A method of allocating actual and incremental losses is shown in Rau (2001), and Rau (2000)—the latter uses a radial network equivalent.

In the world of vertically integrated utilities, B coefficients were used to obtain penalty factors for generators in determining the economical dispatch of generators. It can be shown (Wood and Wollenberg, 1984; Stevenson, 1962) that the loss in a n node network can be approximated[11] as a quadratic expression by

$$P_L = \mathbf{P}^T\ \mathbf{B}\ \mathbf{P}, \tag{8.26}$$

where P_L is the network loss, $\mathbf{P}$ is a n vector of real power injections at nodes, and $\mathbf{B}$ is a $n \times n$ matrix of coefficients.

From the above, we get

$$\frac{\partial P_L}{\partial P_i} = 2B_{ii} + 2 \sum_{j=1,\cdots,n,\ j\neq i} B_{ij} P_j. \tag{8.27}$$

The penalty factor L for the ith generator is defined as

$$L_i = \frac{1}{1 - \frac{\partial P_L}{\partial P_i}}. \tag{8.28}$$

For the economic dispatch of the system, each generator cost is multiplied by its penalty factor (always greater than one) and the generators are then dispatched such that the incremental costs of all generators are equal as in Figure 8.7 of Section 8.3.4.

Our object now is to examine how responsibility for incremental network loss can be determined. The question is whether we can identify such responsibility from the Lagrange multipliers of an optimal power flow solution.

Unfortunately, it is not possible to directly do so if the objective of the dispatch (and optimization) is that of minimizing cost. Lagrange multipliers show the sensitivity of the objective function to constraints. The objective function being the cost of dispatch, Lagrange multiplier can only indicate the sensitivity of cost to constraints, such as the desired demand at nodes. Furthermore, the dispatch cost is affected by congestion costs, voltage limits, and so on, as shown in earlier

[11] See Stagg and El-Abiad (1968), Stevenson (1962), Wallach (1986), and Wood and Wollenberg (1984), for other approximations such as the assumptions regarding phase relationships of current.

examples. Thus, the Lagrange multiplier associated with nodal demands will not only have the *cost of losses* embedded in it, but will also contain cost of congestion, voltage constraints, and machine var limit constraints. When the demand in a node is altered, resulting changes in branch flows will change the network loss as well as the pattern of congestion. The sensitivity of the objective function for changes in branch flow limits, voltage limits, or other constraints can be obtained from the Lagrange multipliers associated with those constraints. However, since there is no constraint associated with losses, the sensitivity of losses cannot be identified directly. It is intertwined into other constraints.

An artifice, though not completely satisfactory, is to obtain an approximate loss allocation by using a reformulated OPF to minimize the cost of loss. Thus we obtain Lagrange multipliers for changes in demand at nodes, which give an indication of the sensitivity of network loss (incremental) to changes in demand at different locations of the system. The following example illustrates the procedure.

lossmin.Xls ***Example.*** Consider the network of the previous example in Section 8.4. Observe that the network losses based on minimization of dispatch cost (of only active power) was 0.0204, without any transmission constraints. Now, our objective, instead of minimizing dispatch cost, is that of minimizing i^2r losses.[12]

It is important to recognize the difference between cost of losses and losses itself. When the cost of loss is minimized, the cost associated with loss is imputed at the cost of generation. Hence, when production cost is minimized, the cost of network losses is automatically minimized. When loss is minimized, the cost of losses, irrespective of the source that delivers that energy, is assumed to be the same from all sources. Thus one minimizes the i^2r loss. The concern about the reduction of losses is of interest to policy makers to study global warming and the reduction of CO_2 emissions. We shall soon see that minimization of losses can be used as an artifice to allocate responsibility for incremental losses.

Results. Figure 8.21 shows the spreadsheet "lossmin.Xls." It is identical to the previous spreadsheet "AC OPF-1.Xls" except for one difference: The objective function now is to minimize the i^2r losses in cell B69. All other constraints are identical to the previous example.

Figure 8.21 shows the final results. Since the objective is to reduce losses, although the offer costs are shown in the figure, they are irrelevant to the optimization procedure. The losses are reduced to 0.01154 from 0.0204. The dispatch cost in Section 8.4 was $295.8158. Assuming that Generators are compensated at their offer prices, the dispatch cost has now increased to $330.976 (cell E26). The distribution of power outputs from machines is quite different from the example in Section 8.4. This can be confirmed by comparing generator outputs in row 19 of Figure 8.11 with the generator outputs in row 16 of Figure 8.21.

Of interest are the Lagrange multipliers. The Lagrange multipliers from the sensitivity sheet are shown in Figure 8.22.

[12] In the competitive markets of today, a system is dispatched to minimize cost and not losses.

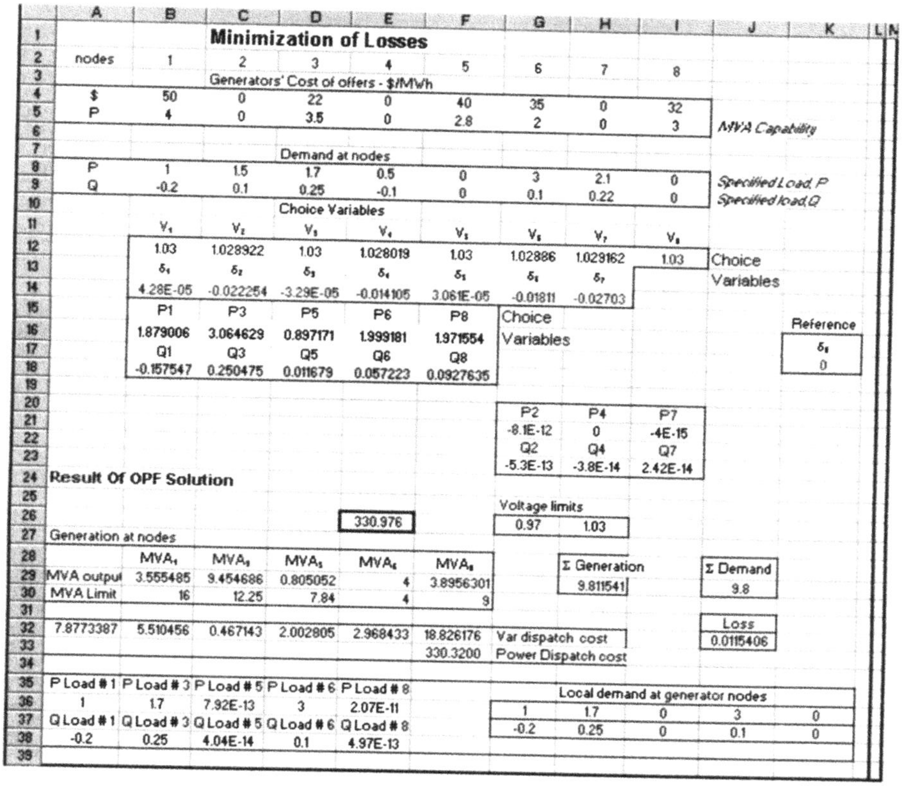

Figure 8.21. Minimization of network losses.

Consider the Lagrange multiplier of −0.00444 associated with P2 in cell G21. The reason for the negative sign is this: The constraint in the Solver requires G21 = 0. The value of G21 is the sum of specified load at node #2 (cell C8) and the flow into the network (negative) in D49. Because of this convention, the value of Lagrange multiplier is negative.

The interpretation of the multiplier is that if P_2 is increased by 1, the loss in the network increases by 0.00444. Similarly, losses associated with P_4 and P_7 can be found in Figure 8.22 to be 0.002838 and 0.005385. Therefore, marginal loss increase is the highest for the load at node #7.

What about incremental losses associated with loads at generator nodes #1, #3, #5, and #8? Generators #1, #3, #5, and #8 are not fully loaded. Hence these generators can supply any increase in loads at these nodes without causing any additional loss (since no resistance is modeled between the loads and the perfect generators at these nodes). Consequently, the Lagrange multipliers associated with cells A36, B36, C36, and E36 are equal to zero, which implies zero incremental loss. However, generator #6 is at its full output. Hence any increase in the load at

Constraints			
Cell	Name	Final Value	Lagrange Multiplier
$A$38	Q Load #1	-0.2	0
$B$38	Q Load #3	0.25	0
$C$38	Q Load #5	4.04E-14	0
$D$38	Q Load #6	0.1	0.00010361
$E$38	Q Load #8	4.97E-13	0
$A$36	P Load #1	1	0
$B$36	P Load #3	1.7	0
$C$36	P Load #5	7.92E-13	0
$D$36	P Load #6	3	0.00361573
$E$36	P Load #8	2.07E-11	0
$B$29	MVA output	3.555485	0
$C$29	MVA output	9.454686	0
$D$29	MVA output	0.805052	0
$E$29	MVA output	4	-0.0009043
$F$29	MVA output	3.89563	0
$G$21	P2	-8.1E-12	-0.0044412
$H$21	P4	0	-0.0028385
$I$21	P7	-4E-15	-0.0053852
$G$23	Q2	-5.3E-13	-0.0001874
$H$23	Q4	-3.8E-14	0.00014395
$I$23	Q7	2.42E-14	-0.0003072

Figure 8.22. Lagrange multipliers for marginal losses.

this node cannot be supplied from this generator; it has to be supplied from other generators in the network. Therefore, the Lagrange multiplier associated with cell D36 is 0.003615, indicating this as the marginal loss for this load.

We obtain the penalty factors for generators using (8.28). For example, for generators #2, #4, and #7, the penalty factors are 1.0044, 1.00284, and 1.0054, respectively.

However, note from (8.27) that the incremental losses is a function of nodal injections of power. The injections in Figure 8.21 are quite different from those that result under actual system dispatch as in Figure 8.11. Consequently, the above penalty factors do not reflect the incremental losses correctly.

In order to compute marginal losses at injection levels corresponding to system dispatch, open sheet 2 of the spreadsheet "lossmin.Xls."

In this case, we increase the load demand at all nodes by a small amount, 0.01 in this example, as indicated in cells B8:I8 in sheet 2. As before, we minimize the network loss, but with an additional constraint. The the constraint is that the output levels of generators should be greater than or equal to those obtained from an OPF to minimize cost. Hence, in cells B24:F24 we have embedded the generation that would result from an OPF if dispatch cost was minimized (see Figure 8.11 in Section 8.4.1). The Solver now includes a stipulation that the choice variables for active power generation in cells B16:F16 is greater than or equal to cells B24:F24.

The spreadsheet shows the results of optimization. Sensitivity 2 shows corresponding Lagrange multipliers which is reproduced in Figure 8.23. Because of the constraint that the network loading level is close to that of cost-minimizing dispatch,

Constraints			
Cell	Name	Final Value	Lagrange Multiplier
$A$36	P Load #1	1.01	0
$B$36	P Load #3	1.71	-0.008936
$C$36	P Load #5	0.01	-0.005825
$D$36	P Load #6	3.01	-0.00526
$E$36	P Load #8	0.01	-0.010438
$A$38	Q Load #1	-0.200001	0
$B$38	Q Load #3	0.25	0.0018455
$C$38	Q Load #5	-5.16E-07	0
$D$38	Q Load #6	0.1	0.001294
$E$38	Q Load #8	-5.5E-11	0.0010281
$B$29	MVA output	0.009422	0
$C$29	MVA output	12.25	-0.063192
$D$29	MVA output	1.774121	0
$E$29	MVA output	4	-0.031195
$F$29	MVA output	9	-0.032442
$H$21	P2	-8.38E-09	0.0011849
$H$24	Q2	-1.68E-09	-0.000783
$I$21	P4	-1.37E-08	0.0050955
$I$24	Q4	-6.9E-10	-0.000948
$J$21	P7	5.64E-09	0.0026078
$J$24	Q7	-4.54E-10	-0.001125

Figure 8.23. Incremental losses at OPF loading level.

the loss has increased to 0.020274 from 0.01154 that resulted under minimum-loss dispatch in Figure 8.21. Additionally, note from sensitivity sheet 2 that the values for $\partial P_L/\partial P_i$ reflected in Lagrange multipliers is different for the two modes of dispatch. For example, at node 4, the incremental loss in Figure 8.22 is 0.00284, whereas in Figure 8.23 it has a value of 0.00509.

Allocation of Responsibility. We now examine how the cost of losses can be allocated to Loads and Generators. Clearly, the allocation can be made in many ways depending on market rules. For example, the total loss can be apportioned to nodes in the ratio of incremental loss factors[13] of Figure 8.23. After apportioning losses, the next question is what the price of loss should be. Again, there are several ways in which loss can be priced depending on rules accepted by participants in a market.

Another way to account for losses is to use incremental loss factors to compute additional costs of generation as penalty factors as in (8.28). Needless to say, there are several other ways in which losses can be allocated and priced. The areas of allocating loss responsibility in the deregulated markets for electricity and its price has attracted intense debate and research at the time of writing this.

[13]Note that the sum of incremental loss factors of Figure 8.23 is not equal to the network loss (0.02024). This is because 0.02024 is the actual i^2r loss while the Lagrange multipliers indicate incremental values, namely, $\partial P_L/\partial P_i$. A matter to note is that these incremental values do not contain any approximations or assumptions as contained in the calculation of B coefficients.

The above example has shown a possible method to compute sensitivity of losses to nodal demands. Problem 10 in Section 8.11 requires the reader to work out exercises regarding losses and cost of losses in order to reinforce the above ideas. The problem asks the reader to examine some additional ways of allocating loss responsibility as well.

8.9 ELEMENTS OF STATE ESTIMATION

8.9.1 Background

In a power system, many measurements are made and telemetered to the central operator. The operator uses these data to take operating decisions. Besides, many measurements are logged on to a digital computer file for future use in analysis and settlements. In many deregulated power systems, the declaration of the locational marginal prices (LMP) is an important task. Most dispatch their systems based on a dc load flow algorithm embedded into the EMS. Therefore, the output of the state estimator is taken to represent the actual state of the network. From this the LMP is synthesized. Of course, to represent the physics of the network properly, the state estimator logic embodies ac network equations.

Several measurements such as MW and MVA line flows, MW and MVA injection by generators into the network, and voltages at several points in a network are taken in a power grid. Depending on the level of instrumentation in a system, a network may have redundant measurements or insufficient measurements.

To understand redundant measurements, consider a node in a network at which voltage, current, and power (active and reactive) are measured. Clearly, from measured power and either of the other two measurements, the third quantity can be computed. Assume that we compute current from power and voltage. The computed value may not agree with the measured current value because of instrument errors. The error may be in either the power or voltage measurement, or possibly in the current measurement. Thus, the measurement of current is redundant, because it can be synthesized. The question is, What is a fair estimate for the voltage and current, and hence for power?

Note in the above that from the measured current and voltage we cannot estimate power, because we do not know the phase angle. On the other side of this spectrum, there may be a lack of measurements. For example, the power flowing into a line at the sending end may be measured, while the power received at the other end may not be measured. Consequently, one cannot compute line loss unless the received power is synthesized. Such a synthesis, or estimation, is possible if the voltage and current measurements and their phase angle at the receiving end are available and if the line impedance parameters are known. As the above implies, there may be certain redundant measurements in some sections and lack of measurements in some other sections of a power system.

Another reason to estimate the data arises from bad measurements. One of the meters may be defective or the telemetry may be bad, or may not send any data at all. It is possible that some instruments are wrongly connected. Thus, estimation of data not measured (or sent) and identification of bad data is an important part of state estimation. Another source of bad data is incorrect representation of network topology arising from incorrect telemetering of circuit breaker status.

The subject of state estimation is vast and has been investigated by several researchers. Schweppe and Handschin (1974) and Clements and Wollenberg (1975) are two examples of this rich field. The reader will benefit from reading an exposé of the basics of state estimation in Wood and Wollenberg (1984).

Our intent in this example is to show how the optimization techniques involved in the state estimation process can be solved practically for small problems. The "real-world" state estimation algorithms are more complex and perform many other computations and estimations than indicated here. Nonetheless, the example gives a basic idea of the process involved.

8.9.2 Simple Example of Estimation

The following is a simple example of estimating acceptable values for measured quantities when there are several measurements with associated accuracies and, possibly, redundancy.

An example in Wood and Wollenberg (1984) considers a six-node 11-branch network. Twenty-six-branch flow measurements (at either end of 13 branches), power injections by generators at two nodes, and voltage measurements at six nodes are estimated in a least-square sense.

In our illustration, we consider the eight-node network of Figure 4.9. If this were to be an actual network in existence, there would be several measurements such as real and reactive power injections, line flows, bus voltages, and so on. As a result, the estimation process would be site-specific depending on which measurements were taken.

In order to illustrate the principle of least-square estimation, which is the basis of estimation algorithms, we consider first a situation of estimating only branch flow measurements. Subsequently, we include voltage measurements as well. After an understanding of these simple examples, an extension of the spreadsheets to include other measurements should be obvious to the reader.

Problem. Table 8.1 shows the power and voltage measurements in the eight-node network of Figure 4.9. Assuming that these were the only available measurements, estimate the state of the system—(What are the estimated voltages and branch power flows?) for the assumed variances associated with the measurements.

Solution. The discussion in Section 3.3.3 and Appendix C has addressed least-square error approximations and maximum likelihood estimators. It is shown Wood and Wollenberg (1984) that the idea behind estimation is to

$$\min_{\mathbf{x}} E(\mathbf{x}) = \sum_{i=1}^{N_m} \frac{[z_i^{meas} - f_i(\mathbf{x})]^2}{\sigma_i^2}, \tag{8.29}$$

Table 8.1. Power measurements and associated variances at nodes

Node	P_{15}	σ	P_{12}	σ	P_{17}	σ	—	
1	−0.63	0.0001	−0.149	0.00008	−0.213	0.0002		
	P_{25}	σ	P_{21}	σ	P_{23}	σ	P_{26}	σ
2	−0.615	0.00015	0.149	0.0001	−0.63	0.00008	−0.4	0.00015
	P_{32}	σ	P_{34}	σ	P_{36}	σ	P_{38}	σ
3	−0.625	0.0002	0.83	0.00015	0.417	0.00008	−0.09	0.00015
	P_{47}	σ	P_{43}	σ	P_{45}	σ		
4	0.385	0.0001	−0.83	0.0001	−0.06	0.0002		
	P_{51}	σ	P_{52}	σ	P_{54}	σ		
5	0.642	0.00015	0.62	0.0001	0.06	0.00008		
	P_{62}	σ	P_{63}	σ	P_{68}	σ		
6	0.4	0.0001	−0.41	0.0002	−1.0	0.0001		
	P_{71}	σ	P_{74}	σ	P_{78}	σ		
7	0.217	0.0001	−0.40	0.0002	−1.98	0.00008		
	P_{83}	σ	P_{86}	σ	P_{87}	σ		
8	0.078	0.0002	−0.95	0.00015	1.93	0.0001		

where **x** represents measured parameters (such as voltage, current, and power), N_m is the number of independent measurements, z_i^{meas} is the ith measured value of the parameter, σ_i is the expected variance in the ith measurement, f_i is the functional relation used to calculate the ith parameter from **x**, and $E(\mathbf{x})$ is the total measurement residual.

Wood and Wollenberg (1984) develop the above expression in terms of covariance matrix and the Jacobian matrix in order to minimize (8.29) by requiring its gradient to be equal to zero. In the following, as a practical first step in understanding state estimation process, we formulate this as a least-square error minimization problem using an Excel spreadsheet.

`state-Estim.Xls`

The spreadsheet "State-Estim.Xls" has two sheets. Sheet 1 shows the simple procedure of least-square error minimization using only flow measurements. Sheet 2 includes voltage measurements to be discussed later.

Sheet 1 is shown in Figure 8.24. As in the earlier spreadsheets, the network parameters are included in the right-hand columns. Cells A10:H35 are network flow equations. The branch-flow equations are embedded into these cells as in the previous spreadsheets. Branch flows are computed in these cells for estimated values of voltages in cells B4:I4 and angles in B6:H6 cells. For example, cell B12 shows the computed value of real power flow from node 1 to node 7 for estimated voltages and angles. Since its sign is negative, the flow of power is from node 7 to node 1.

Cells J10:J35 indicate the measured values given in Table 8.1. The corresponding variances are in column K. The objective function corresponding to (8.29) is in

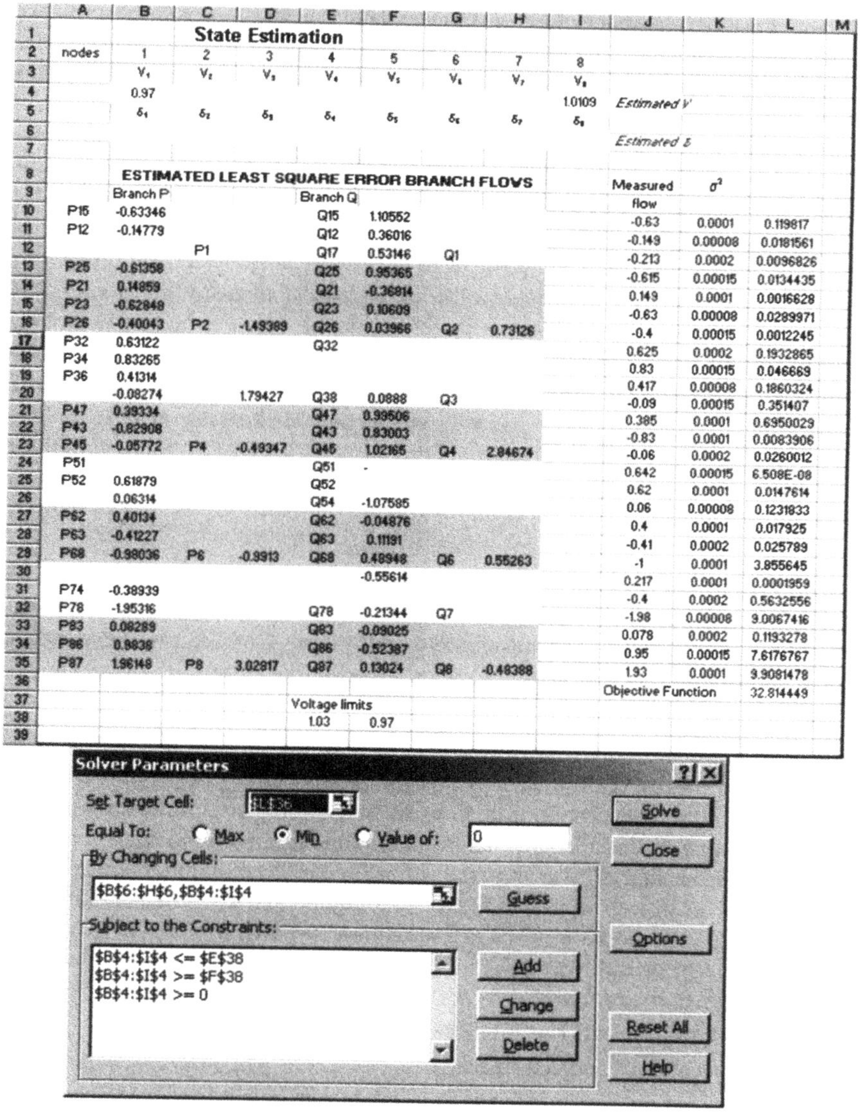

Figure 8.24. Estimation of states for branch power flows.

cell L36. The Solver procedure is to estimate the choice for voltages and angles in order to minimize the objective. Constraints are that we want the voltages to be nonnegative and that the voltage limits should be within some bounds. We need the latter limit because we do not have any guidance as to what a reasonable voltage is. Later, we see in sheet 2 that when voltage measurements are included, this limit is not required. Figure 8.24 shows the result, which obtains the objective as 32.8144.

We encourage the reader to perturb the input data and examine the results. For example, assume that the meter measuring the flow from node 8 to node 7 is connected with reverse polarity. Then the value in cell J35 would be −1.93. Note that the problem still converges but the objective function will be much larger in magnitude than in the above case (90676.9). Consequently, the magnitude of the objective function is an indication of bad or incorrect data. That said, the question is how one can identify which data is bad given a larger objective function.

Another question is how to synthesize missing data. For example, if the telemetry from the location of node 4 were to fail, we can still run the minimization program.[14] Again, the error in estimation increases. If many more channels of telemetry fail, the estimation process breaks down or gives results that are not at all reliable. These matters are subjects of research. Some discussion of these and other related advanced topics can be found in Wood and Wollenberg (1984), Schweppe and Handschin (1974), and Clements and Wollenberg (1975). In particular Abur, Kim and Celik (1995), Clements and Costa (1998), and Singh and Glavitsch (1991) discuss errors in network topological representation that arise from bad data regarding the open or close status of circuit brakers and switches.

Voltage Measurements

Problem: In sheet 2, the voltage measurements at the eight nodes along with the expected variances in their measurements are given. Considering these measurements in addition to those of Figure 8.24, estimate the state of the system.

Solution: We now minimize the least-square sum of not only the power flow measurements but of the voltages as well. Consequently, we do not need the constraint associated with the voltage bounds as in sheet 1. The objective function is now in cell L43. Note that this is a case of having redundant measurements.

The result in sheet 2 shows that the attained objective function has a value of 48.264. The objective function has two components: a component of value 6.398 corresponding to the voltage measurements, and a component of 41.866 corresponding to branch flows. This later component is more than that of sheet 1 when the voltage measurements were not considered. Compare the results of sheet 1 and sheet 2 and observe the differences in estimated parameters.

8.10 SUMMARY

This chapter has introduced the reader to the use of the Solver option of Microsoft Excel to solve nonlinear programs. It is important first to conceive physical problems as those amenable to formulation as nonlinear programs. Once that is done, in these days of availability of excellent commercial software, one can use any suitable software to obtain a solution.

[14] The reader is encouraged to enter zero in cells J21:J23 and observe the resulting estimates for branch flows and compare them to the measured flows.

Chosen example problems were predominantly from the power industry and are those pertaining to the dispatch and control of the power system. There are other activities of the power industry such as reservoir management and hydrothermal coordination in which nonlinear programs are frequently used. Obviously, we cannot discuss them all. However, we hope that the above examples have aided the reader to recognize problems in other areas as those that can be structured as nonlinear programs.

8.11 EXERCISE PROBLEMS

1. Consider the example of optimizing transmission voltage in Section 8.2.1. Solve the problem in the associated spreadsheet, assuming the following:

 (a) The cost of conductor is related to its resistance by the relation cost = $\$\frac{3.0\times 10^8}{R}$.

 (b) The cost of conductor is the same as in Section 8.2.1, but the cost of towers is 300,000 V.

 In the above two cases, what observations can you make regarding the Hessian matrix of (8.2)?

2. In the problem of Section 8.3.1, examine the effect of different expected market prices for stored coal on mining strategy. Study the answer sheet and the sensitivity sheets for different assumed values and make observations regarding possible expansion of mining and/or storage capabilities.

3. For the problem in Section 8.3.2, simulate the effect of ordinances that require the building to be setback from the boundary of the property by at least 30 feet. Study the sensitivity of the problem in Section 8.3.2. Which ordinance requirement (height or floor space) should be reduced to decrease construction cost. What is the sensitivity of the construction cost to the ordinance?

4. In problem 5 of Section 6.10, a solution to an unconstrained problem was sought. We now include some constraints.

 (a) Reconsider the same problem with the condition that the angle θ cannot exceed a value of $22\frac{1}{2}$ degrees. What is the optimum value of d. What is the value and the meaning of the Lagrange multiplier associated with this constraint?

 (b) If it was required that the value of shielding performance index should not be less than 0.32, one has to vary the horizontal distance x of the ground wire from the tower in Figure 6.10. Solve the problem to maximize the shielding performance index subject to it being better than 0.32, and θ not to exceed $22\frac{1}{2}$. Study the sensitivity sheet.

5. For the dispatch problem in Section 8.3.3 with generator costs as given in (8.3):

(a) Increase lower limit of generator #4 to must run at an output of 1.0. Solve the problem using Excel Solver. From the sensitivity sheets, debate the results vis a vis Lagrange multipliers and the slack variables x_i and y_i.

(b) What is the effect of total demand on Lagrange multipliers and x_i and y_i? Study this by requiring the total demand to be 14.0 in the spreadsheet "Dispatch KT equations.Xls" and analyze the results.

6. For the problem of Section 8.4.4, add the following additional branch flow constraints. The suggested values are: magnitude of flow in branch 4–3 $\leq \mp 0.6$ and in branch 2–3 $\leq \mp 0.5$. Examine the sensitivity sheets to obtain the LMP at each node. Compute the congestion rent if all injections into the network were paid at the rate of LMP at the point of injection and all loads were charged at the rate of LMP. Perturb branch flow limits from the above values and examine the effect on congestion rents.

7. Alter the network in the OPF example of Section 8.4 to 10 nodes by adding nonradial branches in any configuration you choose. Assume any reasonable admittance values for the added branches as well as injections and withdrawals at nodes. Solve the resultant OPF problem.

8. In problem 7, alter the admittances of appropriate branches (very high or very low value) and the new injections such that the network collapses to the earlier one of eight nodes in 8.4. Solve the OPF problem of this network "doctored" to eight nodes with original injections to confirm that the same solution as in 8.4 can be obtained.

9. Obtain the ac OPF problem from the spreadsheet of Section 8.4.4. Set the flow limits on branches 2–3, 3–4, and 7–8 to 0.4, 0.4, and 1.5, respectively. Set the flow limits on all other branches high enough such that they are not binding. Obtain the solution to the OPF problem in the spreadsheet to minimize the objective function (production cost).

The goal now is to compute by how much the transfer capabilities of the above confining branches should be upgraded. Making the transfer capabilities greater than required would be wasteful. Therefore, the upgraded capability must be a minimum amount in order to realize all the savings in dispatch costs that are available. Thus the objective now is to minimize the sum of production cost and the sum of increased transfer capabilities in the three branches.

(a) Modify the OPF problem to compute the needed increased capacity of the three restricting branches (individually and collectively) and the corresponding reductions in dispatch cost.

(b) How would the answer change if costs for transfer capacity upgrades were assigned weighting factors of 10, 20, and 40 to branches 2–3, 3–4, and 7–8, respectively?

(c) Examine the sensitivity sheet. What observations can you make from the Lagrange multipliers associated with constraints pertaining to the three transfer capabilities, given that the procedure minimizes a multiobjective function?

Hint: See Rau (2002).

10. This exercise is designed to illustrate ways of computing incremental loss associated with changes in generation.

(a) Consider the example in Section 8.4. Find the optimal dispatch to minimize generation cost:

i. Assume that branch flow transmission capabilities are unlimited.
ii. Flow in branch 7–8 is restricted to 1.6 in either direction as in Section 8.4.4.

For both cases, compute (from sensitivity sheet) and compare the LMP given by $\frac{\partial \pi}{\partial g_i}$, $i = 1, 8$, where g_i is the demand or generation at node i.

(b) In the above, set all the branch resistances to zero and compute the LMP. Compute the difference between payments to injections into the grid (LMP times quantity injected) and the collection from withdrawals (quantity times LMP). Do the differences between these LMP and those of 10a above represent the loss component of LMP?

(c) In the above network with unlimited and limited flow capability, set the objective π to minimize total network loss. Perturb p_1 to p_8 one by one, and note the change in network loss. Compute $\partial\pi/\partial g_i$, $i = 1, 8$ therefrom. Also compute $\partial\pi/\partial g_i$, $i = 1, 8$, from the sensitivity sheet. Compare the two answers obtained.

(d) Find the optimum dispatch to minimize network loss, but increase all nodal demands by a very small amount, and add an additional set of constraint requiring that the generator outputs be greater than or equal to those of Section 8.4. From the Lagrange multipliers, compute $\partial\pi/\partial g_i$, $i = 1, 8$.

Can you suggest a method of pricing and allocating responsibility for losses to generators, loads, or generators and loads? Compute the revenue if the loss responsibility at a node is deemed to be equal to consumption times the marginal loss factor, and is priced at LMP. Compare revenue from any other proposed method with the above value.

Study and compare all the above sensitivity coefficients with a view to understand the reasons for differences. Study the changes in these coefficients if the dispatch of reactive power is also optimized as in Section 8.6.

APPENDIX A

BASIC PRINCIPLES OF ELECTRICITY

A.1 SCOPE

The following description of basic principles is neither rigorous nor complete. It is directed to readers who have not had training in engineering. We start out from basic principles of physics to move quickly to focus on the development of active and reactive power concepts. Those desirous of a more rigorous study of these principles may consult Duckworth (1960).

Evidently, this section of the appendix is redundant to engineers, particularly to those who are electrical engineers.

A.2 CHARGE, ELECTRIC, AND MAGNETIC FIELDS

Matter consists of atoms. The accepted model for an atom has the nucleus at its center and electrons orbiting the nucleus in elliptical orbits at some distances away in what is called shells. The nucleus consists of protons, neutrons, and other elementary particles. Protons have a positive charge, and electrons have a negative charge. The combined charges of these elementary particles cancel, rendering matter that has no net charge in its neutral state.

In certain elements, particularly metals and chemical solutions, a charge separation can be created by the expenditure of energy. For example, in a chemical solution such as in a battery, the negative electrode has an excess of electrons whereas the positive electrode has a deficit of electrons. In a metallic conductor

Optimization Principles, by Narayan S. Rau
ISBN 0-471-45130-4

moving in a magnetic field, a potential difference is created. By potential difference we mean an excess of electrons in one location and a deficit in another so that there is a charge difference between these two locations. This potential difference, or charge density difference, between locations is called the voltage. To produce charge separation or voltage difference requires the expenditure of energy.[1]

Charge has the property that like charges repel and unlike charges attract. Consequently, if there is a net potential difference between two points, any other charge between these two points experiences a net force. A movement or flow of charges results from such a force. Such a flow of charges is called the flow of current, measured in amperes.

In a metallic conductor that is subjected to a charge density difference between two points (voltage difference), a force exists that propels electrons through the atoms of the conductor causing current flow. For example, in a battery or a generator, if there is an external circuit or connection between points that have a voltage difference, a current flow results such that the excess or deficit of charges may be neutralized.

An electric field surrounds a charge. The meaning of the field is that any other charge in the field space experiences a force. Just as in the gravitational field, the force exerted on a charge q_1 by another charge q_2 is the product of the two charges divided by the square of the distance between them as $F = \frac{q_1 q_2}{d^2}$.

Magnetic field is created by current flow. When there is a rate of change of magnetic field, it induces a voltage in objects in the field. If the object is metallic and if the circuit is completed externally, then a time-varying magnetic field produces a current in the circuit. For example, in the case of the coil with N turns as in Figure A.1, the voltage induced by varying magnetic field is given by $N\frac{d\phi}{dt}$, where $\frac{d\phi}{dt}$ represents the time rate of change of the magnetic field. Evidently, if the coil is

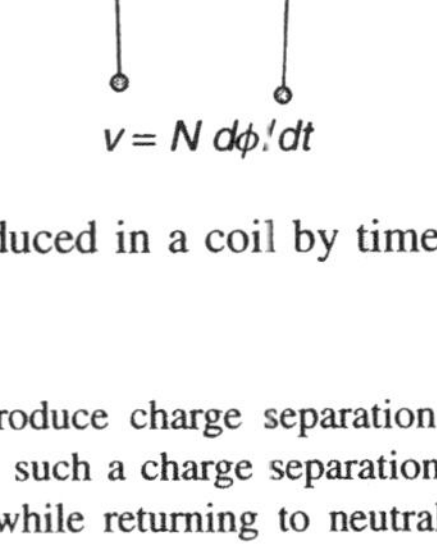

Figure A.1. Voltage induced in a coil by time-varying magnetic field.

[1]This observation is axiomatic. To produce charge separation from the equilibrium state of matter requires energy, because by producing such a charge separation, under appropriate conditions, current and resulting work can be produced while returning to neutral state. Hence, law of conservation of energy requires that to disturb matter from its natural state requires expenditure of energy.

connected suitably, the induced voltage can be made to result in a flow of current in an external path or circuit to do useful work.

Since magnetic field can be created by a flow of current, by varying the current in a coil, a varying magnetic field can be produced. The varying magnetic field can, in turn, induce a voltage in the coil producing the magnetic field as well as in any another conductor or coil linking its field. Consider another coil in the field produced by a coil (or a winding). The induced voltage in such a coil is given by $v = M\frac{di}{dt}$, where v is the voltage, M is called the mutual inductance between the two coils, the value of which is related to their physical dimensions and placement in the field, and $\frac{di}{dt}$ is the rate of change of current in the coil that produces the magnetic field.

If we now consider the coil producing the magnetic field, a voltage opposing the current that is trying to establish the magnetic field is induced in it. This is called Lenz's law, which gives the value of self-induced voltage as $v = -L_s\frac{di}{dt}$, where v is the self-induced voltage, L_s is called the self-inductance of the coil which is related to its physical dimensions and orientation in the field, and $\frac{di}{dt}$ is the rate of change of current in it. Note that the induced voltage, because of the negative sign, opposes the applied voltage that is trying to establish the current.

In a similar vein, time variation of electric charge induces a voltage in an object located in its electric field. The induced voltage v is proportional to the time rate of change of charge given by $\frac{dv}{dt} = S\frac{dq}{dt}$, where S is a constant (reciprocal of capacitance) related to the geometry of the object, and $\frac{dq}{dt}$ is the time rate of charge variation. The object in which the voltage is induced can be metallic or nonmetallic; but only if an external circuit is completed, a current can flow through metallic objects.

A.3 VOLTAGE, CURRENT, AND POWER

The unit of measurement for charge is the coulomb. A force is exerted on a unit charge that experiences an electric field. Movement of charge, if against this force, requires energy. If the energy expended in moving a unit charge between two points against such a force is one joule, then the potential difference between the two points is said to be one volt. Hence, two points differ in potential by v volts if v joules of energy is required to move one coulomb from one point to another.

Current flow results from the flow of positive or negative charges. A flow of charge of one coulomb per second is the unit of current called ampere. From the definition of voltage, it is clear that if more charges (coulombs) move in a field, more energy is expended. Hence the energy per second is the product of voltage times the number of coulombs moved per second. Since the latter is the definition of current (amperes), the energy (or work done) per second is the product of voltage and current. Since this work is done in one second, it represents rate of work done per second. Therefore it is the power (work done per second), which is given by the product of voltage and current. The unit of measurement for power is watts.

Energy in Electric and Magnetic Field. The magnetic and electric fields are created by charge separation, which requires energy to do so. Hence, energy is stored in these fields. The energy in the magnetic field is proportional to the square of the current producing it. It is given by $\xi = \frac{1}{2}Li^2$, where ξ is the energy.

Similarly, energy is stored in electric field. Electric field represents force per unit charge and is designated by E. The energy stored in an electric field is given by $\xi = \frac{\varepsilon E^2}{2}$ joules/m^3, where ε is called the dielectric constant, which is a property of the medium in which the field exists. This expression for energy gives the volume density of energy occupied by an electric field. The energy in a fixed and bounded region of space can be expressed as $\xi = \frac{Cv^2}{2}$ joules, where C is the capacitance defined by the geometry of the region of field and v is the voltage producing the field. The latter expression is generally used for capacitors and transmission lines, the geometry of which is defined.

Kirchhoff's Laws and Circuits. Kirchhoff's laws arise from the concept of continuum mechanics; that is, charges cannot be destroyed. Hence, in an electrical circuit, at any junction (node), the moving charges divide and flow along different paths. The fact that charges are not destroyed is stated as follows: The sum of currents in a node (flowing into it and out of it) is zero. This is not different from the flow of materials or fluids in a pipeline (assuming no leakage!).

The second point is that given that voltage represents the force on charges, the flow of charges is resisted by the internal atomic structure of metallic conductors. This resistance to flow is measured in units of ohms. Hence, for a given voltage (force), the flow of charges is inversely proportional to the resistance. This is expressed as $i = \frac{v}{R}$, where i is the current in amperes, v is the voltage, and R is the resistance.

By a mere application of the above two principles, one can obtain a solution to any direct current (dc) electrical circuit. The parameters are voltages at nodes and currents in branches. The simulation involves a system of linear equations with some specified parameters of voltages and currents. The solution requires a determination of remaining parameters.

A.4 VOLTAGE CURRENT AND POWER IN AC CIRCUITS

We now briefly extend the above concepts to alternating current (ac) circuits.

Generators in ac transmission systems generate a sinusoidally time-varying voltage expressed as

$$v(t) = V_{\max} \sin(\omega t),$$

where ω is the angular frequency and $V_{\max}$ is the instantaneous maximum value of the voltage.[2] The angular frequency of 60-Hz voltage is 376.99 radians/second. Figure A.2 shows a voltage of magnitude $V_{\max} = 10$.

[2] In describing networks, we follow the convention that lowercase letters signify the instantaneous values. Uppercase letters indicate constants (such as R or X), or other quantities identified by appropriate

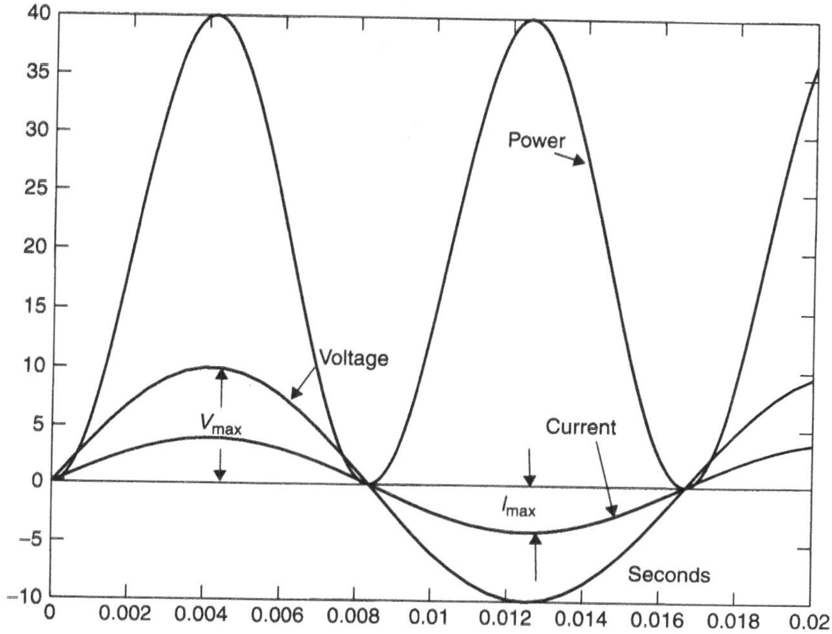

Figure A.2. Instantaneous voltage, current, and power in purely resistive circuit.

If one has an electric circuit, which is purely resistive, since current by Kirchhoff's law is given by voltage divided by resistance, the instantaneous current is also sinusoidal in phase with the voltage. A current of magnitude $I_{max} = 4$ is shown in the figure which corresponds to a resistance of 2.5 ohms. Note in this purely resistive circuit that Kirchhoff's laws are obeyed at every instant. Therefore, instantaneous current is given by instantaneous voltage divided by the resistance.

Since power is the product of voltage and current, the expenditure of energy in each instant is also shown in the figure. The power at all instants is positive. The energy expended in a second is the time integral, or the area under the power curve over a duration of one second.

Given the magnitudes of voltage and current, how do we compute this time integral which is the energy expended in a second? It is not $V_{max} I_{max}$. But it is given by $\frac{I_{max}}{\sqrt{2}} \frac{V_{max}}{\sqrt{2}}$.[3] The quantities $\frac{I_{max}}{\sqrt{2}}$ and $\frac{V_{max}}{\sqrt{2}}$ are called the RMS value or the effective values of sinusoidal current and voltage. The former terminology RMS stands for the *root mean square* value since this value ($\frac{I_{max}}{\sqrt{2}}$ or $\frac{V_{max}}{\sqrt{2}}$ is obtained by evaluating the square root of the mean of the instantaneous sums of the square of voltage or current. In the electric power industry, when one says that the current or voltage is

subscripts such as I_{max} or I_{rms}. However, for RMS quantities of voltage and current, we omit the subscript and express them simply as uppercase V and I.

[3]The confirmation of this by performing necessary integration is left to the reader.

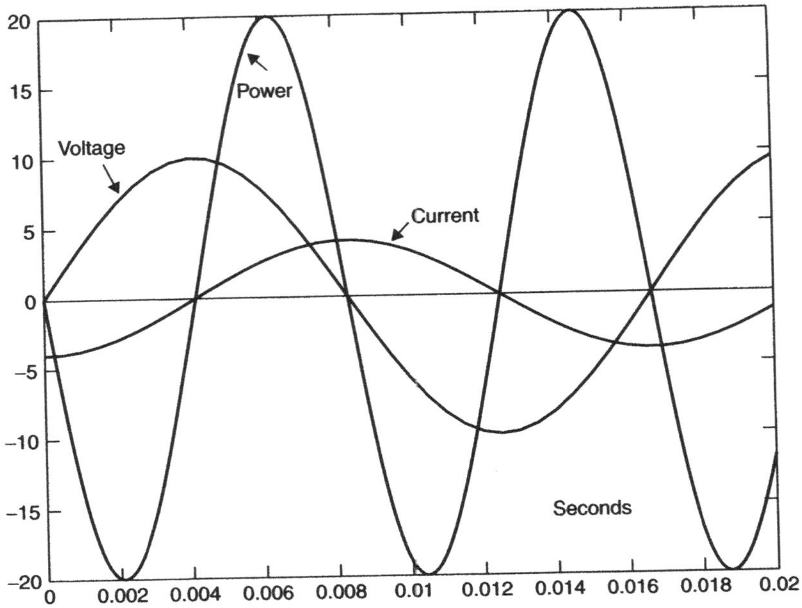

Figure A.3. Instantaneous voltage, current, and power in purely inductive circuit.

some value, it is implied that it is the RMS or effective value. Consequently, for a resistive load, if the applied voltage is 50 kV and the current is 10 kA, we say that the energy consumed is 500 kW per second.

It can be shown that power lost as heat in a conductor or consumed in a resistive load is the product of current square times the resistance. Hence $p = i^2R$.

Now consider a circuit which consists only of inductance in the form of a coil.[4]

Figure A.3 shows that the current lags the voltage by $\frac{\pi}{2}$ radians. A proof of this is omitted. Conceptually, however, it can be explained as follows. Because of the relations in Section A.2, the applied voltage has to be equal to $L\frac{di}{dt}$. Consequently, $\frac{di}{dt}$ or the slope of the current (rate of change) is a maximum when the voltage is at a maximum. A careful examination of the figure observing the voltage magnitude and the related rate of current increase will explain why the current lags the voltage in phase angle by $\frac{\pi}{2}$. Recall that the rate of change of a sinusoid is at its maximum when its value is zero.

The instantaneous power curve in Figure A.3 is obtained by multiplying the voltage and current plots. It has positive and negative components. The positive components indicate that the energy is supplied from the voltage source to the coil to establish the magnetic field. The negative component indicates that the energy

[4]It is a theoretical concept to think of only inductance and zero resistance. A conductor has zero resistance only in a super conducting state when it is cooled to cryogenic temperatures. Nevertheless, this example illustrates the effect of a very small value of resistance.

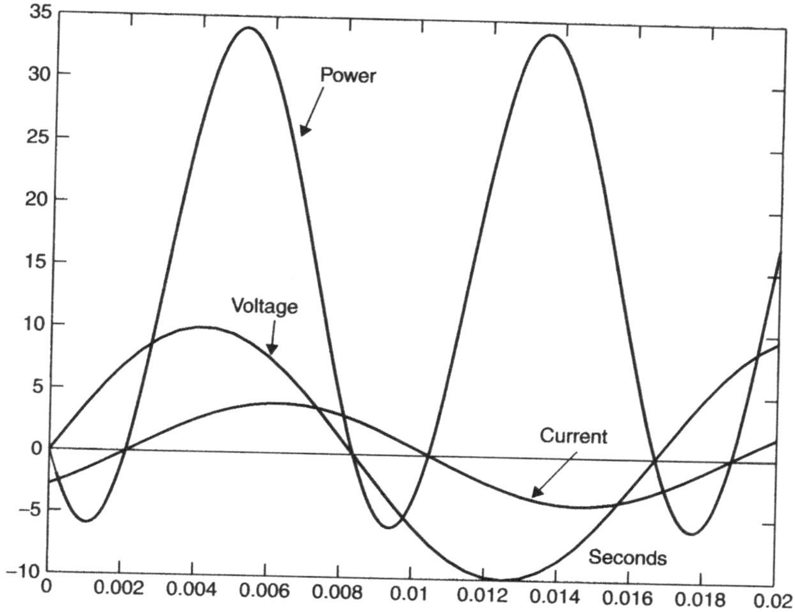

Figure A.4. Instantaneous voltage, current, and power in resistive-inductive circuit.

stored in the magnetic field is returned to the source. Since there is no resistance in the circuit, no net energy is expended; hence the average value of energy consumed (area under the power curve) is zero.

However, in a circuit where there is resistance in addition to inductance, the current lags the voltage by an angle less than $\frac{\pi}{2}$. The angle of lag is given by $\tan^{-1}\frac{\omega L}{R}$, where w is the angular frequency and L and R are the inductance and resistance of the circuit. This angle of lag is called the phase angle.Then the instantaneous power is as indicated in Figure A.4. It has positive and negative components and an average value that is positive representing the power consumption in the resistance. The negative component is indicative of the energy stored in the magnetic field that is returned to the source periodically.

A similar phenomenon of interchange of energy occurs from the electric field established in the electric circuits with capacitance. In circuits with capacitance, the energy stored in the electric field is returned to the source for parts of the cycle. We omit a detailed description of this except to say that capacitance makes the current lead the voltage waveform in phase angle.

A.4.1 Active and Reactive Power

From Figures A.3 and A.4 we see that energy is not delivered into the circuit (load) during the time when the stored energy in the electric and magnetic field is returned to the source. Notice that the instantaneous direction of current is not necessarily negative (that is, from the circuit to the source) when energy is being returned to the source. However, when energy is returned to the source, the instantaneous direction of the current is opposite to that of the voltage.

As said earlier, the current flowing in the circuit expends energy that is given by i^2R. Since there is resistance in transmission lines and other conductors, energy is expended not only in the consumers' apparatus called the load, but also as heat in conductors. Because of the squaring of current in the expression for energy, irrespective of the direction of current, energy is expanded as heat in the resistance of conductors even when the energy from magnetic or electric field is being returned to the source.

One would like to express the energy delivered into the circuit (to end use as well as losses in the conductors) in terms of the effective values of voltage and current. If we examine the instantaneous power curve in the above figures, we need an expression that indicates how much energy was delivered in one second into the circuit for end use (and heat in conductors called losses) and how much was returned to the source. The latter being the energy stored in electric and magnetic fields of the circuit is a mere periodic exchange between fields setup in the circuit and the source of supply. This means that for part of the time, energy is derived from the source to establish the fields; subsequently for the remaining part of time (in one cycle), energy is returned. Hence, no active work is done in the circuit. Since this energy is in reaction to the cyclic establishment of electric and magnetic fields, it is called reactive power. Reactive power is represented as var, which stands for volt-amperes reactive. Useful power delivered to the circuit (losses and end use) is called active power. Active power is represented as watts.

Interested readers may consult books on electric circuits such as Howatson (1996) for details that show the calculus of integration of the instantaneous power curves to derive an expression for active and reactive power. The end results are important. They are

$$\text{Active power (watts)} = VI\cos(\phi)$$

and

$$\text{Reactive power (var)} = VI\sin(\phi),$$

where V and I are the effective values of voltage and current, and ϕ is the phase angle of the current in relation to the voltage.

Furthermore, the Pythagorean addition (square root of sum of squares) of active and reactive power is called the complex power **S** and is designated in volt-amperes (VA) since it is nothing but a product of the effective values of voltage and current.

A.5 APPLICATION TO NETWORKS

The current flow in a branch is given by the difference in voltage in phasor form at the two ends of the branch divided by the impedance of the branch. Analogous to the resistance in dc circuits we use the value of impedance, a complex number, because division of voltage by impedance gives not only the magnitude of the current but its phase relationship as well. In Section B.1, equations that describe the relation between voltages and currents are derived for general networks. Then, voltages and currents (assumed to be effective values) are multiplied to compute complex power **S**. Subsequently, using the above concepts, active and reactive power is calculated. As said earlier, the symbol for active power is P, and the symbol for reactive power is Q; complex power, which is a Pythagorean addition of P and Q, is represented as **S**. In the notation of complex algebra, $\mathbf{S} = p + jQ$, where $j = \sqrt{-1}$.

APPENDIX B

NETWORK EQUATIONS

In the following, we derive equation to the power flow in a general network. There are several books dealing exclusively with network analysis and power flow (Howatson, 1996; Stagg and El-Abiad, 1968; Wallach, 1986; Wood and Wollenberg, 1984). Our intent here is only to outline the rudiments of network analysis to the extent necessary to formulate and solve optimal power flow (OPF) problems using Excel Solver. It is assumed that the reader is familiar with complex algebra, as well as with the notations in common usage by electric power engineering professionals. Those desirous of delving deep into the subject matter of networks are advised to consult Howatson (1996), Stagg and El-Abiad (1968), Wallach (1986), Wood and Wollenberg (1984).

B.1 FLOW IN BRANCHES

In the following development, complex RMS quantities are represented as capital letters. For example, V_i represents complex voltage at node i. The angle of the complex quantity is represented by $\angle$ and its magnitude by $|V_i|$. Similarly, Y represents complex admittance.

Consider a single branch connecting two nodes i and j as portrayed in Figure B.1. By Kirchhoff's laws, the current I_{ij} (complex, RMS value) from i to j is given by

$$I_{ij} = \frac{V_i - V_j}{Z_{ij}} = Y_{ij}\left(V_i - V_j\right). \tag{B.1}$$

Optimization Principles, by Narayan S. Rau
ISBN 0-471-45130-4

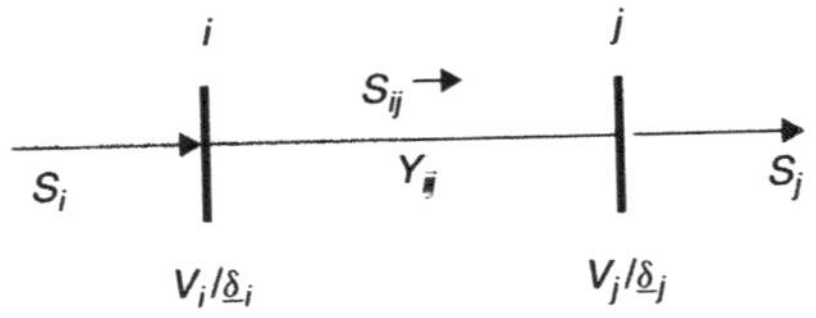

Figure B.1. Complex power flow in a branch.

In (B.1), V_i, V_j represent complex voltages at nodes i and j, and Z_{ij}, Y_{ij} the impedance and admittance between these nodes. The complex power S_{ij} flowing in the branch form i to j (this is the same as the injected power at node i) are given by the product of the voltage conjugate and current[1] as

$$S_{ij} = S_i = V_i^* I_i = V_i^* Y_{ij}(V_i - V_j), \tag{B.2}$$

which can be expanded as

$$S_i = S_{ij} = Y_{ij}|V_i|^2 - Y_{ij} V_i^* V_j, \tag{B.3}$$

which when written in its complex form of magnitude and angle gives

$$S_i = S_{ij} = |Y_{ij}||V_i^2|\angle\theta ij - |Y_{ij}||V_i||V_j|\angle(\theta_{ij} + \delta_j - \delta_i), \tag{B.4}$$

where θ_{ij} represents the angle of the admittance Y_{ij}, and δ_i, δ_j represent the angle of the voltages at nodes i and j.

By similar reasoning, one obtains

$$S_j = -S_{ij} = |Y_{ij}||V_j^2|\angle\theta_{ij} - |Y_{ij}||V_i||V_j|\angle(\theta_{ij} + \delta_i - \delta_j). \tag{B.5}$$

If there are several branches emanating from a node—for instance, three branches as in Figure B.2—the total power injected into the node (positive by convention if injected into the node) is simply given by the sum of branch power flows obtained from (B.4) for each branch. These branch power flows represent flows away from the node, which are negative according to the adopted convention. Consequently, the negative of the sum of flows away from the node is the injected power into the node given by

$$\begin{aligned} S_i = -(S_{ij} - S_{ik} - S_{il}) = {} & |Y_{ij}||V_i^2|\angle\theta_{ij} - |Y_{ij}||V_i||V_j|\angle(\theta_{ij} + \delta_j - \delta_i) \\ & +|Y_{ik}||V_i^2|\angle\theta_{ik} - |Y_{ij}||V_i||V_k|\angle(\theta_{ik} + \delta_k - \delta_i) \\ & +|Y_{il}||V_i^2|\angle\theta_{il} - |Y_{il}||V_i||V_l|\angle(\theta_{il} + \delta_l - \delta_i). \end{aligned} \tag{B.6}$$

[1]Complex power S can be expressed either as $S = VI^*$ or $S = V^*I$. In either case, the sign of the real power is the same. The sign of the reactive power, however, is different for the two definitions. For the former—that is, $S = VI^*$—the sign of lagging reactive power (current lags the voltage) is positive. For the latter definition, sign of lagging reactive power is negative. In this book, we use the definition $Z = V^*I$ consistent with the convention adopted in Stagg and El-Abiad (1968) and Stevenson (1962). All the Excel spreadsheet calculations have been programmed according to this convention.

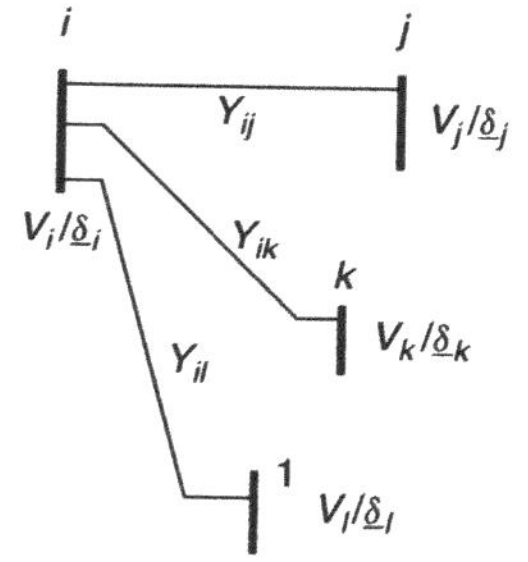

Figure B.2. Illustration of three branches at a node.

By collecting terms containing $|V_i^2|$ in the above, the final expression for complex power injected into the network at a node can be written as

$$S_i = |Y_{ii}||V_i^2|\angle\theta_{ii} - |Y_{ij}||V_i||V_j|\angle(\theta_{ij} + \delta_j - \delta_i)$$
$$- Y_{ik}||V_i||V_k|\angle(\theta_{ik} + \delta_k - \delta_i) - |Y_{il}||V_i||V_l|\angle(\theta_{il} + \delta_l - \delta_i), \quad \text{(B.7)}$$

where Y_{ii} represents the nodal admittance at node i whose value is equal to the sum of admittances of all branches that terminate at this node.

In a similar manner, injected power at any other node to which many branches are connected can be written as the sum of all branch flows out of that node.

The total injected current into the network at node i is

$$I_i = Y_{ij}(V_i - V_j) + Y_{ik}(V_i - V_k) + Y_{il}(V_i - V_l), \quad \text{(B.8)}$$

where terms can be combined to obtain

$$I_i = (Y_{ij} + Y_{ik} + Y_{il})V_i - Y_{ij}V_j - Y_{ik}V_k - Y_{il}V_l. \quad \text{(B.9)}$$

If we consider all nodes $i = 1, n$ of the network, the set of equations for the current is

$$\begin{bmatrix} I_1 \\ I_2 \\ \vdots \\ I_n \end{bmatrix} = \begin{bmatrix} \sum\limits_{n=1,\text{all br.}} Y_{1n} & -Y_{12} & -Y_{13} & \cdots & -Y_{1n} \\ -Y_{21} & \sum\limits_{n=1,\text{all br.}} Y_{2n} & -Y_{23} & \cdots & Y_{2n} \\ \vdots & \vdots & \vdots & \vdots & \vdots \\ -Y_{n1} & -Y_{n2} & \cdots & \cdots & \sum\limits_{n=1,\text{all br.}} Y_{nn} \end{bmatrix} \begin{bmatrix} V_1 \\ V_2 \\ \vdots \\ V_n \end{bmatrix}. \quad \text{(B.10)}$$

The equation for injected current at nodes in terms of the nodal voltages is written in compact form as

$$\mathbf{I} = [\mathbf{Y}]\,\mathbf{V}, \quad \text{(B.11)}$$

where [**Y**] is called the network admittance matrix. The pattern of the admittance matrix is as follows. Its diagonal entries are the sum of admittances of all branches terminating at that node. This is called the driving point admittance of the network at that node. Its off-diagonal terms are negative and their values are equal to the branch admittances. Therefore, if there is no branch connecting two nodes i and j, the value of the entry in the ijth term of the admittance matrix is zero. Clearly, in a large network, only a few branches, say four or five, emanate from a node. Consequently, for an n node network, the $n \times n$ network admittance will be sparse (many zero terms). As a consequence of the above, the value of the diagonal entries is equal to the negative sum of all the off-diagonal entries along that row.

Since the row sums of the network admittance matrix is zero, it is a singular matrix and cannot be inverted. However, as we shall see later, the choice of a reference bus allows us to eliminate a corresponding row and column of the admittance matrix. Such a reduced matrix is nonsingular. Furthermore, if line capacitance is considered as in Figure B.3, where it is represented as a "pi equivalent," the admittance matrix will not be singular. This is because the diagonal terms, in addition to the sum of branch admittances will have the total line capacitance at the node added to them. In this instance, since there are three branches emanating from node i, the diagonal term for node i will be equal to $Y_{ij} + Y_{ik} + Y_{il} + 0.5(C_{ij} + C_{ik} + C_{il})$, where C_{ij} is the admittance due to the capacitance of the branch ij. Thus, the row sum of the admittance matrix will no longer be zero, rendering the matrix nonsingular.[2]

Equations for Power. As said earlier, complex power S is represented by $S = V^*I$. Therefore, from (B.11), the expression for power is obtained as

$$\mathbf{S} = \text{Diag}\left[\mathbf{V}^*\right] \mathbf{I}, \tag{B.12}$$

where **S** represents a vector of the total complex power injected into the network at nodes,[3] Diag [**V**] is a diagonal matrix of voltage at nodes, and **I** is a vector of currents injected into the network at nodes given by (B.10). A substitution of (B.10) for **I** into this gives the expression of (B.7) for injected complex power. The real and imaginary parts of complex power, real and reactive powers, are obtained

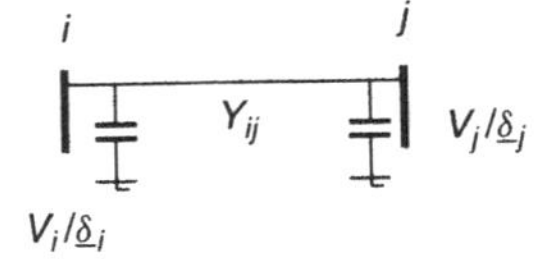

Figure B.3. Representation of line capacitance.

[2]In the examples shown in this book, line capacitance is not included.

[3]It was shown earlier that the total injected power into a node is the sum of flows in all branches out of that node, plus (of course) any local consumption (load) at that node.

from (B.7) as

$$P_i = |Y_{ii}||V_i^2|\cos(\theta_{ii}) - |Y_{ij}||V_i||V_j|\cos(\theta_{ij} + \delta_j - \delta_i) \\ - |Y_{ik}||V_i||V_k|\cos(\theta_{ik} + \delta_k - \delta_i) \\ - |Y_{il}||V_i||V_l|\cos(\theta_{il} + \delta_l - \delta_i), \tag{B.13}$$

and

$$Q_i = |Y_{ii}||V_i^2| sin(\theta_{ii}) - |Y_{ij}||V_i||V_j|\sin(\theta_{ij} + \delta_j - \delta_i) \\ - Y_{ik}||V_i||V_k|\sin(\theta_{ik} + \delta_k - \delta_i) \\ - |Y_{il}||V_i||V_l|\sin(\theta_{il} + \delta_l - \delta_i). \tag{B.14}$$

Observe the pattern of these equations; they contain the square of the voltage of the node under consideration and the product of voltage of that node and of all other nodes connected to it. The equations are nonlinear. The nonlinearity is introduced on two accounts: (a) the product of nodal voltages and (b) the sine and cosine of voltage angles. A more detailed discussion of the network equations can be found in several references such as Stagg and El-Abiad (1968).

In all, there are two equations per node, one for power and the other for reactive power. There are four variables at each node: P_i, Q_i, $|V_i|$, and δ_i. In an n node network, there will be $2n$ equations for $4n$ variables. Therefore, in order to obtain a solution, we specify $2n$ variables such that we can solve for remaining $2n$ variables using $2n$ equations.

In power engineering practice, at nodes (buses) where there is a demand for power (load bus), the values for P and Q are specified. At nodes where a generator is connected (generator bus), since a generator can normally control the voltage at its terminals via an automatic voltage controller, the terminal voltage and the real power output of the generator are specified. Consequently, the solution of the power flow equation obtains values for other $2n$ unknowns: (a) the voltage and angle at load buses and (b) the angle and reactive power at generator buses.

Before a discussion of the solution of the power flow equation for the remaining $2n$ unknowns, we digress to examine the formulation and solution of the simplified linear version of the power flow equations.

B.2 THE DC LOAD FLOW

The linearized version of the power flow equations (B.13) and (B.14) is called the dc load flow. Here, it is assumed that the admittances are purely imaginary—that is, that there is no resistance in the transmission network. Consequently, a lossless network is presumed. Then, the angle of the admittances in (B.13) will be $-\frac{\pi}{2}$. With this assumption, $\cos(\theta_{ij} + \delta_j - \delta_i) = \sin(\delta_i - \delta_j)$, which transforms (B.13) to

$$P_i = -Y_{ij}|V_i|V_j|\sin(\delta_j - \delta_i) - Y_{ik}|V_i|V_k|\sin(\delta_k - \delta_i) - Y_{il}|V_i|V_l|\sin(\delta_l - \delta_i). \tag{B.15}$$

If the angle differences are assumed to be small (i.e., $\delta_i - \delta_j \simeq 0$), we have $\sin(\delta_i - \delta_j) = (\delta_i - \delta_j)$. Furthermore, if all voltages are equal[4], and generally assumed to have a value of one per unit (i.e., $|V_i| = |V_j| = \cdots = |V_n| = 1$), equation (B.15) further simplifies to

$$P_i = -Y_{ij}|V_i||V_j|(\delta_j - \delta_i) - Y_{ik}|V_i||V_k|(\delta_k - \delta_i) - Y_{il}|V_i||V_l|(\delta_l - \delta_i), \quad \text{(B.16)}$$

or

$$\begin{aligned} P_i &= Y_{ij}|V_i||V_j|\delta_i + Y_{ik}|V_i||V_k|\delta_i + Y_{il}|V_i||V_l|\delta_i \\ &\quad -Y_{ij}|V_i||V_j|\delta_j - Y_{ik}|V_i||V_k|\delta_k - Y_{il}|V_i||V_l|\delta_l, \end{aligned} \quad \text{(B.17)}$$

and equation (B.14) vanishes because of the assumption that the admittance angle is $-\frac{\pi}{2}$ and that $\delta_i - \delta_j \simeq 0$.

Since $|V_i| = |V_j| = 1$, equation (B.17) reduces to

$$P_i = (Y_{ij} + Y_{ik} + Y_{il})\delta_i - Y_{ij}\delta_j - Y_{ik}\delta_k - Y_{il}\delta_l. \quad \text{(B.18)}$$

The first term in this equation represents the nodal or driving point admittance. Equation (B.18), written for all nodes, represents a set of linear equations, which can be written as

$$\mathbf{P} = [\mathbf{Y}]\,\boldsymbol{\delta}, \quad \text{(B.19)}$$

which gives

$$\boldsymbol{\delta} = [\mathbf{Y}]^{-1}\,\mathbf{P}. \quad \text{(B.20)}$$

In the above, $[\mathbf{Y}]$ is the network admittance matrix with nodal admittance in its diagonal and possessing other properties described earlier, $\boldsymbol{\delta}$ is a vector of angles $[\delta_1, \delta_2 \cdots \delta_n]$, and $\mathbf{P}$ is a vector of injected power into the network at nodes given by $[P_1, P_2, \ldots, P_n]$. The solution of (B.20) gives angles $\boldsymbol{\delta}$ of node voltages from which line flows can be calculated using the magnitude of branch admittance as

$$P_{ij} = \left|Y_{ij}\right| (\delta_i - \delta_j). \quad \text{(B.21)}$$

Hence, in a dc load flow, line power flows are dependent only on the differences between the angles of voltage at modes.

Net Injected Power. The above equations represent the net injected power into the network at a node. At some nodes, both generation and load can be present. Consequently, the net injected complex power into the network at node i is given by

$$S_i = S_{ig} - S_{iL}, \quad \text{(B.22)}$$

where the subscripts g and L stand for generation and load. The real and imaginary components of this, as before, give real and reactive power injections.

[4]This assumption is called a "flat start," meaning that a plot of voltages at all nodes is flat.

B.3 SOLVED PROBLEMS

B.3.1 The dc Load Flow

Figure B.4 shows a printout of the program "APP-dcflow.Xls." The network is also indicated in this figure and is identical to that of Figure 4.9. This network is used in many illustrations found in this book. **App-dcflow.Xls**

The admittance matrix is shown in cells B8:I15. The admittance is purely imaginary (no resistance component). Study the values of branch admittances and the network topology of connections in relation to the admittance matrix. For example, in the first row, there are off-diagonal entries only in cells C8, F8, and H8 since node 1 is connected only to nodes 2, 5, and 7. As indicated for the structure of the admittance matrix, the diagonal entrees representing driving point admittance at that node is the negative sum of off-diagonal entries.

Node 8 is arbitrarily chosen as the reference. The angle δ_8 at his node is the reference (normally zero) against which the angles of all other nodes are computed. Consequently, the eighth row and column of this matrix is eliminated. While matrix [**Y**] is singular, the reduced matrix by this elimination of column and row is not singular. Matrix $\left[\mathbf{Z}'\right]$ is the inverse of the reduced matrix shown in cells B18:H24. The angles of nodes are computed per (B.20) as a product of $\left[\mathbf{Z}'\right]$ and $\mathbf{P}_{inj}$ vector that is shown in cells M18:25.

The required generation and load demand at nodes are shown in B3 to I4. Row 5 which is a difference between the generation and demand at each node per (B.22) is the net injection into the network. This is also shown as the injection vector $\mathbf{p}_{inj}$ in cells K18:K25.

The branch flows can be computed once the angles are known using (B.21). From (B.19) we observe that flow in branches—for example, branch 1–2—is given by $P_{12} = |Y_{12}|\,(\delta_1 - \delta_2)$. This is computed in cell M9 as 0.6945. The positive sign associated with the flow 1–2 indicates that the flow is from node 1 to 2. Had it been the other way, its sign would have been negative.

We now introduce the concept of ***shift factor matrix*** to calculate all branch flows. We search for a shift factor matrix such that the branch flows are given by

$$\mathbf{P}_{br} = [\mathbf{S}_F]\mathbf{P}_{inj}, \tag{B.23}$$

where $\mathbf{P}_{inj}$ is a vector of nodal power injections into the network, and $\mathbf{P}_{br}$ is a vector of power flows in branches.

In the spreadsheet under consideration, $[\mathbf{Y_{br}}]$ is the branch admittance matrix which has diagonal entries of admittance values for all the branches (12 in this example) of the network. Matrix [**A**] is a node-branch incidence matrix with entries of 1 for the starting node and −1 for the terminating node. For example, for branch 1–5, we write 1 at node 1 and −1 at node 5. Then it can be easily shown that the matrix $\mathbf{S}_F$ is given by

$$[\mathbf{S}_F] = [\mathbf{Y_{br}}]\,[\mathbf{A}]\left[\mathbf{Z}'\right]. \tag{B.24}$$

DC Load flow formulation

nodes	1	2	3	4	5	6	7	8	
P	2	0	3.5	0	2.8	2	0	3	*Specified Generation*
L	1	2.5	1.7	3.5	0	2.5	2.1	0	Specified Demand
	1	-2.5	1.8	-3.5	2.8	-0.5	-2.1	3	Net injection

Y Matrix - B (Imgy.)COMPONENT OF Y

	1	2	3	4	5	6	7	8
1	54	-20	0	0	-20	0	-14	0
2	-20	78	-15	0	-25	-18	0	0
3	0	-15	86	-40	0	-21	0	-10
4	0	0	-40	90	-20	0	-30	0
5	-20	-25	0	-20	65	0	0	0
6	0	-18	-21	0	0	74	0	-35
7	-14	0	0	-30	0	0	89	-45
8	0	0	-10	0	0	-35	-45	90

Branch 1-2 Flow

0.694

Y Inverse =Z'

0.04	0.02	0.01	0.02	0.03	0.01	0.01	0
0.02	0.03	0.02	0.02	0.02	0.01	0.01	0
0.01	0.02	0.03	0.02	0.02	0.01	0.01	0
0.02	0.02	0.02	0.03	0.02	0.01	0.01	0
0.03	0.02	0.02	0.02	0.04	0.01	0.01	0
0.01	0.01	0.01	0.01	0.01	0.02	0	0
0.01	0.01	0.01	0.01	0.01	0	0.02	0
0	0	0	0	0	0	0	0

Pinj	δ
1	-0.005
-2.5	-0.04
1.8	-0.02
-3.5	-0.061
2.8	0.007
-0.5	-0.022
-2.1	-0.045
3	0

Branch flow

1-2	0.6945
1-5	-0.251
1-7	0.5564
2-3	-0.302
2-5	-1.182
2-6	-0.322
3-4	1.6497
3-6	0.0472
3-8	-0.199
4-5	-1.368
4-7	-0.483
6-8	-0.775
7-8	-2.026

Y_{br} Branch admittance martix

1-2	20	0	0	0	0	0	0	0	0	0	0	0	0
1-5	0	20	0	0	0	0	0	0	0	0	0	0	0
1-7	0	0	14	0	0	0	0	0	0	0	0	0	0
2-3	0	0	0	15	0	0	0	0	0	0	0	0	0
2-5	0	0	0	0	25	0	0	0	0	0	0	0	0
2-6	0	0	0	0	0	18	0	0	0	0	0	0	0
3-4	0	0	0	0	0	0	40	0	0	0	0	0	0
3-6	0	0	0	0	0	0	0	21	0	0	0	0	0
3-8	0	0	0	0	0	0	0	0	10	0	0	0	0
4-5	0	0	0	0	0	0	0	0	0	20	0	0	0
4-7	0	0	0	0	0	0	0	0	0	0	30	0	0
6-8	0	0	0	0	0	0	0	0	0	0	0	35	0
7-8	0	0	0	0	0	0	0	0	0	0	0	0	45

A Node Branch Incidence Matrix

	1	2	3	4	5	6	7	8
1-2	1	-1	0	0	0	0	0	0
1-5	1	0	0	0	-1	0	0	0
1-7	1	0	0	0	0	0	-1	0
2-3	0	1	-1	0	0	0	0	0
2-5	0	1	0	0	-1	0	0	0
2-6	0	1	0	0	0	-1	0	0
3-4	0	0	1	-1	0	0	0	0
3-6	0	0	1	0	0	-1	0	0
3-8	0	0	1	0	0	0	0	-1
4-5	0	0	0	1	-1	0	0	0
4-7	0	0	0	1	0	0	-1	0
6-8	0	0	0	0	0	1	0	-1
7-8	0	0	0	0	0	0	1	-1

Y br * A

1-2	20	-20	0	0	0	0	0	0
1-5	20	0	0	0	-20	0	0	0
1-7	14	0	0	0	0	0	-14	0
2-3	0	15	-15	0	0	0	0	0
2-5	0	25	0	0	-25	0	0	0
2-6	0	18	0	0	0	-18	0	0
3-4	0	0	40	-40	0	0	0	0
3-6	0	0	21	0	0	-21	0	0
3-8	0	0	10	0	0	0	0	-10
4-5	0	0	0	20	-20	0	0	0
4-7	0	0	0	30	0	0	-30	0
6-8	0	0	0	0	0	35	0	-35
7-8	0	0	0	0	0	0	45	-45

SF SHIFT FACTOR MATRIX -- = Ybr * A * Z

1-2	0.33	-0.18	-0.04	0.01	0.03	-0.06	0.05	0
1-5	0.27	-0.02	-0.04	-0.07	-0.25	-0.02	0.02	0
1-7	0.4	0.2	0.08	0.06	0.22	0.07	-0.07	0
2-3	0.15	0.25	-0.14	-0.03	0.13	0.02	0.01	0
2-5	-0.07	0.2	0.01	-0.09	-0.36	0.05	-0.04	0
2-6	0.25	0.36	0.09	0.12	0.25	-0.13	0.08	0
3-4		0.02	0.31	-0.38	-0.14	0.09	-0.14	0
3-6	0.09	0.07	0.3	0.18	0.11	-0.18	0.08	0
3-8	0.14	0.16	0.25	0.17	0.16	0.11	0.08	0
4-5	-0.21	-0.18	0.03	0.16	-0.39	-0.03	0.02	0
4-7	0.13	0.2	0.28	0.46	0.26	0.13	-0.16	0
6-8	0.34	0.44	0.39	0.31	0.37	0.69	0.16	0
7-8	0.52	0.4	0.36	0.52	0.48	0.2	0.76	0

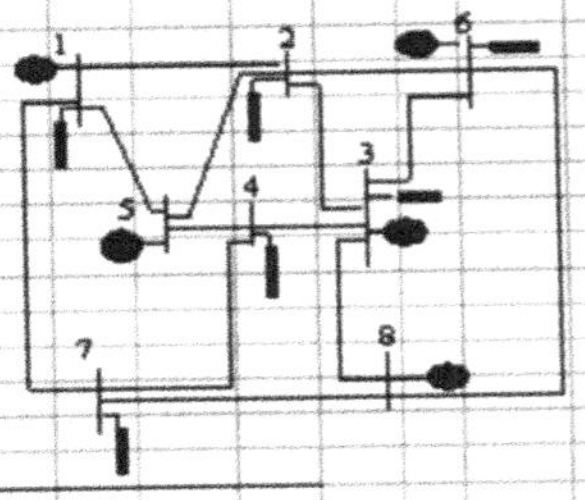

Figure B.4. Solution of dc load flow.

The computed value for the $[\mathbf{S}_F]$ matrix are shown in cells B59:I71. Next the flow in all branches resulting from (B.23) is shown in cells Q18:Q30.

In conclusion, all flows in the network are computed.

B.3.2 The ac Load Flow

The solution of the nonlinear equations of (B.13), (B.14) is normally obtained in the power industry using the Newton–Raphson (NR) method.[5] The reader can find the details of the NR method in several books (Stagg and El-Abiad, 1968, Wallach, 1986). Essentially, the method improves on an initial guess for a solution by searching on the gradient of an objective. The objective is to minimize the mismatch between the desired nodal power and reactive power injections and the computed injections. The essence of NR method is shown under Section 5.3.2 as a general case of solving nonlinear equations.

In the example of the attached Excel Solver solution (Open "App-acflow.Xls") shown in Figure B.5, instead of the conventional NR method, the square of the error between real and reactive power mismatches is minimized to obtain the solution. After all, in order to obtain the solution, the procedure of Solver does indeed searches on the gradient of the objective function as well. In that sense, the solution by using the Excel Solver is similar to the process of the NR method in converging to a solution. However, the mechanics of the NR method is not explicitly obvious to the reader by using Excel Solver. Nonetheless, this example serves the purpose of understanding a setup procedure to obtain a solution to nonlinear power flow equations using the Solver.

The spreadsheet contains the network diagram which is identical to the one used for the dc load flow solution. However, the resistance of branches is also included (cells B30: I37) from which the magnitude and angles of branch admittances are computed. **App-acflow.Xls**

Recall that the real power injection, voltages at generator bus, and real and reactive power demand in load buses are generally specified in the ac load flow problem formulation. The goal is to solve for remaining voltages and all angles (except the reference). In rows 4, 6, and 7, specified real and reactive powers are indicated. Nodes 2, 5, and 7 have no connected generation but have loads. Consequently, the net injection into the network at these nodes must equal this demand (negative by convention). Correspondingly, the real power injected into the network at nodes where there are generators, nodes 1, 3, 5, 6, and 8, must be equal to the specified generation in row 4. The specified voltages for nodes with generators have been specified as 1, 1.03, 1, 1.02, and 1.03 pu in row 9. Since the network loss is not known *a priori*, we cannot specify all the required generation. The generation at one bus is left floating to take on any value (presumably within the limits of the generator output) to balance total generation with demand plus

[5] Prior to this method becoming popular, an iterative solution using the Gauss–Sidel method was a common practice.

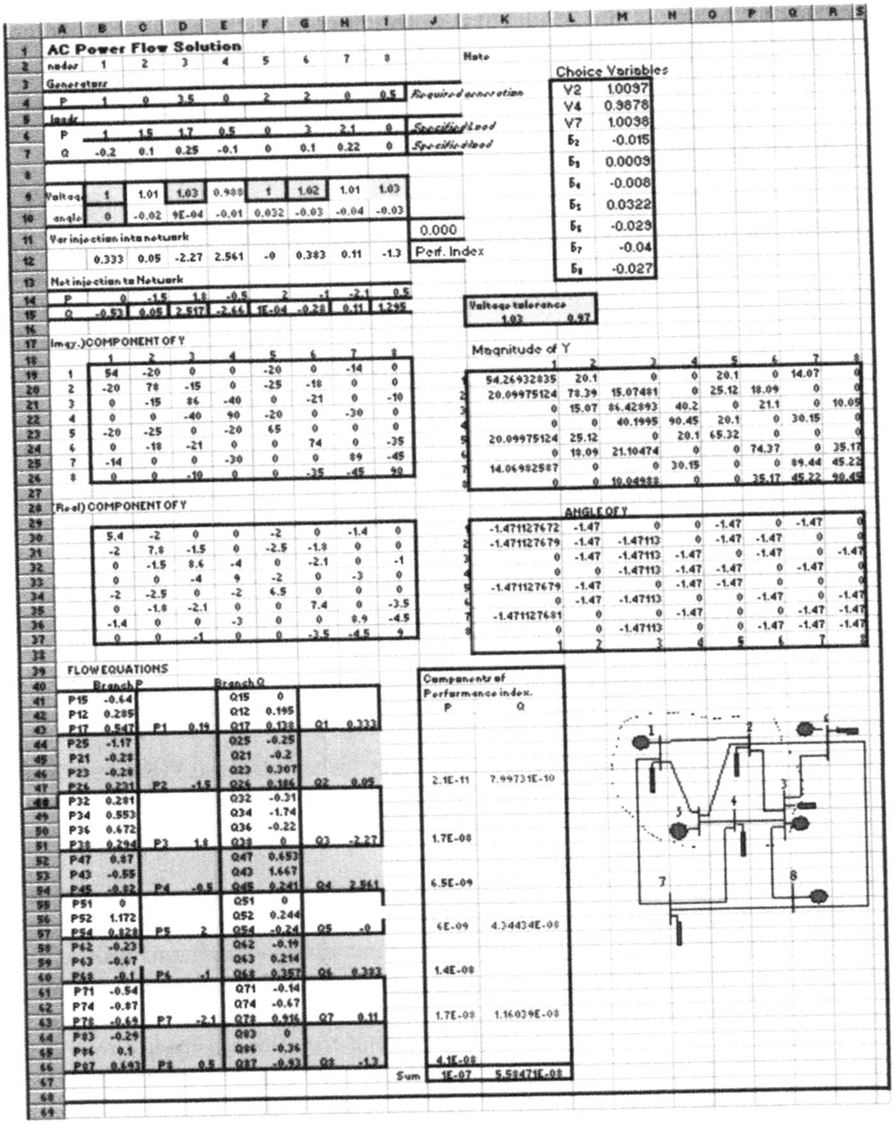

Figure B.5. Solution of ac load flow.

loss in the network. Such a bus, called the slack bus, is also used as the reference node[6] for angles. The voltage at the reference node is kept at a chosen value, 1 pu generally. In this example, node one has been chosen as the reference node.

[6]The purpose of the slack bus is to accommodate unspecified generation. The reference node for angle can be a different node. These two are generally taken to be the same node for convenience and simplicity.

Choice variables for the problem are the remaining voltages and angles. They are: $|V_2|$, $|V_4|$, and $|V_7|$ at the load buses, and the angles δ_2 to δ_8 at all nodes except the slack or reference node. In rows 41 to 66, formulas describing branch flows have been entered. These entries correspond to the relations shown in (B.13) and (B.14). The formulas are in terms of specified voltages and angles as well as the voltages and angles to be obtained by the solution process. Amidst these, column D and H contain the injection of real and reactive power into the network derived as a sum of appropriate branch flows.

In cells J47:K66, components of the performance index, the square of the difference between the computed values for P and Q and the desired values have been entered. For generator buses, the square of the difference only between the desired (real power) generation and the actual generation is entered. We are not concerned at present about the reactive power output of generators. It is assumed that any resulting value of reactive power necessary to maintain the specified voltage at the generator terminals is acceptable. Therefore, the objective function is a sum of all these squares as represented in cell J11. Finally, entries in rows 12, 14, and 15 are an echo of some selected values for convenience in programming the problem.

The entry for the Solver is to minimize cell J11 with M4:M13 as choice variables. However, in order to ensure that very large (e.g., 10 pu!) or negative voltages are not selected by the optimization procedure, we stipulate a constraint that $|V_2|$, $|V_4|$ and $|V_7|$ should be within a range, 0.97 to 1.03 pu. Similarly, in order to obtain values for angles only in the first quadrant, we require δ_2 to δ_8 to be within reasonable bounds.

Starting with any assumed values for choice variables, the Solver converges to the final solution shown in Figure B.5. Note that all voltage limits are honored and that we obtain the required injection at nodes. Branch flows are also computed and are indicated in cells B41:B66 and F41:F66. This information will be used in an example in Section 8.4 to ensure that branch flows are within certain limits.

APPENDIX C

RELATION BETWEEN PSEUDO-INVERSE AND LEAST-SQUARE ERROR FIT

C.1 MINIMUM SUM-OF-SQUARES ERROR FIT

Consider an observation of a physical phenomenon where observations are made for two parameters.[1] Let the parameters be x and y. For example, in a simplified missile tracking system, these two parameters can be the azimuth and velocity. Our interest is in finding a relation between x and y. One can assume any relationship between these parameters and try to estimate constants in such a relational formula. However, one is generally armed with a knowledge of the process involved in order to choose a relationship that makes sense. For example, if it is known that the relationship between x and y is linear, if one chooses a sinusoidal relationship, one may get the ensuing mathematical procedure to work, but the obtained results may not make much sense. Consequently, care has to be exercised in choosing a sensible relationship.

In the following, the mathematical process of the parameter estimation process is outlined. Here, our purpose is to show that the least-square error fit, the pseudo-inverse matrix, and the maximum likelihood estimators are all the same. In this example, we choose a liner relationship between x and y to make our point.[2]

[1] An extension of this mathematics to more than two parameters can be found in Lawson and Hanson (1974).

[2] There is a wealth of literature in statistics and estimation theory that discusses a nonlinear relationship between several variables. Because they are too numerous to mention, the reader interested in studying this field further may benefit from a literature search with appropriate key words.

Optimization Principles, by Narayan S. Rau
ISBN 0-471-45130-4

Let k observations be made from which we want to establish a linear relationship between y and x of the form

$$y = a_0 + a_1 x. \tag{C.1}$$

The computed value of y'_k from the estimated values of a_0 and a_1 will differ from the actually observed value y_k. This error can be written as

$$(y_k - y'_k) = \varepsilon_k = y_k - (a_0 + a_1 x_k). \tag{C.2}$$

Our intent is to minimize the sum of the squares of errors of all the observations. The square is taken to ensure that positive and negative errors do not cancel. The objective for minimization of error is as follows;

Minimize

$$E = \sum_{i=1}^{k} \varepsilon_i = \sum_{i=1}^{k} \left[y_i - (a_0 + a_1 x_i)\right]^2. \tag{C.3}$$

This criterion for error minimization is generally referred to as a *"least-square error"* (LSQ) fit.

In order to minimize the error, the partial derivative of E with respect to a_0 and a_1 must vanish. Hence,

$$\frac{\partial E}{\partial a_0} = -2 \sum_{i=1}^{k} \left[y_i - (a_0 + a_1 x_i)\right] = 0, \tag{C.4}$$

$$\frac{\partial E}{\partial a_1} = -2 \sum_{i=1}^{k} \left[y_i - (a_0 + a_1 x_i) x_i\right] = 0. \tag{C.5}$$

Elimination of the constant 2, along with a rearrangement of (C.4) and (C.5), gives

$$k a_0 + a_1 \sum_{i=1}^{k} x_i = \sum_{i=1}^{k} y_i, \tag{C.6}$$

$$a_0 \sum_{i=1}^{k} x_i + a_1 \sum_{i=1}^{k} x_i^2 = \sum_{i=1}^{k} x_i y_i \tag{C.7}$$

Equations (C.6) and (C.7) represent two simultaneous equations for parameters a_0 and a_1 (with coefficients computed from observed x and y values) which can be written as

$$[\mathbf{A}]\,\mathbf{a} = \mathbf{b}, \tag{C.8}$$

in which

$$[\mathbf{A}] = \begin{bmatrix} k & \sum_{i=1}^{k} x_i \\ \sum_{i=1}^{k} x_i & \sum_{i=1}^{k} x_i^2 \end{bmatrix}, \tag{C.9}$$

$$\mathbf{a} = \begin{bmatrix} a_0 \\ a_1 \end{bmatrix}, \tag{C.10}$$

and

$$\mathbf{b} = \begin{bmatrix} \sum_{i=1}^{k} y_i \\ \sum_{i=1}^{k} x_i y_i \end{bmatrix}.$$

The solution to (C.8) is

$$\mathbf{a} = [\mathbf{A}]^{-1}\,\mathbf{b}.$$

In the above, from an observation of the physical process, all the data necessary to compute [**A**] and **b** are obtained. Therefore, the estimation parameters vector **a** can be computed.

An extension of this straight-line fit to an nth-order polynomial is straightforward. As shown in Pipes and Harvill (1970), the corresponding [**A**], **a**, and **b** are

$$\mathbf{A} = \begin{bmatrix} k & \sum_{i=1}^{k} x_i & \sum_{i=1}^{k} x_i^2 & \cdots & \sum_{i=1}^{k} x_i^n \\ \sum_{i=1}^{k} x_i & \sum_{i=1}^{k} x_i^2 & \sum_{i=1}^{k} x_i^3 & \cdots & \sum_{i=1}^{k} x_i^{n+1} \\ \vdots & \vdots & \vdots & \vdots & \vdots \\ \sum_{i=1}^{k} x_i^n & \sum_{i=1}^{k} x_i^{n+1} & \sum_{i=1}^{k} x_i^{n+2} & \cdots & \sum_{i=1}^{k} x_i^{2n} \end{bmatrix}, \tag{C.11}$$

$$\mathbf{a} = \begin{bmatrix} a_o \\ a_1 \\ \vdots \\ a_n \end{bmatrix}, \tag{C.12}$$

and

$$\mathbf{b} = \begin{bmatrix} \sum_{i=1}^{k} y_i \\ \sum_{i=1}^{k} y_i x_i \\ \vdots \\ \sum_{i=1}^{k} y_i x_i^n \end{bmatrix}. \tag{C.13}$$

C.2 PSEUDO-INVERSE

The above procedure has shown a method to fit a linear relationship to a given set of data points. Extending this to a higher order, a quadratic equation for each of the k observations can be written for terms up to x^2 as follows:

$$y_1 = a_0 + a_1 x_1 + a_2 x_1^2 \tag{C.14}$$

$$y_2 = a_0 + a_1 x_2 + a_2 x_2^2 \tag{C.15}$$

$$\vdots = \vdots \tag{C.16}$$

$$y_k = a_0 + a_1 x_k + a_2 x_k^2. \tag{C.17}$$

The goal now is to estimate values for a_0, a_1, and a_2. Clearly, all these equations are exactly satisfied only if all observations fall exactly on the curve to be desired. Nonetheless, if we assume that all the equations in (C.14) to (C.17) are exactly satisfied, they can be written using matrix notation as follows.

$$[\mathbf{X}]^T \mathbf{a} = \mathbf{y}, \tag{C.18}$$

where

$$[\mathbf{X}] = \begin{bmatrix} 1 & 1 & 1 & \cdots & 1 \\ x_1 & x_2 & x_3 & \cdots & x_k \\ x_1^2 & x_2^2 & x_3^2 & \cdots & x_k^2 \end{bmatrix}, \tag{C.19}$$

$$\mathbf{y} = \begin{bmatrix} y_1 \\ y_2 \\ \vdots \\ y_k \end{bmatrix}, \tag{C.20}$$

and

$$\mathbf{a} = \begin{bmatrix} a_o \\ a_1 \\ a_2 \end{bmatrix}. \tag{C.21}$$

In the above, [**X**] is not a square matrix. Its dimensions are $n \times k$, where k represents the number of observations (equations), and n is the order of the polynomial plus one that is used as a fit. Therefore, equation (C.18) cannot be solved by taking the inverse of the matrix [**X**] because it is nonsquare. However, (C.18) can be made square by premultiplying it through by [**X**]:

$$[\mathbf{X}][\mathbf{X}]^T\mathbf{a} = [\mathbf{X}]\,\mathbf{y}. \tag{C.22}$$

Since $[\mathbf{X}][\mathbf{X}]^T$ is square, a solution can be obtained for **a**. If we now denote $[\mathbf{X}]\,[\mathbf{X}]^T$ by [**D**], (C.22) can be written as

$$\mathbf{a} = [\mathbf{D}]^{-1}[\mathbf{X}]\,\mathbf{y}. \tag{C.23}$$

Furthermore, if we compare the terms of matrix [**D**] with matrix [**A**] in (C.11), we see that the two matrices are identical. Also, the product [**X**]**y** in (C.23) can be seen to be identical to (C.13). Therefore, the above procedure of computing $[\mathbf{D}]^{-1}$ has indeed generated equations to minimize the sum of square of the errors.

C.3 MAXIMUM LIKELIHOOD ESTIMATORS

Suppose a measurement is made by two independent observers.[3] Each observation might have an inherent error in it either due to different instruments being used or because observations are made at different locations influenced by the environment. In any event, we associate an error associated with different measurements. If there are only two observations, we might say that the first one is likely to have an error of ±5%, and the second ±6%. Furthermore, let these errors be normally distributed.

Assume that, behind these independent measurements, there is a true value τ that we are trying to estimate from the measurements. If m_1 and m_2 are the two measured values, the probability density function $f(m)$ of the two measurements

[3]There are several examples of such independent observations. For instance, two observers at different locations might be measuring the position of a celestial body or the velocity of a missile. In electrical engineering terms, it can be the measurement of current or power by two separate instruments or measurements. By measurements, we also include synthesis. Power can be synthesized as a product of current and voltage. Consequently, one measurement of power may be by a watt-meter and another a synthesis as the product of current and voltage.

are

$$f(m_1) = \frac{1}{\sigma_1\sqrt{2\pi}} \exp\left(\frac{-(m_1 - \tau)^2}{2\sigma_1^2}\right), \tag{C.24}$$

$$f(m_2) = \frac{1}{\sigma_2\sqrt{2\pi}} \exp\left(\frac{-(m_2 - \tau)^2}{2\sigma_2^2}\right), \tag{C.25}$$

where the assumption is that the two measurements are normally distributed around a mean equal to the true value of τ with a standard deviation of σ_1 and σ_2.

The likelihood function is the probability of obtaining the two measurements m_1 and m_2. Since the random errors and the observations are assumed to be independent, the probability density of obtaining m_1 and m_2 as measurements is the product of (C.24) and (C.25) given by

$$f(m_1, m_2) = \left[\frac{1}{\sigma_1\sqrt{2\pi}} \exp\left(\frac{-(m_1 - \tau)^2}{2\sigma_1^2}\right)\right]\left[\frac{1}{\sigma_2\sqrt{2\pi}} \exp\left(\frac{-(m_2 - \tau)^2}{2\sigma_2^2}\right)\right]. \tag{C.26}$$

The probability associated with m_1 and m_2 as outcomes is obtained from the density function as $f(m_1)dm_1$ and $f(m_2)dm_2$, where $f(m_1)$ and $f(m_2)$ are as in (C.24) and (C.25). This probability is expressed as

$$\begin{aligned} \text{Prob}\ \{m_1, m_2\} &= \left[\frac{1}{\sigma_1\sqrt{2\pi}} \exp\left(\frac{-(m_1 - \tau)^2}{2\sigma_1^2}\right)\right] \\ &\times \left[\frac{1}{\sigma_2\sqrt{2\pi}} \exp\left(\frac{-(m_2 - \tau)^2}{2\sigma_2^2}\right)\right] dm_1 dm_2, \\ &\quad dm_1, dm_2 \to 0. \end{aligned} \tag{C.27}$$

The maximum likelihood of obtaining these outcomes requires the maximization of (C.27) with respect to τ, which is the same as maximizing the natural logarithm of (C.27). Hence we get

$$\begin{aligned} \text{Ln}\ [\ \text{Prob}\ .\{m_1, m_2\}] &= \text{Ln}\ \left[\frac{1}{\sigma_1\sqrt{2\pi}}\right] + \text{Ln}\ \left[\exp\left(\frac{-(m_1 - \tau)^2}{2\sigma_1^2}\right)\right] \\ &+ \ln\left[\frac{1}{\sigma_2\sqrt{2\pi}}\right] + \text{Ln}\ \left[\exp\frac{-(m_2 - \tau)^2}{2\sigma_2^2}\right], \end{aligned} \tag{C.28}$$

or

$$\begin{aligned}\text{Ln [Prob }.\{m_1, m_2\}] = &\left[\text{Ln }(1) - \text{ Ln }(\sigma_1\sqrt{2\pi})\right] + \frac{-(m_1-\tau)^2}{2\sigma_1^2} \\ &+ \left[\ln (1) - \text{ Ln }(\sigma_2\sqrt{2\pi})\right] + \frac{-(m_2-\tau)^2}{2\sigma_2^2}.\end{aligned} \tag{C.29}$$

Hence

$$\begin{aligned}\max [\text{ Prob }.\{m_1, m_2\}] = \max_{\tau} &\left[-\ln(\sigma_1\sqrt{2\pi}) - \left(\frac{(m_1-\tau)^2}{2\sigma_1^2}\right)\right. \\ &\left. - \ln (\sigma_2\sqrt{2\pi}) - \left(\frac{(m_2-\tau)^2}{2\sigma_2^2}\right)\right],\end{aligned} \tag{C.30}$$

which is identical to

$$= \min_{\tau} \left[\left(\frac{(m_1-\tau)^2}{2\sigma_1^2}\right) + \left(\frac{(m_2-\tau)^2}{2\sigma_2^2}\right)\right]. \tag{C.31}$$

The minimum of the above equation is found by

$$\frac{d}{d\tau}\left[\left(\frac{-(m_1-\tau)^2}{2\sigma_1^2}\right) + \left(\frac{-(m_2-\tau)^2}{2\sigma_2^2}\right)\right] = 0, \tag{C.32}$$

resulting in

$$\frac{m_1-\tau}{\sigma_1^2} + \frac{m_2-\tau}{\sigma_2^2} = 0. \tag{C.33}$$

From (C.33), we get the relation between two observations and the true value as

$$\frac{m_1-\tau}{\sigma_1^2} = -\frac{m_2-\tau}{\sigma_2^2}, \tag{C.34}$$

giving a value for τ as

$$\tau = \frac{(m_1\sigma_2^2 + m_2\sigma_1^2)}{(\sigma_1^2 + \sigma_2^2)}. \tag{C.35}$$

Clearly, if the standard deviations of the errors of measurements are the same, as expected, we get

$$\tau = \frac{(m_1 + m_2)}{2}. \tag{C.36}$$

In particular, it is important to see that (C.31) is nothing but a statement requiring the minimization of the sum of squares of errors in the measurements.

In view of the above formulations of the pseudo-inverse, least-square minimization, and the maximum likelihood, we conclude that they all lead to the same end result.

C.4 SOLVED PROBLEMS

The pseudo-inverse and LSQ techniques are very powerful methods to obtain solutions to practical problems. The following examples illustrate this.

C.4.1 Computation of Pseudo-Inverse Iteratively

Pseudo. inv. eg. Xls

This example shows how computer routines can be used in a "brute force" way to obtain results. For an overdetermined system of equations $[\mathbf{A}]\mathbf{x} = \mathbf{b}$, the coefficient matrix $[\mathbf{A}]$ of rank three (cells C6:E10) is given, along with the $\mathbf{x}$ (cells G6:G8) vector and the right-hand side $\mathbf{b}$ (cells I6:I10). It is required to compute $[\mathbf{A}^{+}]$.

Figure C.1 shows the results of the spreadsheet. The procedure is that cells C16:G18 are set up as choice variables to obtain all the elements of the matrix

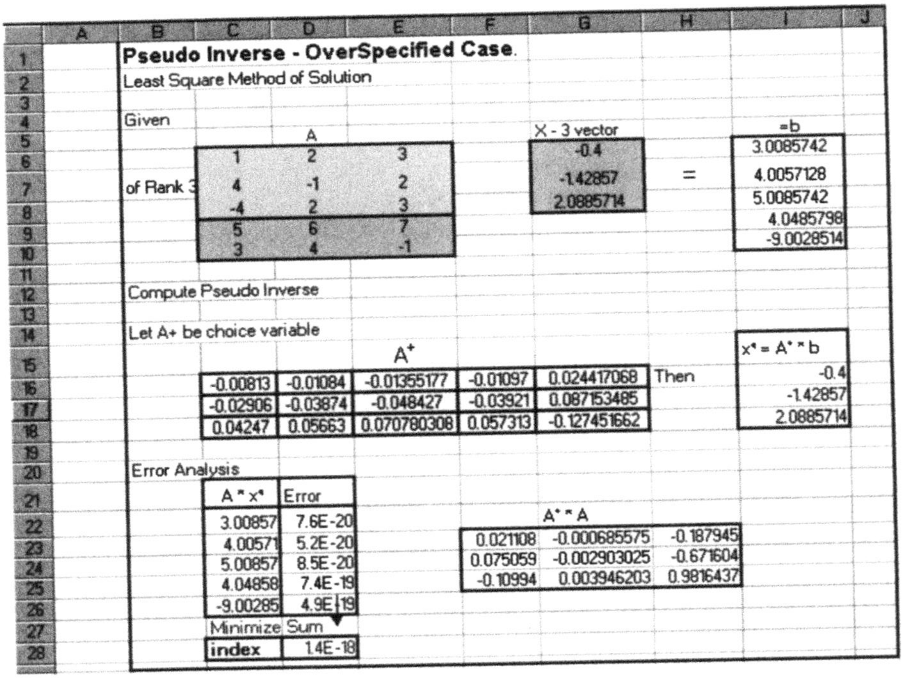

Figure C.1. Computation of pseudo-inverse iteratively.

$[A^+]$. Then, knowing **b**, we compute the approximate $\mathbf{x}^q$ as the product of [**A**] and **b** in cells I16:I18. From this $[\mathbf{x}^q]$, we compute vector **b** as the product of [**A**] and $[\mathbf{x}^q]$. Obviously, since the initial guess by the Solver for the pseudo-inverse is very likely to be incorrect, we compute the square of the error of differences between such computed values of **b** (in cells C22:C26) and the actual value of **b** (in cells I6:I10). The sum of the error squares (cell D28) is chosen as the index to be minimized.

The above iterative process converges to a satisfactory solution.

C.4.2 Least-Square Error Polynomial Fit

Problem:

Fit a third-order polynomial to the following set of data.

poly-nomial.fit.Xls

X:	0	1	10	15	20	25	30
Y:	0	18.033	24.2925	10.9307	4.1848	1.5529	0.5724

Solution: Figure C.2 indicates the Excel spreadsheet of the solution. The given data are entered in rows 3 and 4. The coefficients of the polynomial, a_0 to a_3, are set as choice variables in row 7, cells 1 to 4. Then, the value of the function for selected choice variables at different data points is computed in cells F6:16. The corresponding function values are below this with the square of the differences in cells F8:L8. Finally, the objective function of sum-of-squares errors is in cell L9. The Solver is asked to minimize this value.

As in the previous case, a nonlinear model is chosen for the Solver because of squares. The resulting coefficients are seen in the figure.

Note that the fit is not very good. Ideally, we like the objective to be as close to zero as possible. Also, a comparison of rows 6 and 7 and columns F and L reveals that the computed function is somewhat different from the data points. Why is this so? How can the accuracy be improved? Is there any other function or a higher-order polynomial that can give a better fit?

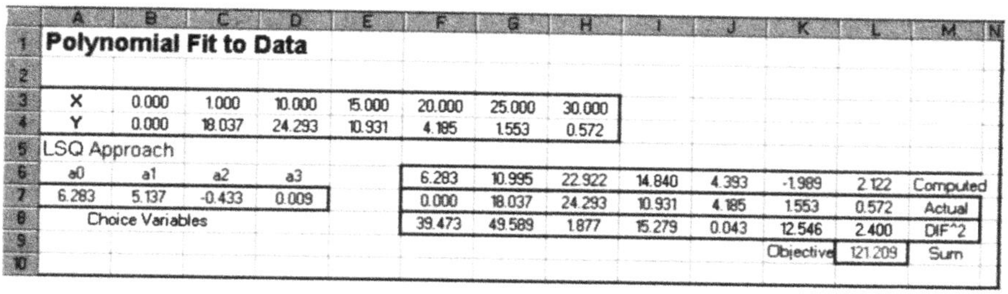

	A	B	C	D	E	F	G	H	I	J	K	L	M
1	**Polynomial Fit to Data**												
2													
3	X	0.000	1.000	10.000	15.000	20.000	25.000	30.000					
4	Y	0.000	18.037	24.293	10.931	4.185	1.553	0.572					
5	LSQ Approach												
6	a0	a1	a2	a3		6.283	10.995	22.922	14.840	4.393	-1.989	2.122	Computed
7	6.283	5.137	-0.433	0.009		0.000	18.037	24.293	10.931	4.185	1.553	0.572	Actual
8		Choice Variables				39.473	49.589	1.877	15.279	0.043	12.546	2.400	DIF^2
9											Objective	121.209	Sum
10													

Figure C.2. Polynomial fit to data.

An important observation is that the above least-square procedure can be used to fit any other function to a set of data. For example, one can fit an exponential function, or any nonlinear function, to a set of data using this procedure.

C.5 EXERCISE PROBLEMS

1. Given a matrix **A**

$$[\mathbf{A}] \begin{bmatrix} 1 & 2 & 3 \\ 1 & -1 & 2 \\ -4 & 2 & 3 \\ 5 & 6 & 7 \\ 3 & 4 & -1 \end{bmatrix},$$

compute the pseudo-inverse by the procedure described by (3.23) in Section 3.3.3. Check the product of the pseudo-inverse obtained thus and the matrix [**A**]. Observe the difference, if any, between that and the corresponding product from iteratively obtained pseudo-inverse in cells F23:H25 of spreadsheet Pseudo.inv.eg. Xls. What conclusions can you draw?

2. For the problem of Section C.4.2, attempt to fit other functions (such as higher-order polynomials or sum of exponential functions) in order to reduce the least-square error. Is there an improvement in the fit by using such other functions?

BIBLIOGRAPHY

Abur, A., Kim, H., and Celik, M., Identifying Unknown Circuit Breaker Statuses in Power Networks, *IEEE Transactions on Power Systems*, Vol. 10, No. 4, November 1995.

Bazaraa, Mokhtar, S., Sherali Hanif, D., and Shetty, C. M., *Nonlinear Programming, Theory and Algorithms*, John Wiley and Sons, 1979.

Bellman, R., *Introduction to Matrix Analysis*, McGraw-Hill, New York, 1970.

Caramanis, M. C., Bohn, R. E., and Schweppe, F. C., *Optimal Spot Pricing: Practice and Theory*, *IEEE Transactions on Power Systems*, Vol. PAS101, No. 9, September 1982.

Chiang, Alpha C., *Fundamental Methods of Mathematical Economics*, third edition, McGraw-Hill, 1984.

Clements, K. A., and Wollenberg, B. F., *An Algorithm for Observability Determination in Power System State Estimation*, IEEE Summer Power Meeting, 1975, Paper A75-447-3.

Clements, K. A., and Costa, A. S., Topology Error Identification Using Normalized Lagrange Multipliers, *IEEE Transactions on Power Systems*, Vol. 13, No. 2, May 1998.

Crew, M. A., and Kleindorfer, P. R., *The Economics of Public Utility Regulation*, The MIT Press, Cambridge, MA, 1986.

Dantzig, George D., Maximization of a Linear Function of Variables Subject to Linear Inequalities, in Activity Analysis of Production and Allocation, T. C. Koopmans (ed.), John Wiley and Sons, New York, 1951.

Duckworth, H. E., *Electricity and Magnetism*, Holt, Rinehart and Winston, New York, 1960.

Faddeeva, V. N., *Computational Methods of Linear Algebra*, Dover Publications, New York, 1959.

Finney, Ross L., and Thomas, George Brinton, *Calculus*, Addison-Wesley, Reading, MA, 1993.

Optimization Principles, by Narayan S. Rau
ISBN 0-471-45130-4

Gantmacher, F. R., *The Theory of Matrices*, Vols. I and II, Chelsea Publishing Co., New York, 1950.

Gill, Philip E., Murray, Walter, and Wright, Margaret, H. *Practical Optimization*, Academic Press, New York, 1981.

Hagen, Martin T., Demuth, Howard B., and Beale, Mark, *Neural Network Design*, PWS Publishing Company, 1999.

Hillier, Frederick S., and Liberman, Gerald J., *Operations Research*, Second Edition Holden-Day, San Francisco, 1974.

Hitchcock, F. L., The Distribution of Product from Several Sources to Numerous Localities, *Journal of Mathematical Physics*, Vol. 20, 1941, pp. 224–230.

Howatson, A. M., *Electrical Circuits and Systems: An Introduction for Engineers and Physical Scientists*, Oxford University Press, New York, 1996.

Hu, T. C., *Integer Programming and Network Flows*, Addison-Wesley, Reading, MA, 1969.

Hull, John C., *Options, Futures, and Other Derivatives*, fourth edition, Prentice-Hall, Englewood Cliffs, NJ, 1999.

Karmarker, N., A New Polynomial-Time Algorithm for Linear Programming, *Combinatorics*, Vol. 4, 1984, pp. 373–395.

Koopmans, T. C., *Optimum Utilization of The Transportation System, Proceedings of the International Statistical Conference*, Washington, D.C., 1947.

Lawson, Charles S., and Hanson, Richard J. *Solving Least Square Problems*, Prentice-Hall, Englewood cliffs, NJ, 1974.

Mamoh, James A., *Electric Power System Applications of Optimization*, Marcel Dekker, New York, 2000.

Pierre, Donald A., *Optimization Theory with Applications*, John Wiley and Sons, New York, 1969.

Pipes, Louis A., and Harvill, Lawrence R., *Applied Mathematics for Engineers and Physicists*, McGraw-Hill Book, New York, 1970.

Rao, Singiresu S., *Engineering Optimization: Theory and Practice*, third edition, John Wiley and Sons, New York, 1996.

Rau, N. S., Transmission Losses and Congestion Costs Allocation—An Approach Based on Responsibility, *IEEE Transactions on Power Systems*, Vol. 15, No. 4, November 2000, pp. 1401–1409.

Rau, N. S., Radial Equivalents to Map Networks to Market Formats—An approach Using Quadratic Programming, *IEEE Transactions on Power Systems*, Vol. 16, No. 4, November 2001, pp. 856–861.

Rau, N. S., Transmission Congestion and Expansion Under Regional Transmission Organizations, *IEEE Power Engineering Review, Power Engineering Letters*, Vol. 22, No. 9, September 2002.

Samuelson, P. A., *Foundations of Economic Analysis*, Harvard University Press, Cambridge, MA, 1947.

Schweppe, F. C., and Handschin, E., Static State Estimation in Power Systems, *IEEE Proceedings*, Vol. 62, July 1974.

Singh, N., and Glavitsch, H., Detection and Identification of Topological Errors in Online Power System Analysis, *IEEE Transactions on Power Systems*, Vol. 6, No. 1, February 1991.

Song, Yong-Hua, *Modern Optimization Techniques in Power Systems*, Kluwer Academic Publishers, Hingham, MA, 2002.

Stagg, Glen, W., El-Abiad, Ahmed H., *Computer Methods in Power System Analysis* McGraw-Hill, New York, 1968.

Stevenson, William D., Jr., *Elements of Power System Analysis*, McGraw-Hill Book, New York, 1962.

Stewart, James, *Calculus*, Brooks/Cole Publishing Co., Monterey, CA, 1999.

Wallach Y., *Calculations and Programs for Power System Networks*, Prentice-Hall, Englewood Cliffs, NJ, 1986.

Wood, Allen J., and Wollenberg, Bruce F., *Power Generation, Operation, and Control*, John Wiley and Sons, New York, 1984.

INDEX

Optimization Principles, by Narayan S. Rau
ISBN 0-471-45130-4

ABOUT THE AUTHOR

Narayan S. Rau received his Bachelor's degree in Electrical Engineering from the University of Mysore, India in 1957, his Master of Applied Science degree for Electrical Engineering from the University of Saskatchewan Saskatoon, Canada in 1965 and his Ph.D. in Electrical Engineering: Power Control Systems from the University of Toronto, Canada in 1973. Dr. Rau currently works as a principal engineer in the Planning department at ISO New England (formerly New England Power Pool). He is a Fellow of the IEEE and a member of the Power Engineering Society.

Printed and bound by CPI Group (UK) Ltd, Croydon, CR0 4YY
16/09/2026
14963442-0002